70 Advances in Polymer Science

Fortschritte der Hochpolymeren-Forschung

Key Polymers
Properties and Performance

With Contributions by
R. M. Aseeva, M. Biswas, S. Packirisamy,
S. Rostami, D. J. Walsh, I. M. Ward,
G. E. Zaikov

With 117 Figures and 22 Tables

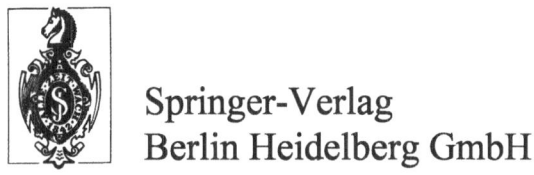

Springer-Verlag
Berlin Heidelberg GmbH

ISBN 978-3-662-15213-3   ISBN 978-3-540-39434-1 (eBook)
DOI 10.1007/978-3-540-39434-1

Library of Congress Catalog Card Number 61-642

© Springer-Verlag Berlin Heidelberg 1985
Originally published by Springer-Verlag Berlin Heidelberg New York Tokyo in 1985
Softcover reprint of the hardcover 1st edition 1985

2154/3020-543210

Editors

Editorial

With the publication of Vol. 51 the editors and the publisher would like to take this opportunity to thank authors and readers for their collaboration and their efforts to meet the scientific requirements of this series. We appreciate the concern of our authors for the progress of "Advances in Polymer Science" and we also welcome the advice and critical comments of our readers.

With the publication of Vol. 51 we would also like to refer to a editorial policy: *this series publishes invited, critical review articles of new developments in all areas of polymer science in English (authors may naturally also include workes of their own).* The responsible editor, that means the editor who has invited the author, discusses the scope of the review with the author on the basis of a tentative outline which the author is asked to provide. The author and editor are responsible for the scientific quality of the contribution.

Manuscripts must be submitted in content, language, and form satisfactory to Springer-Verlag. Figures and formulas should be reproducible. To meet the convenience of our readers, the publisher will include "volume index" which characterizes the content of the volume.

The editors and the publisher will make all efforts to publish the manuscripts as rapidly as possible, i.e., at the maximum six months after the submission of an accepted paper. Contributions from diverse areas of polymer science must occasionally be united in one volume. In such cases a "volume index" cannot meet all expectations, but will nevertheless provide more information than a mere volume number.

Starting with Vol. 51, each volume will contain a subject index.

Editors   Publisher

Table of Contents

The Preparation, Structure and Properties of Ultra-High Modulus Flexible Polymers

I. M. Ward
Department of Physics,
University of Leeds,
Leeds LS2 9JT, UK

In this review the preparation, structure and properties of ultra-high modulus polyolefins are discussed. First, detailed consideration is given to the preparation of ultra-high modulus polyethylene and polypropylene fibres by spinning from dilute solution and by gel-spinning and hot drawing. This is followed by an account of the tensile drawing procedures which have led to the production of high modulus fibres and films in polyethylene, polypropylene and polyoxymethylene. Finally, consideration is given in the production of high modulus solid section products (rod, sheet and tube) in all these polymers by ram extrusion, hydrostatic extrusion and die-drawing.

It is concluded that in these different processing routes the overriding consideration is to achieve very high extensional deformation at a molecular level. The structure of the initial isotropic polymer is therefore of key importance, the essential requirement being an adequate but not restrictive number of molecular entanglements, so that the molecular network can be stretched effectively to very high draw ratios and give very high molecular orientation. Molecular weight and morphology are therefore important, together with the drawing conditions which must permit sufficient mobility for the chains to move freely between network entanglements.

Following a review of present structural understanding of these highly oriented polymers, a detailed account is presented of mechanical behaviour, including dynamic mechanical relaxations, creep and strength. This is followed by discussion of thermal properties (melting behaviour, thermal conductivity, thermal expansion, shrinkage) and barrier properties (permeability to liquids and gases, solubility). It is of some practical importance that the improvements in stiffness and stength are accompanied by substantial improvements in thermal stability, in barrier properties and in chemical resistance.

List of Symbols

B	Constant
c	Surface crack length
$\overset{*}{c}$	Fitted constant
d	Fibre diameter
d_f	Die exit diameter
E	Young's modulus
E_c	Crystal modulus
E_m	Modulus of Maxwell element
E_v	Modulus of Voigt element
G_c	Strain energy release rate
k	Boltzmann's constant
$K_{\parallel}$	Thermal conductivity parallel to orientation direction
$K_{\perp}$	Thermal conductivity perpendicular to orientation direction
L	Long period
$\overline{L}_{002}$	Average crystal thickness
m	Constant
m_s	Molecular weight of a statistical segment
M_e	Molecular weight between entanglements
$\overline{M}_n$	Number average molecular weight
$\overline{M}_w$	Weight average molecular weight
N_c	Number of statistical chain segments between crosslinks
N_e	Number of statistical chain segments between entanglements
p	Probability of crystalline sequence linking two adjacent lamellae
P	Hydrostatic pressure
$\overset{*}{S}$	Amorphous phase solubility
R	Deformation ratio at distance x from die cone apex
R_A	Actual deformation ratio
R_N	Nominal deformation ratio
t	Time
T	Absolute temperature
ΔU	Activation energy
v	Shear activation volume
v_f	Die exit velocity
x	Distance from die cone apex
Y	Geometric factor
α	Die semi-angle
α_0	Intrinsic thermal expansivity
$\alpha_{\parallel}$	Thermal expansivity parallel to orientation direction
ε	Strain
$\dot{\varepsilon}$	Strain rate
$\dot{\varepsilon}_0$	Pre-exponential factor
$\dot{\varepsilon}_p$	Plateau creep rate
$\varepsilon/\overline{k}$	Lennard-Jones force constant
λ	Draw ratio

λ_{max} Maximum draw ratio
Φ Polymer volume fraction
σ Stress
σ_B Breaking stress
σ_0 Applied stress
σ_T Tensile stress
τ Retardation time of Voigt element
τ_f Shear flow stress
Ω Pressure activation volume
η_m Viscosity of Maxwell element
η_v Viscosity of Voigt element

1 Introduction

Although it had been appreciated since the early studies of Meyer and Lotmar [1] in the 1930's that there was a substantial gap between the theoretical stiffness of the chain in several commercially available polymers and the achievement of stiffness by existing processes, it was not until the 1970's that this gap was bridged. It is in polyethylene that the results have been most dramatically realised, and oriented fibres and rods have been produced with room temperature Young's moduli in the range from 50–100 GPa [2-7]. Solution spun fibres have even been prepared with a Young's modulus at low temperatures of 288 GPa [8], which is very close to the highest theoretical estimate of 324 GPa [9].

In this review article, an account will be presented of the different methods by which high modulus materials have been produced from flexible polymers. Much of the discussion will be concerned with polyethylene, although comparable results have been obtained for polypropylene and polyoxymethylene, and these will also be considered. The initial stimulus to this research came from the quest for high stiffness, but other properties have also been enhanced, including strength, thermal and chemical stability, and barrier properties. The present article updates and extends previous reviews [10-12] of progress in this exciting new area of polymer science.

2 Solution and Melt Processes

2.1 Solution Spinning and Drawing

A remarkable development in polymer science was the observation by A. J. Pennings [13-15] that, when dilute solutions of polyethylene were cooled under conditions of continuous stirring, very fine fibres were precipitated on the stirrer. These fibres possessed a remarkable morphology consisting of a fine central core of extended chain polyethylene with an outer sheath of folded chain polymer material, so that electron microscopy revealed a beautiful shish-kebab structure. The possible significance of this result was recognised by Frank [16], who emphasised the importance of extensional as distinct from shear deformation in achieving high molecular extension and hence high modulus. This idea was developed in elegant series of experiments by Frank, Keller, Mackley [17-24] and their colleagues at Bristol University where converging jets or rollers were used to create elongational flow fields in both flowing solutions and melts. It did not prove possible, however, to produce high modulus material. In the case of the solutions the proportion of central core extended chain material to outer sheath chain-folded material was comparatively low. In melts, it appeared that if the conditions were chosen to reduce the rate of chain-folded crystallisation the overall rate of solidification became so fast that flow ceased.

A further breakthrough in this area was then achieved by Zwijnenberg and Pennings [6] in 1976, with the discovery that ultra high modulus polyethylene fibres could be produced by "seeded crystallisation" of fine fibres, winding up at very high temperatures. In the optimum arrangements for this process shown in the schematic diagram of Fig. 1, the seed crystal is located between the inner and outer surface of a Couette viscometer. With the seed crystal in the middle of the gap "free growth" was obtained similar to that observed previously by Zwijenberg and Pennings [25] under laminar flow conditions. When the seed crystal comes into contact with the rotor surface, the so-called 'surface growth' occurs where continuous production of a very fine fibre can be maintained for several days. Moreover, by growing fibres at high tempe-rature, materials with truly remarkable properties were obtained. A set of stress-strain curves is shown in Fig. 2, from which it can be seen that a modulus of ~ 100 GPa and a tensile strength of 3.5 GPa can be achieved. Detailed studies by Pennings and his collaborators [26], and by Keller and his collaborators [24, 27] have shown that the formation of a very high molecular weight gel layer by absorption on the rotor surface plays a vital role. The molecular network of the gel layer is then stretched in the elongational flow field to give the high modulus product. The limitation of this approach is the comparatively low production rate, which was about 150 cm/min, limited by fibre fracture at the highest haul-off rates.

The next major development takes up the idea of the gel layer, and combines this with the tensile drawing of fibres at high temperature, hopefully to give a preparation route which is more acceptable in terms of production rates. We therefore now have a two stage process, in which a fibre of suitable initial structure is first produced, followed by a hot stretching process, and historically two parallel accounts have been given more or less contemporaneously. Smith and Lemstra [7] describe the

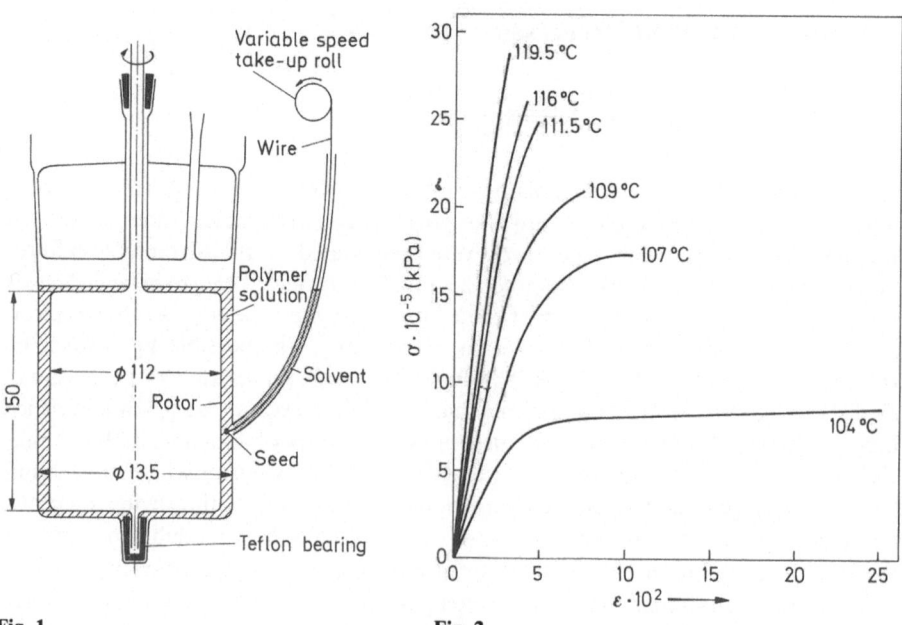

Fig. 1   Fig. 2

Fig. 1. Couette apparatus (A) for continuous growth of polyethylene macrofibres from dilute p-xylene solutions. The macrofibre grows on the surface of the Teflon rotor, leaves the vessel through a Teflon pipe, and is wound up on a take-up roll.

Fig. 2. Stress (σ)-strain (ε) curves for some polyethylene macrofibres grown from a 0.5 % xylene so-lution by the surface growth technique at various temperatures

production of a gel fibre by spinning a 2% solution of high molecular weight poly-ethylene in decalin into cold water. The gel fibre is dried and then drawn in a hot air oven at 120 °C. At the highest draw ratio ~30, a fibre with 90 GPa modulus and 3 GPa tensile strength was obtained. Kalb and Pennings [28] describe the production of a "porous" fibre by spinning a 5% solution of polyethylene in paraffin oil. Again the fibre is dried, and in this case drawn in a temperature gradient from 100–148 °C to a draw ratio of 16. This gave a product with 106 GPa modulus and 3 GPa tensile strength.

The preparation of polyethylene fibres with strengths in the range 3–4 GPa by the gel spinning route described by Pennings as discussed above, required spinning ex-trusion speeds of about 1 m/min. Although higher spinning speeds could be achieved by stretching the filaments in the molten state, this led to a dramatic reduction in the final strengths after drawing. For example a strength of less than 0.5 GPa was obtained for a winding up speed of 20 m/min. Pennings and his colleagues [29] have speculated that this relates to recoiling of the molecules leading to the generation of elastic turbulences which then disrupt the entanglement network so that the molecules are not elongated in the drawing process. Smook and Pennings [30] have shown that the addition of 1% (weight) of Aluminium stearate ($Al(OOC_{17}C_{35})_3$) permits the tensile properties after drawing to be retained to comparatively high wind up speeds (Fig. 3). It has been proposed that an absorbed aluminium stearate layer will be formed between the flowing polymer solution and the die wall. This layer then inhibits the occurrence of elastic flow instabilities. The effectiveness of adding aluminium stearate is limited to low concentrations. At higher concentrations, the aluminium stearate provides the formation of more intramolecular entaglements which are not effective in stretching the molecules.

Pennings and his collaborators [29] have examined the hot drawing behaviour of the range of spun fibres. The fibres were stretched between moving rollers, and the draw ratio determined as the ratio of the take-up speed on the second roller to the

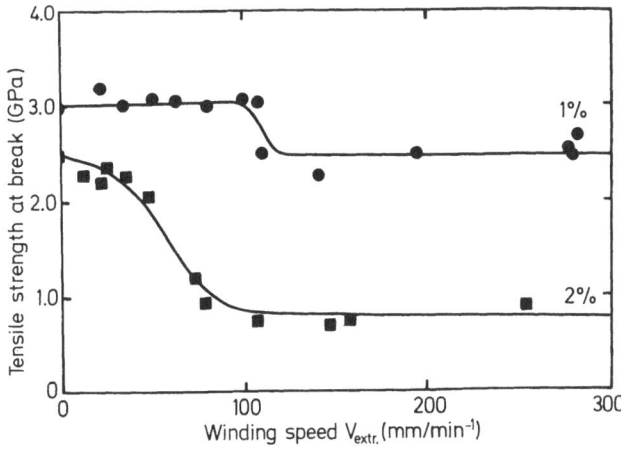

Fig. 3. Tensile strength at break, σ_b, after hot-drawing against take-up speed during gel-spinning of UHMWPE at different concentrations of aluminium stearate (wt.-%) in the spinning solution (Die: e = 14.5 cm, α = 6° and D = 1.8 mm). With permission of the publishers Chapman & Hall Ltd, (C)

input speed on the first roller, which corresponded to a constant velocity of 2.65 cm/min. The importance of a parallel oriented shish kebab morphology at the spinning stage, as well as the topology of the entanglement network, has been emphasised. Smook [31] has envisaged the hot drawing process as involving the deformation of a structure consisting of alternating bundles of elongated molecules and clusters of unoriented molecules, which are connected to each other by entanglement couplings. Pennings and his collaborators [29, 31, 32] have measured the axial tensile force as a function of time for stretching the fibres from a draw ratio of 3.5 to the maximum draw ratio over a wide temperature range. The results were analysed on the basis that stress is generated overcoming viscous flow. An apparent elongational viscosity is then derived as a function of temperature and draw ratio. It is suggested that there are three distinct temperature regions identified by different activation energies. There is a low temperature regime associated with the flow of separate fibrils and the unfolding of lamellar crystals, an intermediate regime where the lamellae melt and the fibrils aggregate, and a high temperature regime of very high chain mobility where slippage of individual chains occur. Although these speculations are attractive, it is clear that there is still room for a more comprehensive phenomenological treatment.

Some further studies of the gel spinning and drawing process have been undertaken by Manley and co-workers [33-35]. In polyethylene, their efforts were concentrated on two aspects. First, there was the drawing of dried gel films at ambient temperatures. These gel films had a structure similar to single crystal mats where the lamellar crystals are oriented parallel to the film surface. Draw ratios of 20 could be obtained, much greater than for cold drawing of high molecular weight bulk polymer. Moreover, by annealing these gel films at 110 °C, the maximum draw ratio could be increased to 30. This improvement was attributed to the increase in entanglements to act as interlamellar crosslinks and give more effective drawing.

In a further development Peguy and Manley [36] report the drawing of polypropylene gel films. The key result is summarised in Fig. 4, which shows the modulus/draw ratio

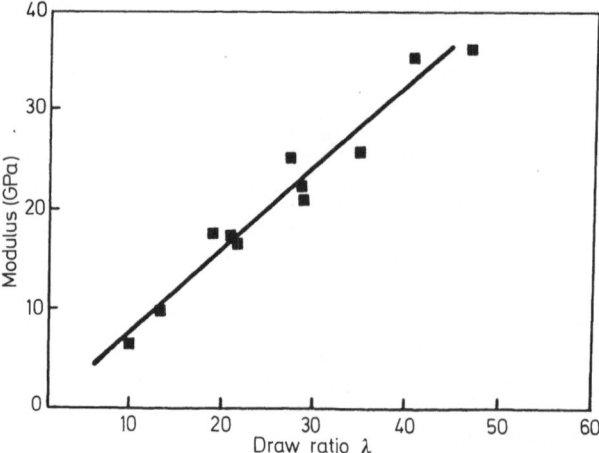

Fig. 4. Young's modulus/draw ratio relationship for dry polypropylene gel films at 140 °C. With permission of the publishers Butterworth & Co. (Publishers) Ltd. (C)

relationship for drawing at 140 °C. It is remarkable to note that values of draw ratio as high as 57 were recorded, and modulus results were obtained for draw ratio 47.5, where a value of 36 GPa was obtained. This is very close to the theoretical estimates and crystal strain values of $\sim 40\text{–}45$ GPa.

2.2 Molecular Entanglements and the Stretching of a Molecular Network

In the one-stage solution surface-growth technique as well as the two-stage gel spinning and drawing route, it has been established that the absorption of the very high molecular weight polyethylene on the rotor surface and the formation of a gel layer are key features. In the original method of Pennings and his co-workers the entanglement network of the gel is stretched in the elongational flow field of a Couette or Poiseuille flow apparatus, as originally postulated by Frank [16], and clearly illustrated in subsequent experiments by Frank, Keller and Mackley [17–23].

It can therefore be concluded that the production of high modulus polyethylene occurs by the stretching of a molecular network in the solid phase deformation routes of tensile drawing, hydrostatic and ram extrusion and die drawing, and for the solution routes of one-stage surface growth and gel spinning and drawing. In all cases the limiting draw ratio λ_{max} relates to the chain length between entanglements.

For a simple rubber-like network, the maximum draw ratio varies with the number N_c of statistical chain segments between crosslinks as

$$\lambda_{max} = (N_c)^{1/2} \tag{1}$$

In the solid phase, support for a similar contention has come from the drawing of polyethylene terephthalate, combined with stress optical measurements, and also for the behaviour of craze fibrils in a range of glassy polymers.

Smith, Lemstra and Booij [37] developed a similar argument for a gel-spun fibre. The molecular weight between entanglements in the undiluted polymer melt is denoted by M_e. The number of statistical chain segments between entanglements is given by

$$N_e = \frac{M_e}{m_s},$$

where m_s is the molecular weight of a statistical chain segment. In a solution with a polymer concentration c exceeding the value for onset of coil-overlap, the molecular weight between entanglements $(M_e)_{soln}$ is greater and is given approximately by

$$(M_e)_{soln} = (\varrho/c)M_e = M_e/\Phi \tag{2}$$

where ϱ is the bulk density of the polymer and Φ is the polymer volume fraction.

Smith et al. assume that the entanglements are equivalent to crosslinks, so that $N_e = N_c$.

Combining equations (1) and (2) then gives

$$\lambda_{max\,gel} = (N_e)_{soln}^{1/2} = (N_e/\varphi)^{1/2} = \lambda_{max\,solid}\varphi^{-1/2} \tag{3}$$

where $\lambda_{\text{max gel}}$ and $\lambda_{\text{max solid}}$ are the limiting draw ratios for the gel spun and drawn and solid phase drawn polymers respectively.

M_e for high molecular weight PE is reported by Smith et al. as 1900, giving $N_e = 13.6$ and $\lambda_{\text{max solid}} = 3.7$.

The results obtained by drawing gel spun polyethylene fibres at different temperatures are shown in Fig. 5. It can be seen that the relationship between $\lambda_{\text{max gel}}$ and $\Phi^{-1/2}$ predicted by Eq. (3) holds to a good approximation. Furthermore, the intercept at $\Phi = 1$ for a draw temperature of 90 °C was found to be 3.8, in good agreement with the value of 3.7 estimated above. The higher values of $\lambda_{\text{max gel}}$ at higher temperatures were attributed to chain slippage.

Pennings and his colleagues [29] have remarked on the influence of the hot drawing process in reducing the number of entanglements from that existing in a quiescent solution. For example, for a 5 % solution of polyethylene in paraffin oil, with $\Phi \simeq 0.6$, a value for $M_{e\ (\text{soln})}$ of 17×10^3 kg/kmol can be estimated, corresponding to about 12 entanglements per chain. By subjecting fibres to a series of cross-linking treatments by irradiation, it was possible to determine the number of effective network chains in the gel for each irradiation treatment on the basis of swelling measurements. Extrapolation to zero radiation dose then gave a value of 2 entanglements per chain in the unirradiated drawn fibre, a marked reduction from the value of 12 for the solution.

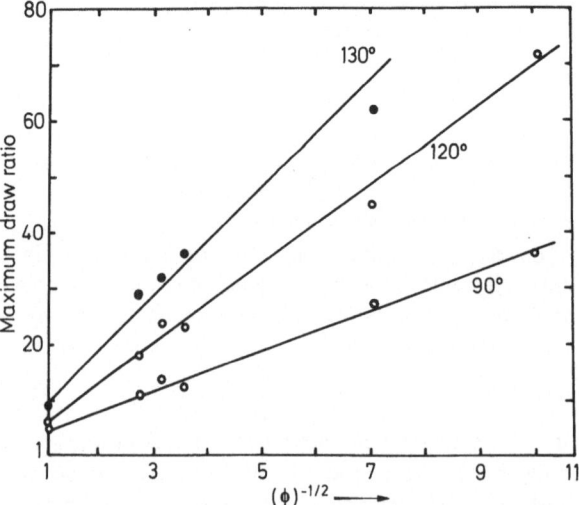

Fig. 5. Maximum draw ratio vs (initial polymer volume fraction)$^{-1/2}$ at the indicated temperatures. With permission of the publishers John Wiley & Sons, Inc. (C).

2.3 Crystallisation During Melt Flow Under Pressure

High modulus polyethylenes have also been produced by the application of pressure to the polymer when it is molten under conditions of extrusion through a narrow capillary. It appears that in one case at least these methods differ in kind from those which involve stretching the molecules either in solution, or in the solid phase.

Van der Vegt and Smit [38] observed that blockage could occur when molten polyethylene is extruded in a capillary rheometer under conditions where the polymer would be expected to remain fluid. This observation was followed up by Southern and Porter [39,40], who examined the effect of a sudden minor reduction in temperature (by 1 °C say) during the extrusion of polyethylene in an Instron rheometer. It appeared that the large increase in pressure which occurred was due to crystallisation in the elongational flow field, which leads to termination of flow. A transparent strand of polyethylene could be extracted from the capillary. This varied in properties across its section and the overall dynamic modulus of 6.6 GPa, measured at 110 Hz, was comparatively low [41]. However DSC measurements indicated a high melting component which was attributed to extended chain material, and later research described the measurement of moduli of a higher order of magnitude. In subsequent work by Porter and his colleagues [42], the pressure was maintained so that the plug of solid crystallised polymer was slowly extruded. Later studies by this group have moved to straight extrusion of a pre-formed billet. This is akin to ram-extrusion and hydrostatic extrusion and is discussed elsewhere in this review.

Southern and Porter's original studies were extended in a more precise fashion by Keller and co-workers [43] in a series of experiments where the flow of polymer in the capillary rheometer was deliberately stopped by sudden insertion of a needle valve. This produced a sudden rise in pressure, and crystallisation of the polymer. After cooling to room temperature, the plug of polymer was removed and was found to possess a very high modulus of ~70 GPa. Its structure was however quite different from other high modulus polyethylenes. A clear SAXS pattern was observed corresponding to a lamellar texture with a long period of ~300 Å. The comparatively low melting point showed that there was no extended chain material present. On the contrary, transmission electron microscopy revealed a structure of interlocking penetrating lamellae. It was suggested that the high modulus is due to the constraints on the rubber-like material between the lamellae, similar to that observed when sheets of steel are laminated with thin layers of rubber. Due to the incompressibility of the rubber and the lateral constraints imposed by bonding to the steel plates, the axial deformation of the rubber is severely restricted.

3 Solid Phase Deformation Processes

3.1 Tensile Drawing

The developments in high draw polyethylene stemmed initially from an exploratory examination by Andrews and Ward [44] of the influence of molecular weight and molecular weight distribution on the drawing of linear polyethylene at ambient temperatures. In spite of considerable scatter in the results, due primarily to difficulties in controlling polymer morphology in the small scale spinning and drawing experiments, it was concluded that the draw ratio was sensitive to $\overline{M}_w$, the weight average molecular weight. A second important observation was the excellent correlation between the modulus of the drawn fibres and the draw ratio, the modulus increasing from 4 to 20 GPa as the draw ratio was increased from 7 to 13.

These preliminary results were confirmed and extended dramatically in subsequent investigations by Capaccio, Ward and co-workers [2, 45 – 52]. A key step was the recognition that the draw temperature should be optimised to a critical temperature range (for each polymer grade) so that the molecular mobility is adequate to permit maximum draw without the onset of flow drawing due to molecular relaxation processes. It is essential that the molecular chains should move through the crystalline regions, giving a link between drawing and the α-relaxation process. This link has been quantified by detailed studies of drawing and creep behaviour which will be described in a subsequent section of this review. Further work on the drawing of monofilaments [46] showed that polymer with $\overline{M}_w$ = 69,000 could be successfully drawn to a draw ratio of 20 with a Young's modulus of 40 GPa.

It was also recognised at this stage that the molecular weight effects would be more readily identified if the morphology of the sample could be controlled. In particular, the work of Fatou and Mandelkern [53] was known to show that differences in the crystallisation rates of different molecular weight species were only marked at low degrees of supercooling. The effect of initial morphology was studied by varying the thermal treatment in the compression moulding of sheets, monitoring the temperature continuously.

The choice of polymer was restricted to commercial available grades of LPE, made available by the courtesy of BP Chemicals Ltd. In a series of investigations [45 – 49] it was shown that the extreme thermal treatments of quenching directly into water at ambient temperatures, and slow cooling from the melt at about 8 °C/min to 110 °C (to permit crystallisation to occur at low supercooling) and then quenching into water, produced marked differences in drawing behaviour for low molecular weight polymers. A spectacular result was the discovery that polymer of $\overline{M}_w$ = 67,800, with the slow cooled thermal treatment, could be drawn in a Instron at 75 °C to give a draw ratio of 30 and a Young's modulus of 70 GPa (Fig. 6). Several important observations were made at this stage by Capaccio and Ward. First, it was confirmed that the Young's modulus related to the draw ratio, so that high draw ratios are essential for the production of high modulus. Secondly, it was clear that the drawing process is sensitive to the molecular weight and molecular weight distribution. In particular, following previous work by Way and Atkinson [54] on polypropylene, the segregation of low molecular weight material in the slow cooled sheets was recognised as an important factor, and for low draw temperatures (75 °C) can act as a plasticiser to assist ease of draw. These results were confirmed and restated in a subsequent investigation by Barham and Keller [55].

At this juncture, the research at Leeds University expanded in two major directions. First, there was the development of a practical melt spinning and drawing process, some of this work being in collaboration with industrial concerns. This research is described largely in patent applications. A number of alternative possibilities were found for the production of fibres at acceptable production rates. One process, devised by Capaccio, Ward and Smith [56], describes the rapid quenching of melt spun fibre followed by high speed drawing at high temperatures. 75 °C is a reasonable temperature for drawing low molecular weight polymer at low extension rates, but temperatures above 100 °C are required for fast processing.

Secondly, it was shown that polypropylene [57] and polyoxymethylene [58] could also be drawn to high draw ratios and high Young's moduli. Essential ingredients were

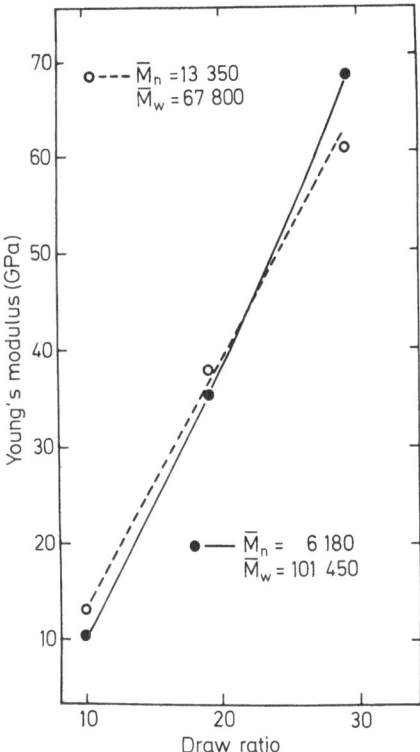

Fig. 6. Modulus-draw ratio relationship for LPE tapes drawn at 75 °C

shown to be a comparatively high draw temperature, and the choice of polymer with suitable molecular weight characteristics. Again, it was shown to be advantageous to use low molecular weight polymer, providing that the molecular weight is not reduced to a level at which the fracture strength of the polymer is unacceptably low. As in polyethylene, excellent correlations were observed between the Young's modulus and the draw ratio, but in contrast with this polymer, effects of initial thermal treatment were not found to be significant. This is probably because of the comparatively narrow molecular weight distributions of the optimum grades of polymer used. In poly-propylene, the narrow distribution occurs because low molecular weight polymer was obtained by degradation of high molecular weight polymer. In POM we are dealing with a condensation polymer, where $\overline{M}_w/\overline{M}_n \sim 2$. Figure 7 shows the room temperature modulus draw ratio relationship for low molecular weight polypropylene samples drawn at various temperatures. It can be seen that maximum moduli of ~ 20 GPa are obtained, which compare favourably with a 'theoretical' modulus ~ 45 GPa. Subsequent studies by Taylor and Clark [59] have confirmed the achievement of high modulus polypropylene, and further work at Leeds University has shown that such materials can be produced at commercially acceptable rates.

The studies of polyoxymethylene at Leeds University and its inclusion in a patent application [60], slightly predate the publication of an independent study by Clark and Scott [61] where high modulus POM was produced by a two-stage drawing process. The latter was regarded by Clark as essential to the successful preparation of high modulus material. Comparative results obtained by Brew and Ward [58] for single stage and two

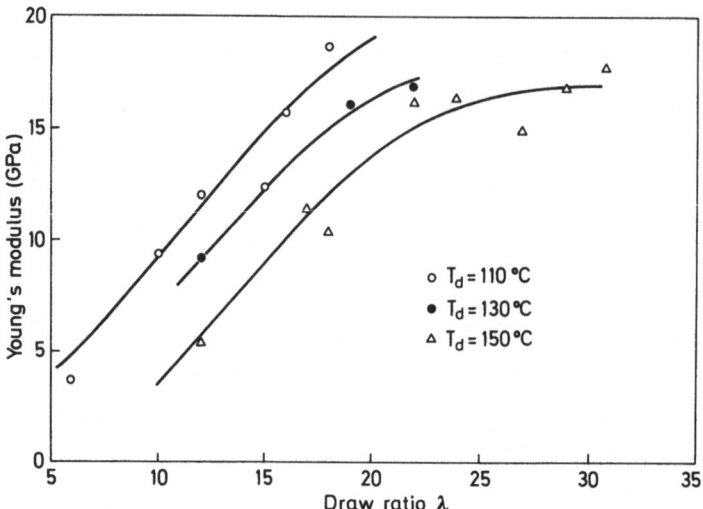

Fig. 7. Modulus-draw ratio relationship for low molecular weight ($\overline{M}_w = 1.8 \times 10^5$) PP samples drawn at different temperatures

stage drawing of POM are shown in Fig. 8. It was found that, contrary to the implications of Clark's ideas, materials produced by single stage drawing were marginally superior to those of two-stage drawing, the latter agreeing well with results reported by Clark and Scott. The superiority of the single state drawing was attributed by Brew and Ward to the deleterious effects of annealing which occur in two-stage drawing as the polymer is being raised to temperature prior to the second stage of drawing. Attempts to produce a viable commercial drawing process for POM have not been successful. It appears from the investigation of Brew and Ward that high draw and high modulus could only be obtained within a very narrow window of temperatures and strain rates. Figure 8 illustrates the situation very clearly. Only for a crosshead speed of 1 cm/min at a temperature ca. 145 °C, can the highest modulus material be obtained. It was tentatively suggested that this is because both the crystalline and the non-crystalline material are involved in the deformation process, and that there must be a suitable coincidence of rates for the two processes to occur simultaneously. This prevents any simple combination of strain rate and temperature being an adequate condition for drawing, so that drawing at high rates and high temperatures is not equivalent to drawing at low rates and low temperatures.

The early studies of Capaccio and Ward [2, 45, 46], which led to the discovery of high modulus polyethylene, were based on drawing low molecular weight polyethylene homopolymer at comparatively low temperatures. Subsequent investigations were addressed at three major extensions of this work:

1) The drawing behaviour of LPE homopolymers with a very wide range of molecular weights.
2) Examination of the effect of temperature on drawing behaviour and Young's modulus of drawn polymers.
3) The drawing behaviour of polyethylene copolymers.

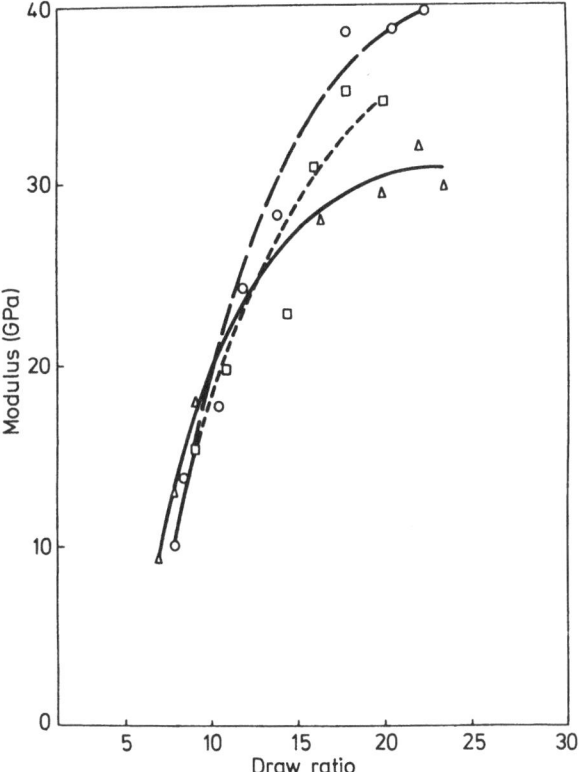

Fig. 8. Modulus-draw ratio relationship for POM samples drawn at 145 °C. Single-stage drawing was performed at 10 cm/min ($\triangle$) and 1 cm/min ($\bigcirc$). In the two-stage drawing experiment ($\square$) the samples were first drawn for 72 s at 10 cm/min and then at 1 cm/min

At this point, it is important to emphasise that the concept of a natural draw ratio, where the polymer draws through a neck so that its initial cross-section is reduced to the final cross-section by localised plastic deformation in the neck only, is not applicable to the drawing of polyethylene, in contrast to poly(ethylene terephthalate) or nylon. Instead, we have the concept of "effective drawing" i. e. drawing which produces a genuine transformation of the structure to give molecular alignment, as distinct from flow drawing, where macroscopic deformation occurs without such alignment. Moreover, in general there is a continuous drawing of material beyond the neck as the time of draw is increased. The neck region passes imperceptably into a region of further draw down and at any instant of time the draw ratio varies along the sample from unity in the undrawn cross-section to a maximum value in the central section. An important finding was that the maximum draw ratio achieved varies with time. It is therefore necessary to compare samples at a fixed draw time.

With these ideas in mind, consider the results of Capaccio, Crompton and Ward [48] for a wide range of LPE samples, drawn in an Instron at 75 °C. Figure 9 shows the maximum draw ratio as a function of time for samples varying in $\overline{M}_w$ and $\overline{M}_n$, where results are presented for quenched and slow-cooled samples. Full details of the molecular weight characteristics of the polyethylene homopolymers used in this investigation, together with other homopolymer grades discussed later, are given in Table 1.

Table 1.

Sample	Polymer Grade	Melt flow index* (MFI)	$\overline{M}_n$	$\overline{M}_w$	$\overline{M}_w/\overline{M}_n$
1	Rigidex 140–60[a]	12.7	13,350	67,800	5.1
2	Rigidex 50[a]	6.0	6,180	101,450	16.4
3	S 50[b]	5.3	12,300	101,000	8.2
4	P 10[b]	4.5	16,800	93,800	5.6
5	Rigidex 25[a]	3.0	12,950	98,800	7.6
6	Rigidex 9[a]	1.1	6,060	126,600	20.9
	006-60	0.6	25,000	135,000	5.4
	HO50-55	5.0[e]	25,000	250,000	10
7	P 40[b]	0.15	26,300	265,000	9.9
8	HO20-54P[a]	<0.05	33,000	312,000	9.5
9	Hostalen-Gur[c]	—	—	3,500,00[d]	—

* BS 2782 Method 105C
[a] BP Chemicals Int. Ltd.
[b] Experimental Polymer
[c] Farbwerke Hoechst A.G.
[d] Light scattering
[e] High load MFI

The major conclusions to be drawn from this study are
(i)   For all except the very highest molecular weight polymer, the draw ratio is time dependent.
(ii) Although draw ratios of 30 and more are obtained in the shortest draw times for polymers with lowest $\overline{M}_w$ and the slow cooled morphology, high draw ratio can also be obtained for higher molecular weight samples and/or quenched morphology provided that the draw time is extended.
(iii) The effect of initial thermal treatment is not apparent for $\overline{M}_w > 200,000$.
(iv) There is a unique modulus/draw ratio relationship, independent of molecular weight and initial thermal treatment. This is shown in Fig. 10.

Detailed structural studies, starting with optical micrographs (Fig. 11) showed that the differences in morphology relate to the crystallisation of low molecular weight polymer and low degrees of supercooling. Under such conditions, the crystallisation takes place above the melting point of the lowest molecular weight species, which are then segregated in the crystallisation process, because initially they do not take part in crystallisation. Furthermore, crystallisation is occurring under conditions of low viscosity where branching due to secondary nucleation is less likely and a more homogeneous growth occurs, hence the regions of homogeneous orientation seen in Fig. 11(b). The more regular lamellar texture of such samples has been confirmed by detailed comparison of small angle X-ray scattering, Raman longitudinal acoustic mode lines, and measurements of crystal thickness from nitric acid etching followed

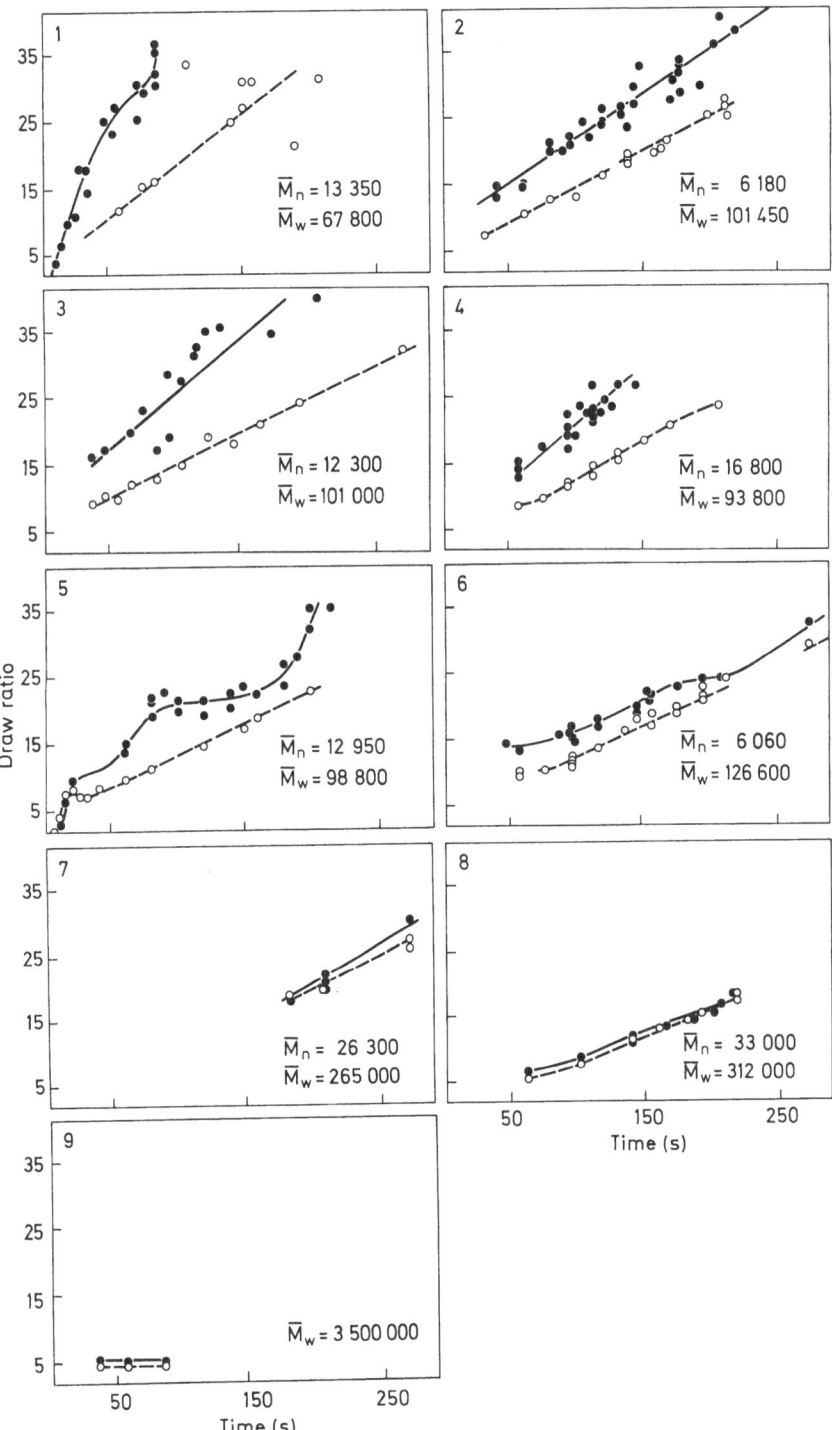

Fig. 9. Draw ratio as a function of time of draw for quenched (open circles) and slow-cooled (solid circles) LPE samples

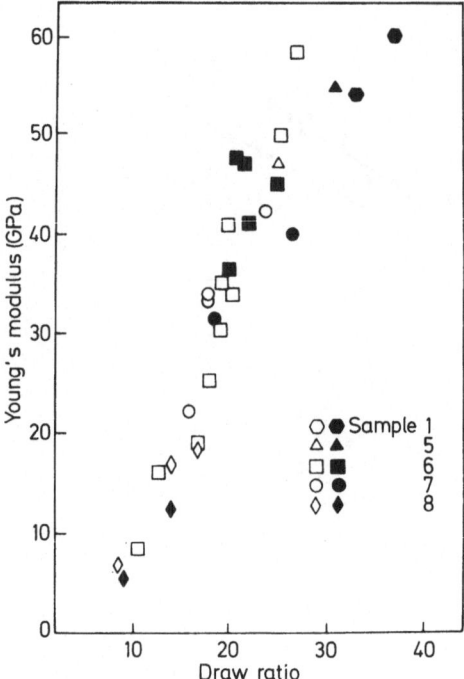

Fig. 10. Modulus versus draw ratio for a variety of quenched (open symbols) and slow-cooled (solid symbols) LPE samples drawn at 75 °C

by gel permeation chromatography analysis of the degradation product [62]. The segregation of non-crystallising low molecular weight material was also confirmed by the appearance of a narrow line component in a composite broad line NMR spectrum [63]. The ease of draw of the slow cooled low molecular weight material was therefore attributed to the unfolding of a more regular lamellar texture, the reduction in tie molecules between lamellae due to optimum thermal treatment and low $\overline{M}_w$, and the segregation of low molecular weight material. The latter was also confirmed by the observation of the effect of $\overline{M}_n$ (as distinct from $\overline{M}_w$) on draw ratio/ time plots.

A general conclusion is that the molecular topology and the deformation of a molecular network are the overriding considerations in determining the drawing behaviour. Key elements in this network are physical entanglements in the amorphous regions and junction points formed by molecular chains being incorporated in the same crystal lamellae, so that there are two sources for semi-permanent network junction points. For high molecular weight polymers the physical entanglements predominate and morphology is therefore not important. For low molecular weight polymers where there are few physical entanglements but higher crystallinity, morphology is very important.

These ideas explain why low molecular weight polymers must either be drawn slowly at low temperatures or fast at high temperatures. High molecular weight polymers, on the other hand, can only be drawn at high temperatures [52]. Figure 12

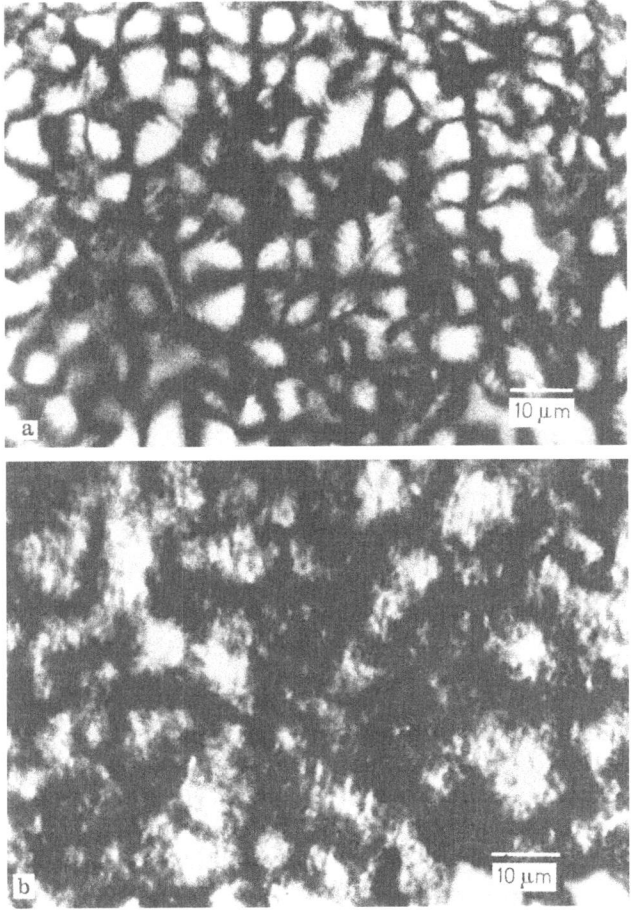

Fig. 11. Photomicrographs of sections from isotropic sheets of low molecular weight LPE: (a) quenched and (b) slow-cooled sheets

shows results for a polymer with $\overline{M}_w \sim 8 \times 10^5$. It is important to note that, whereas the macroscopic draw ratio increases, monotonically to a maximum draw ratio of 40 as the draw temperature is increased to 130 °C, the highest temperature drawing is "ineffective" so that the maximum modulus of 40 GPa is obtained at about 120 °C. Even Hostalen GUR with $M_w \sim 3.5 \times 10^6$ can be drawn to quite high draw ratios, but at temperatures above the usual melting point of the polymer [51]. Jarecki and Meier [64, 65] confirmed that there was an optimum temperature for each grade of LPE and suggested that as high a draw temperature as possible be chosen commensurate with achieving high modulus, so that internal voiding can be reduced and transparent samples produced.

An extremely important result is that the modulus/draw ratio relationship for the high molecular weight samples drawn at high temperature is approximately coincident with the "universal" relationship established for drawing at 75 °C, and again indepen-

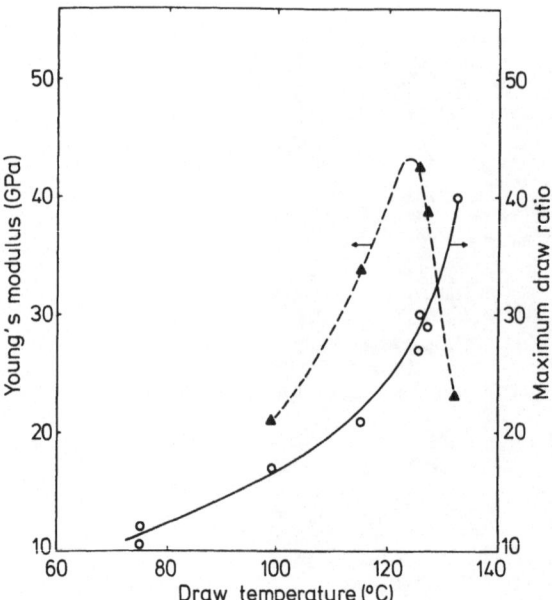

Fig. 12. Effect of draw temperature on the maximum draw ratio attainable (solid line) for high molecular weight ($\overline{M}_w = 8 \times 10^5$) LPE. Also shown (broken line) the room temperature modulus of samples drawn to the maximum draw ratio at each temperature

dent of the initial thermal treatment. However, as will be discussed below, there are important improvements to be obtained in creep behaviour and tensile strength.

In a recent investigation [66] the tensile drawing behaviour of a range of polyethylene copolymers has been studied. Full details of the copolymer composition, including the molecular weight characteristics are given in Table 2, which gives details for all polyethylene copolymers discussed in this paper. Dumbbell samples cut from quenched sheets were drawn at 75 or 115 °C at a standard crosshead speed. It was found that even at the very low concentration of one side branch per 1,000 carbon atoms there was a

Table 2.

Grade	Melt flow index (MFI)	$\overline{M}_n$	$\overline{M}_w$	Branch	Branch concentration (10^{-3}C)
Ethyl/Bu	5.4[a]	—	250,000[b]	Ethyl	c
HO60-45		23,000	283,000	n-butyl	~1.3
R2000	0.2	15,000	170,000	Butyl	~1
002-55		23,000	160,000	n-butyl	~1
002-40	0.16	20,800	135,000	n-butyl	~4
R40	4	—	120,000	Methyl	~6–7
R85	9	10,700	80,800	Methyl	~7

[a] High load MFI;
[b] Estimate from MFI;
[c] Estimated branch content similar to HO60-45

very marked effect on the strain hardening behaviour and the maximum draw ratio which could be achieved. Figure 13 shows results for two ethylene-hexene 1 copolymers with a concentration of n-butyl side branches of 1 and 4 per 1,000 carbon atoms respectively and a homopolymer of comparable molecular weight, from which it can be seen that the strain hardening increases and the maximum draw ratio decreases, with increasing branch concentration. Long branches are more effective than short

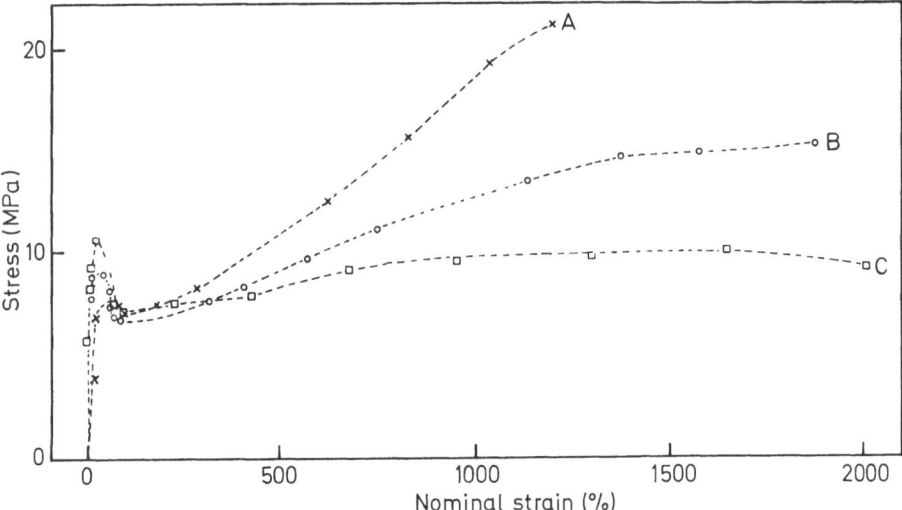

Fig. 13. Load-extension curves for ethylene-Hexene .1 copolymers A 4 butyl/10^3C; B 1 butyl/10^3C; C homopolymer

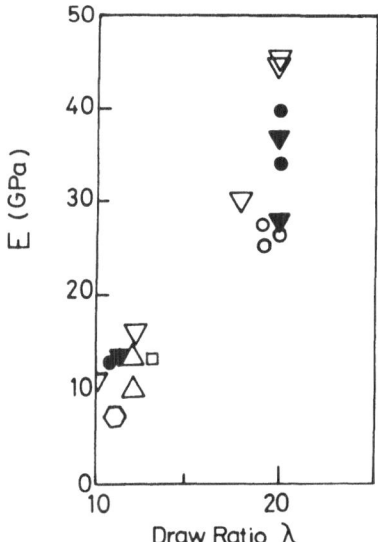

Fig. 14. Room-temperature Young's modulus as a function of draw ratio. Draw temperature $T_d = 75\,°C$: (○) H050/55, (□) Eth/But, (△) H060/45, (▽) 002-55, (○) 002-40, (●) 006-60, (△) R 50

branches in limiting the draw ratios achieved, and methyl branches produce comparatively small effects. The similarity between the introduction of long side branches and the effects of increasing $\overline{M}_w$ or radiation cross-linking is evident, and suggests that the branches significantly reduce the molecular motions required for the process of plastic deformation. This is consistent with recent speculations on the effect of branching on reptation motions.

It is interesting to note that the Young's modulus/draw ratio relationship for these copolymers is virtually identical to that for homopolymers (Fig. 14). However the ratio of average crystal lengths to long period is lower for a given value of modulus than for homopolymers. This result led Clements et al. [67] to suggest that there may be contributions to the stiffness due to taut tie molecules as well as crystalline bridges. It was clear that the simple crystalline bridge model, which fitted the homopolymer results extremely well, was not applicable.

3.2 Hydrostatic Extrusion and Ram Extrusion

In this review article, primary attention will be given to the results of hydrostatic extrusion measurements on polymers by the author and his colleagues at Leeds University. This work will be described in the context of major developments which have taken place on both hydrostatic extrusion and ram extrusion by other groups, notably Takayanagi and his colleagues [68-72] in Japan, Porter and his colleagues in U.S.A. [73,74], Keller and colleagues [75] in U.K. and more limited studies of hydrostatic extrusion elsewhere [76,77].

In ram extrusion, a plug of isotropic polymer is deformed in the solid phase by the application of pressure by a piston which extrudes the polymer through a die of reducing cross-section. Takayanagi and his colleagues [68] examined the behaviour of several polymers with this technique. In polyethylene, reduction ratios in the die of 10–16 were obtained, and the products showed high crystallite orientation and some increase in Young's modulus, values in the range of 10 GPa being reported.

In an important series of experiments, Porter and colleagues [5,39-42,73,74] have also used ram extrusion to produce highly oriented polymers. As discussed above, their earliest studies were, however, different in principle from those of Takayanagi where the polymer is in the solid phase during the whole operation. In contrast, Southern and Porter [39,40] performed ram extrusion under conditions where the polymer was molten before entering the die. As the extrusion progressed the polymer crystallised and solidified within the die and very fine strands of highly oriented polymer were produced Later studies in this area by Porter and his colleagues [73] have followed the Takayanagi technique of extruding in the solid phase, and similar work has also been undertaken by Farrell and Keller [75]. Although very significant enhancement of properties has been obtained by ram extrusion, particularly in polyethylene, the high friction between the polymer and the metal cylinder limits the process to very low extrusion rates. Farrell and Keller also showed that there was an appreciable variation in the structure across the extrudate section, corresponding to large shear deformations near the billet surface due to friction. Porter and his colleagues have shown that these disadvantages of ram extrusion can be reduced by the ingenious technique of co-extrusion [78] where a sheet of polymer is extruded

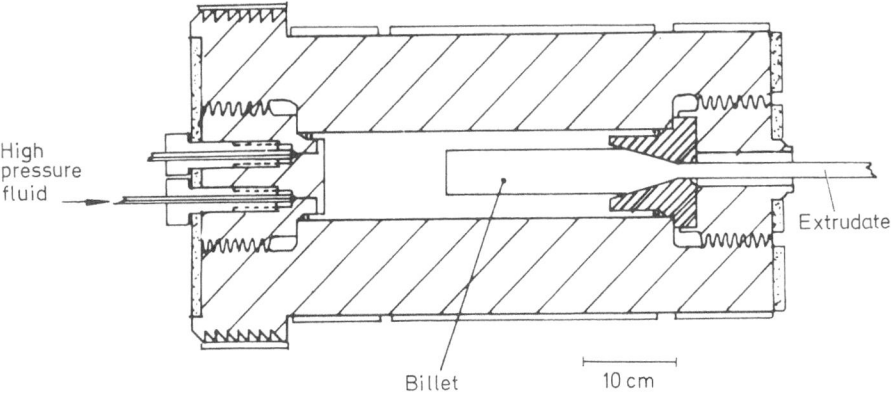

High
pressure
fluid

Extrudate

Billet   10 cm

Fig. 15. Apparatus for hydrostatic extrusion of polymers

between a "split billet" consisting most usually of two hemi cylinders, where a circular billet of a second polymer has been split longitudinally along its diameter.

Many of the disadvantages of ram extrusion are eliminated in hydrostatic extrusion, where the piston is replaced by high pressure fluid. As shown in Fig. 15, the billet stands clear of the cylinder walls, so that this element of friction is eliminated. The pressure transmitting fluid also lubricates the interface between the die and the billet, so that the deformation is essentially plug-flow (i.e. an extensional deformation, identical to drawing polymer with a free surface). Homogeneous oriented sections are therefore produced. A small haul-off force is applied to provide control of the extrusion process, and this also serves to ensure that the extrudates are straight.

Early studies of the hydrostatic extrusion of polymers were undertaken by Buckley and Long [79], Alexander and Wormell [80], Nakayama and Kanetsuma [76], and Williams [81]. Most of the results obtained for enhancement of properties were somewhat disappointing, although Williams obtained oriented polypropylene extrudates with Young's moduli up to ~16 GPa.

The full potential of the hydrostatic extrusion technique became apparent in 1974, when the production of ultra high mudulus polyethylenes with stiffnesses up to 60 GPa were reported [3, 4]. The main process parameter in hydrostatic extrusion is the nominal extrusion ratio R_N, the ratio of the billet cross-sectional area to that of the die exit (assuming deformation occurs at constant volume, which is a very good approximation). Because polymers can exhibit "die swell" in extrusion, it is convenient also to define an "actual" extrusion ratio R_A, based on the ratio of the initial and final billet cross-sections. R_A is, of course, directly comparable to the draw ratio in tensile drawing (assuming plug-flow) and in practice $R_A \approx R_N$ for all but the lowest reduction ratios.

It was considered initially that because there is no net tensile stress in a hydrostatic extrusion process it might be possible to impose very large plastic deformations without incurring fracture. It was indeed shown [3, 4] that R_N of 30 can be imposed in polyethylene comparable to the draw ratios of 30 achieved in a tensile drawing process, and that both processes are limited by the strain hardening behaviour of the material, which is determined solely by the total plastic strain imposed. This led to an important principle, namely that the processing behaviour of each polymer is determined by a mechanical equation of state, the so-called true stress-strain curve, which is a function of strain rate, temperature and pressure. The precise details of this stress-strain curve depend on the chemical composition of the polymer, and factors such as its molecular weight distribution, chemical regularity and initial morphology. Related to this important principle for the processing behaviour, is a second principle which has already been highlighted by the tensile drawing studies. This principle states that under conditions of effective deformation (i.e. effective draw) the properties of the extrudate depend only on the total plastic strain imposed i.e. the extrusion ratio.

The studies of hydrostatic extrusion at Leeds University [3,4,11,82-91] have used two facilities, a small scale extrusion vessel with bore dimensions 11.5 mm diameter and 140 mm length and a large vessel with 83 mm bore diameter and 400 mm length. The smaller vessel was used to examine a wide range of polyethylene polymers, similar to that chosen for the tensile drawing studies of Capaccio, Ward and co-workers [48]. A set of results for Rigidex 50 grade ($\overline{M}_w \sim 100,000$) is shown in Fig. 16, where the hydrostatic pressure required to give a given velocity of extrudate v_f is shown for a series of imposed deformation ratios R_N. In all cases, the semi angle of the conical die was 15°, and R_N determined by changing the initial diameter of the billet, using the same die and hence the same die exit diameter in all cases. It can be

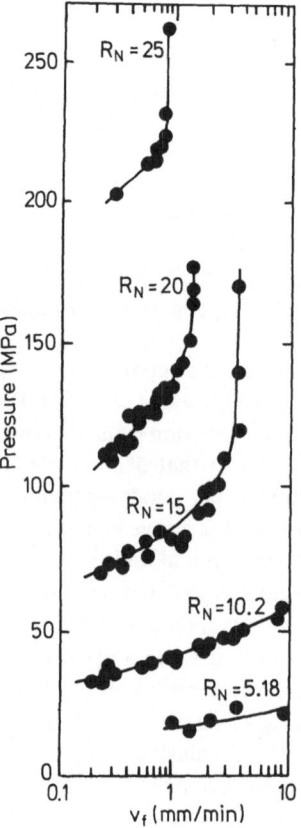

Fig. 16. Relationship between pressure and extrusion velocity at 100 °C for 2.5 mm R50 LPE extrusions at different values of R_N

seen that for low values of R_N, the extrudate velocity increases more or less linearly with pressure, as would be expected for the extrusion of a material of constant viscosity. For high values of R_N, however, there is a sudden increase in pressure if attempts are made to increase the extrudate velocity beyond a critical value. This critical value decreases with increasing R_N. A quantitative analysis of the mechanics of the hydrostatic extrusion process [89, 90] has shown that this very rapid upturn in pressure is due to two main factors:
1) the pressure dependence of the flow stress of the polymer,
2) the very rapid increase in the flow stress of the polymer with increasing plastic strain and, equally important, increasing strain rate.

It has been shown that it is convenient to describe these features by the Eyring equation for a thermally activated flow process. The strain rate $\dot{\varepsilon}$ is then given by

$$\dot{\varepsilon} = \dot{\varepsilon}_0 \exp - \left(\frac{\Delta U - \tau_f v + P\Omega}{kT} \right) \qquad (4)$$

where $\dot{\varepsilon}_0$ is a pre-exponential factor, ΔU is the activation energy, τ_f is the shear flow stress, v is the shear activation volume, P is the hydrostatic pressure, Ω is the pressure activation volume, k is Boltzmann's constant and T is absolute temperature.

With increasing plastic strain, the shear activation colume v decreases, giving increased strain-rate sensitivity, and increased pressure dependence (since the latter depends on Ω/v, where Ω remains constant).

Assuming "plug flow" in the die region, the strain rate at any point x is given by

$$\dot{\varepsilon}_x = \frac{4v_f}{d_f} \left[\frac{R}{R_N} \right]^{3/2} \tan \alpha \qquad (5)$$

where v_f is the die exit velocity, d_f is the die exit diameter, α is the die semi-angle and R is the deformation ratio of material at the distance x from the die cone apex.

Eq. (5) shows that the strain rate increases rapidly near the die exit. There is therefore a very rapid rise in pressure near the die exit where the flow stress of the polymer is increasing with increasing plastic strain and with strain rate, and where the increasing plastic strain also gives rise to a larger effect of the hydrostatic pressure on the flow stress.

For different polymers the results can be more readily appreciated by examining the change in pressure with extrusion ratio R_N for a constant extrudate velocity. Results for different polyethylenes are shown in Fig. 17, where the rapid upturn occurs at comparatively low extrusion ratios. For different polymers results are shown in Fig. 18, together with the best analytical fits based on modified Hoffman-Sachs [92] analysis, which incorporates the strain, strain rate and pressure dependent flow stress according to Eq. (4) and the Avitzur [93] strain rate field of Eq. (5). Figure 17 shows that the extrusion ratio at which the pressure rise occurs is very dependent on molecular weight, and that the results for hydrostatic extrusion parallel those for drawing, in accordance with the underlying principle that both relate to the strain hardening as quantified by Eq. (4) with a strain dependent activation volume. It is also clear that only in LPE, and to a somewhat lesser extent, in PP and POM, can very high extrusion ratios be obtained, so that the behaviour of other polymers such

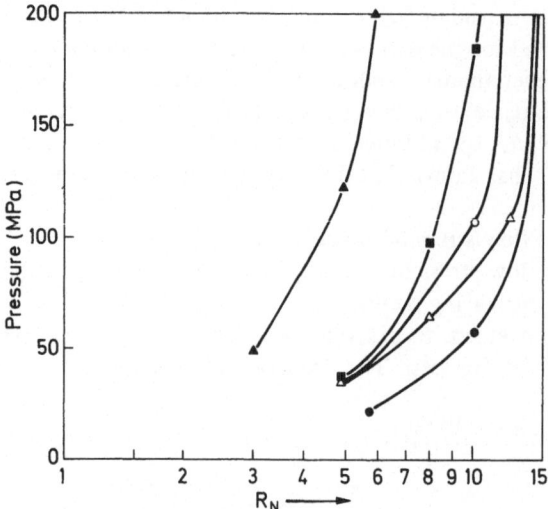

Fig. 17. Comparison of pressure-ln R_N curves for 2.5 mm LPE extrusions at 100 °C (product velocity 10 mm/min): ▲ HGUR, ■ R2000, ● R50, ○ R25, △ R140

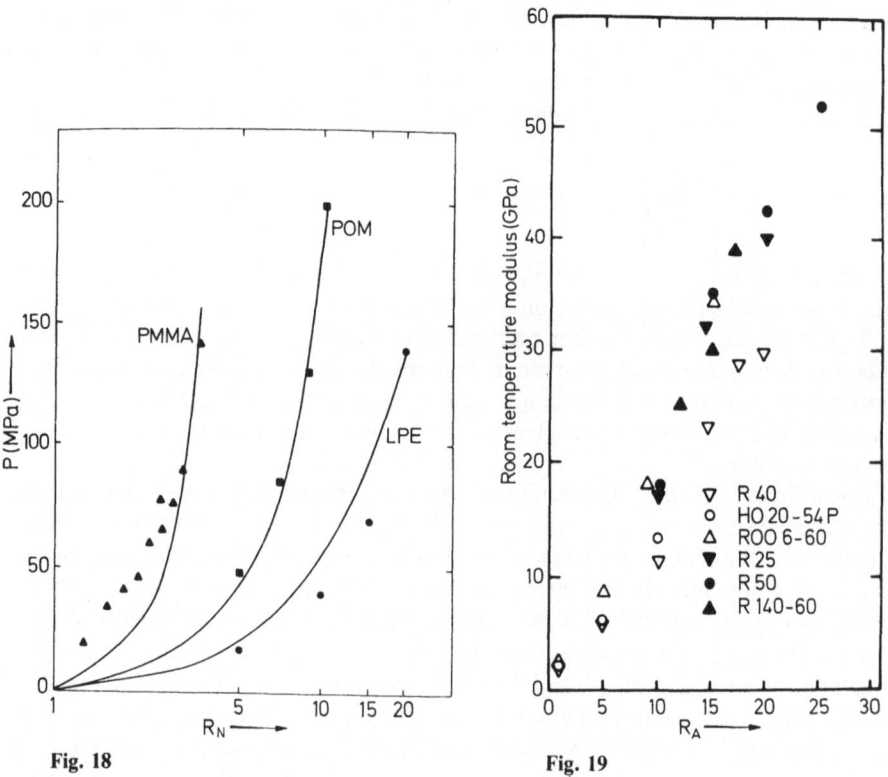

Fig. 18   Fig. 19

Fig. 18. Best analytical fits to experimental extrusion pressure — extrusion ratio R_N data

Fig. 19. Room temperature modulus as a function of extrusion ratio R_A for 2.5 mm diameter polyethylene extrudates ($T_N = 100$ °C)

as PMMA shown in Fig. 18 will be limited to comparatively low degrees of deformation.

An attempt at a quantitative analysis of the ram extrusion process was undertaken by Takayanagi and co-workers [68], also using the Hoffman-Sachs lower bound approach. An empirical true stress-strain relationship of the form

$$\log (\sigma/\overset{*}{\sigma}) \log (\varepsilon/\overset{*}{\varepsilon}) = -\overset{*}{c} \qquad (6)$$

was assumed, where $\overset{*}{\sigma}$, $\overset{*}{\varepsilon}$ and $\overset{*}{c}$ are fitted constants.

The theory did not model accurately the rise in pressure with increasing extrusion ratio, due to the omission of the effects of strain rate and pressure on the flow stress.

An upper bound analysis of the hydrostatic extrusion of linear polyethylene has been given by Gupta and McCormick [77]. This followed the approach outlined above for the flow stress (Eq. (4)) taking results from Coates and Ward [94] to describe the increase in strain rate sensitivity by a reduction in the shear activation volume. Reasonable fits were obtained for experimental data up to comparatively low deformation ratios ($R_N = 9$), giving further support for the more extensive treatments presented by Ward and coworkers.

The Young's moduli of the small-diameter extrudates were uniquely related to the extrusion ratio R_A to a very good approximation. As shown in Fig. 19, this relationship does not depend on the molecular weight of the polymer, consistent with the second principle enunciated above. In fact, it appears from extensive studies of the structure and properties of oriented LPE, PP and POM that comparable materials are produced in large section by hydrostatic extrusion to those produced as fibres or tapes by tensile drawing.

The limitations of hydrostatic extrusion have been brought out by the discussion of the mechanics of the process, from which it is clear that isothermal extrusion of small-diameter material can only proceed very slowly, at rates of ~ 1 cm/min at best. There are two advantages of increasing the scale of operation, which can lead to more practical extrusion rates. First, it can be seen from Eq. (5) that for a given strain rate field the exit velocity increases in proportion to the product diameter. Secondly, as shown by Hope and Parsons [87], extrusion can take place in a stable manner in an *adiabatic* regime as distinct from the isothermal regime considered so far. Results are shown in Fig. 20 for the hydrostatic extrusion of R40 grade polyethylene. In the isothermal regime on the left of the diagram, the effect of increasing the extrusion pressure is to increase the extrudate velocity until the process becomes unstable and a badly deformed extrudate is produced. If, however, the pressure is reduced and the velocity still increased, stable extrusion will continue in an adiabatic regime with velocities up to ~ 50 cm/min. Of course, it is necessary to ensure that extrusion in this adiabatic regime does not lead to a major reduction in properties due to annealing.

It has proved possible to make good quality extrudates in a number of different polymers (notably LPE, PP and POM) and for a range of products including non-circular and tubular sections [88,91]. For the simple non-circular section there is little change in the processing conditions. For tubular and I-sections, however, higher pressures are required due to higher strain rates in the deformation zone, increased die friction and less possibility for adiabatic heating.

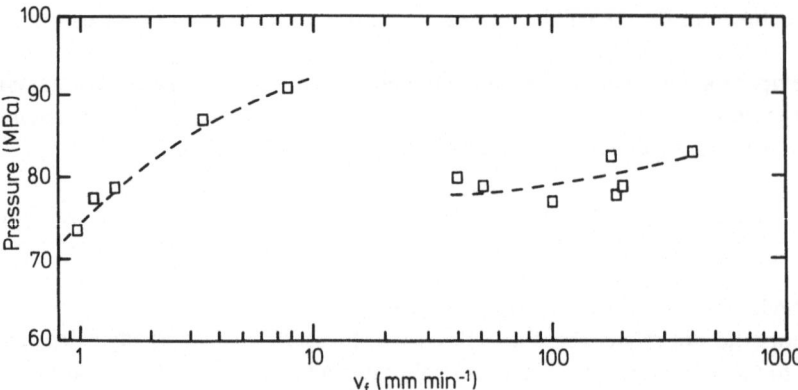

Fig. 20. Extrusion pressure v. extrudate velocity for polyethylene — large scale R40 polyethylene extrusion (T_N = 90 °C, product diameter 15.5 mm)

Hydrostatic extrusion can also be applied to polymers filled with short fibres, so that products are obtained where the fibres are aligned in addition to the polymer matrix. Excellent results have been obtained with glass reinforced POM [95], where the Young's moduli of the extruded products were about twice those of unfilled polymer with a comparable deformation ratio. A theoretical modelling of the development of mechanical anisotropy in these materials [96], using the aggregate model [97,98] was shown to be successful provided it is assumed that

1) the composite consists of a series-coupled array of sub-units, each containing continuous fully oriented fibres in a fully oriented matrix,
2) the orientation of the sub-units develops with deformation in a pseudo-affine manner.

A recent study has been undertaken of the hydrostatic extrusion of linear polyethylene over a wide range of ambient hydrostatic pressures and extrusion temperatures [99]. The basic experiment is a pressure to pressure hydrostatic extrusion of either conventional melt crystallized polyethylene or pressure crystallized chain extended polyethylene to a fixed extrusion ratio. A particular point of interest was to attempt to perform extrusion under conditions where the polyethylene is maintained in the hexagonal phase identified by Bassett and co-workers [100]. In the first instance, it was shown that well controlled pressure to pressure extrusion of melt crystallized polyethylene could be obtained throughout the temperature range 100–200 °C. The long period of such extrudates depended only on the degree of supercooling and the relationship between the Young's modulus and the ratio of the mean crystal length to the long period was maintained throughout, (see Sect. IV below). Only limited success was achieved in extrusion of materials whilst maintaining the material in the hexagonal phase. On the other hand, it was found that the hydrostatic extrusion of pressure crystallized chain extended polyethylene, both with and without back pressure, produced extrudates having considerably enhanced stiffness at much lower deformation ratios than those required for the extrusion of bulk crystallized material. This offers some potential advantages with regard to producing high modulus products with a wider range of molecular weight. It is interesting to note that the relationship between modulus and the ratio of average crystal length to long period was identical for all these materials (see Sect. 4 below).

3.3 Die Drawing

Eq. (5) shows that in hydrostatic extrusion the maximum strain rate is experienced at the die exit, where the plastic strain is greatest. In tensile drawing, on on the other hand, the maximum strain rate is experienced in the initial neck region, and as discussed above, the further plastic deformation takes place in the tapering region beyond the neck. The hydrostatic extrusion and tensile drawing processes can be regarded in terms of different paths across the true-stress-strain rate surface as shown in Fig. 21, where it can be seen that much higher stresses are reached in the hydrostatic extrusion process. This led Ward and co-workers [101–106] to examine die drawing, where the polymer billet is pulled through a heated die of reducing cross-section, as an alternative process for the preparation of oriented polymers in large sections.

A schematic diagram of the die drawing operation is shown in Fig. 22. There are essentially three deformation regimes:

I.   The polymer retains contact with the die wall and is deformed essentially isothermally at the temperature of the containing die.

II. The polymer is not contact with the die wall, and draws as in free tensile drawing, with a neck. The polymer is contained within the heated die, but it is likely that the drawing process is partly controlled by adiabatic heating of the polymer in the necked region.

III. The polymer continues to deform in the region beyond the die, but as it is now cooling rapidly it soon reaches its final deformation ratio.

As in the case of hydrostatic extrusion, the die drawing studies at Leeds University have required the small scale rig shown in Fig. 23 for initial studies and a large scale rig for the production of substantial size rods, tubes and sheets.

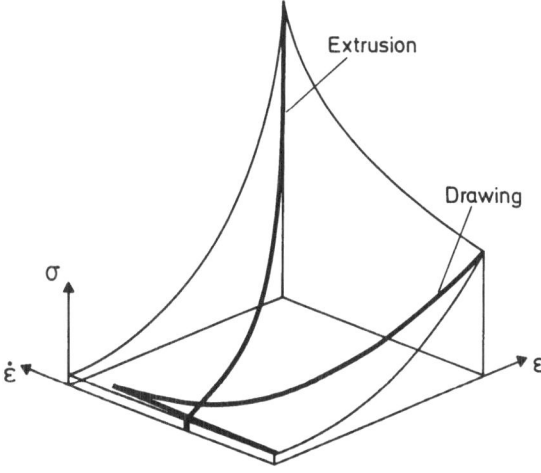

Fig. 21. Schematic diagram of process flow stress paths for elements of material undergoing extrusion through a conical die of small semi-angle, and a uniaxial tensile test

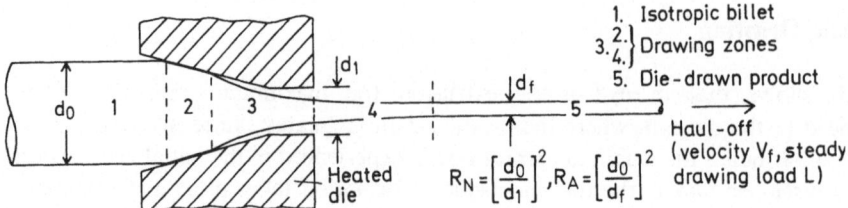

Fig. 22. Schematic diagram of die drawing process

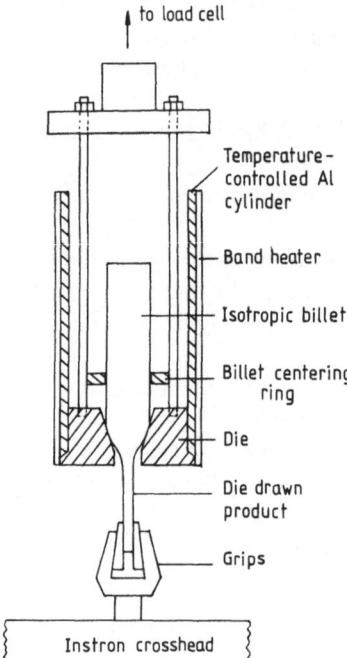

Fig. 23. Experimental die drawing apparatus

The operation of die drawing is illustrated in Fig. 24 which shows results for polypropylene [101]. As for hydrostatic extrusion there is the nominal deformation ratio R_N where

$$R_N = \frac{\text{Original billet cross-sectional area}}{\text{Die exit cross-sectional area}}$$

and the actual deformation ratio R_A where

$$R_A = \frac{\text{Original billet cross-sectional area}}{\text{Final product cross-sectional area}}$$

Fig. 24a shows results for $R_N = 5$ and 7. It can be seen that for low imposed draw speeds $R_A = R_N$, i.e. the polymer remains in contact with the die wall throughout. When the imposed draw speed is increased the polymer leaves the die wall and R_A increases quite dramatically. In contrast to hydrostatic extrusion higher deformation ratios are obtained at *higher* production speeds, in this case reaching 50 cm/min. It

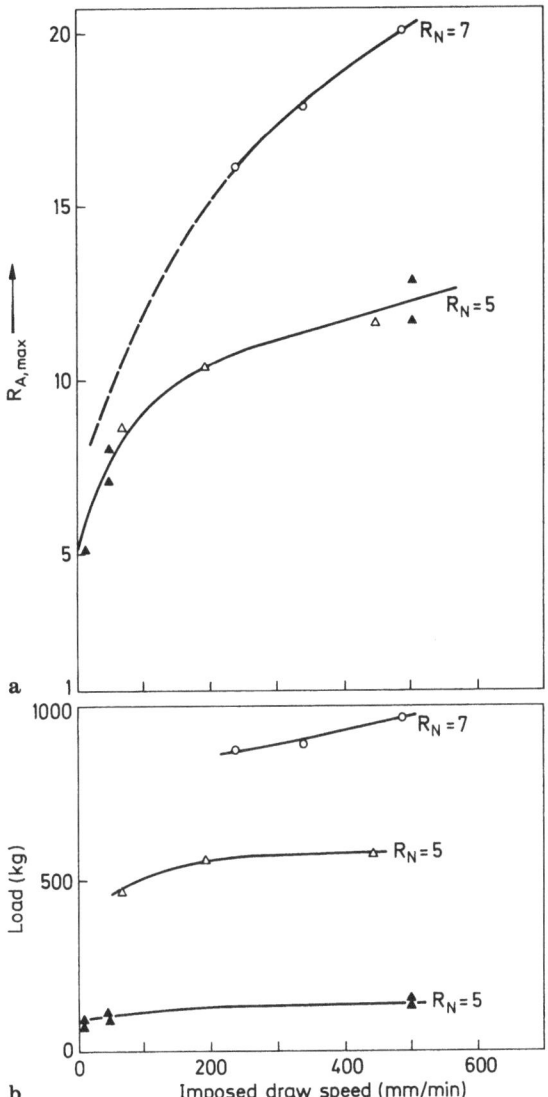

Fig. 24. (a) Dependence of maximum steady deformation ratio R_A on imposed draw speed for PP copolymer die drawn at a nominal temperature of 110 °C. (○, $R_N = 7$, 15.5 mm die; △, $R_N = 5$, 15.5 mm die; ▲ $R_N = 5.7$ mm die). (b) Relationship between steady state draw load and imposed draw speed for PP copolymer nominally at 110 °C. ○, $R_N = 7$, 15.5 mm die; △ $R_N = 5$, 15.5 mm die; ▲ $R_N = 5.7$ mm die)

is also important to note that, as shown in Fig. 24(b), the draw load only increases very slightly with increasing draw speed, because adiabatic heating of the polymer in the neck reduces the flow stress so that the draw tension remains approximately constant. The relationship between modulus and deformation ratio R_A for die drawing is closely similar to that for hydrostatic extrusion and tensile drawing. For the polypropylene

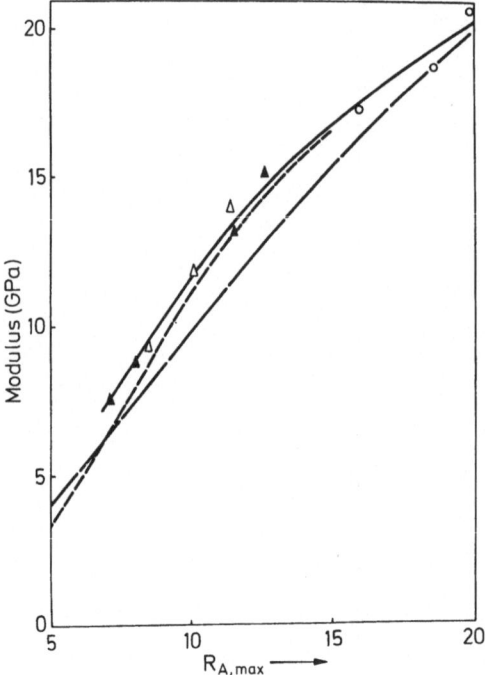

Fig. 25. Axial Young modulus, determined at 0.1 % maximum strain at 21 °C, versus maximum steady draw ratio for PP copolymer die drawn at a nominal temperature of 110 °C, ($\bigcirc$, $R_N = 7$, 15.5 mm die; $\triangle$ $R_N = 5$, 15.5 mm die; $\blacktriangle$ $R_N = 5$, 7 mm die). The results of Cansfield et al. [57] for PP homopolymer fibres drawn at 110 °C, (———) and Williams [81] for PP homopolymers hydrostatically-extruded at 110 °C (------) are included for comparison

samples studied here the results are shown in Fig. 25, and it can be seen that products with stiffnesses ~ 20 GPa are produced at the highest values of $R_A \sim 20$.

Fig. 26 shows results for die-drawing of several grades of linear polyethylene [102], for a constant $R_N = 4$. There is a clear molecular weight effect, the polymers in increasing $\overline{M}_w$ being 006-60, H120 and H020. In this case, the polymer of intermediate molecular weight gives the best performance in terms of attainable product properties and production rate. The relationship between the flexural stiffness and deformation ratio R_A is however still approximately unique, as can be seen from Fig. 27. Small scale die drawing studies have now been undertaken for a number of other polymers, including polyoxymethylene [104], polyvinylidene fluoride [105] and polyetheretherketone [106]. In POM it was possible to obtain products with stiffnesses up to ~ 20 GPa, and in PEEK stiffnesses up to 12 GPa. For PVDF the extent of draw is limited, but die-drawing at comparatively high temperatures (140 °C) enabled material with $R_A \sim 6$ and a room temperature flexural modulus ~ 4 GPa to be obtained. Perhaps more important, such material showed almost 100% conversion from the non-polar Form II to the polar Form I, which is of direct significance for the piezoelectric applications of this polymer.

In addition to the small scale die-drawing, a major activity at Leeds University has been the production of rod, tube and sheet in substantial sections on the large-

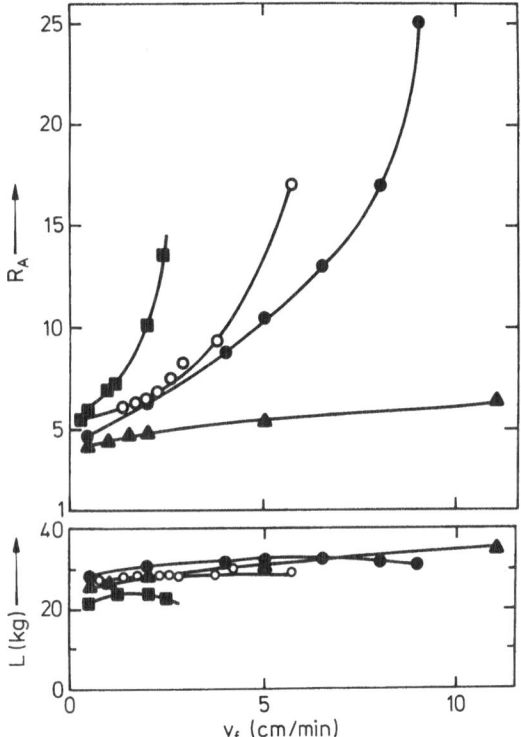

Fig. 26. Comparison of behaviour of different polymer grades. Deformation ratio (R_A) and draw load (L) versus draw velocity (v_f) for $R_N = 4$. ○ 006, ● H120, ▲ H020, ■ 002 co-polymer

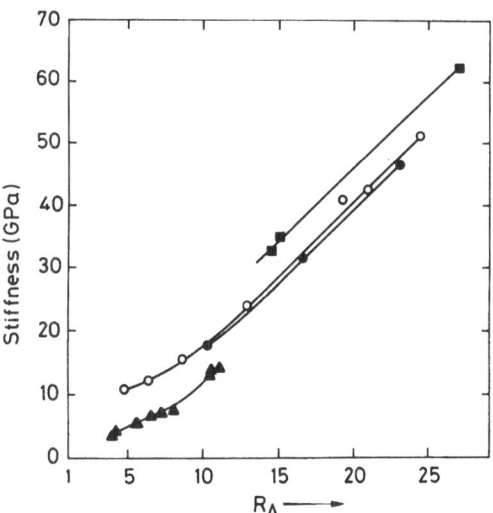

Fig. 27. Relationship between 10 sec flexural stiffness and deformation ratio (R_A) for different polymer grades. ○ 006, ● H120, ▲ H020, ■ 002 co-polymer

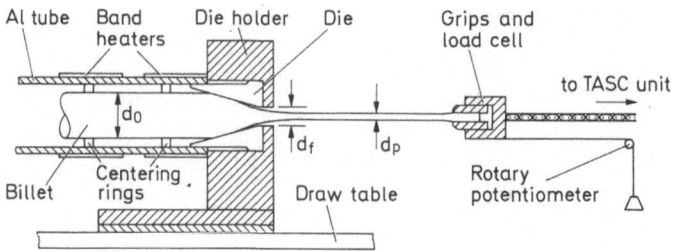

Fig. 28. Large scale die drawing rig

scale die drawing rig [107] shown in Fig. 28. Using this facility 12 m lengths of the appropriate material can be produced. It was found that the guidelines for operating conditions established in the small scale studies could be very readily applied to the large-scale process.

4 The Structure of Ultra-High Modulus Polymers

The structure of conventionally drawn polymers has received considerable attention, and amongst others, Peterlin has contributed to our understanding of the transformation of the original spherulitic texture into a highly aligned fibrillar structure. Peterlin [108] has proposed that the increased modulus of the oriented polymer can be attributed to the presence of taut tie molecules which bridge the crystal blocks of the aligned structure. At this point in time, there is no general agreement regarding the structure of these highly oriented polymers. Several issues, however, can be tentatively resolved. With the exception of the pressure crystallised melt structures of Odell et al. [43], it appears that chain orientation and elongation are key factors i.e. for solution and gel spinning processes, and for tensile drawing, hydrostatic extrusion and die drawing. It is by no means clear, and indeed unlikely on the basis of present structural evidence, that the morphology of the fibres produced by the solution and gel spinning processes are identical to those produced by the solid phase deformation processes. It also seems unlikely that the structures of the high modulus polyethylenes produced by the solid phase deformation processes are identical in nature to those of polypropylene and polyoxymethylene produced in a similar manner. With these broad reservations in mind, the structural studies of these polymers will now be reviewed.

It is only in the case of high modulus polyethylene produced by tensile drawing or hydrostatic extrusion that extensive structural studies have been undertaken. For these materials it was found [109,110], in agreement with previous studies by Peterlin and co-workers [111], that there is a two-point small angle pattern, corresponding to the long period, which is typically ~ 200 Å, and depends only on the extrusion or drawing temperature, being independent of the deformation ratio. Several structural techniques, including quantitative WAXS measurements of the average crystal lengths based on the 002 reflection [110], dark field transmission electron microscopy [112,113], and nitric acid etching followed by gel permeation chromatography [114,115] show that the average crystal lengths increase with draw ratio to reach ~ 500 Å, but that there are comparatively few crystals with lengths of 1000 Å or more. Similar results have been obtained for ram extruded polyethylenes by Thomas and his colleagues [116], although they emphasise that there can be a significant proportion of long crystals in some instances.

It was proposed by Gibson et al. [117] that a structural model for ultra-high modulus polyethylene, which is consistent with the structural data, is a crystalline bridge model, where the crystal blocks characteristic of low draw ratio polyethylene become increasingly linked by crystalline bridges as the deformation ratio is increased (Fig. 29). This model explains the retention of the SAXS two-point pattern, albeit with decreasing intensity at high draw ratios, and the fact that the average crystal length is greater than the long period. It also forms a basis for a quantitative explanation of the increase in stiffness with draw ratio, as will be discussed in detail later. If it is assumed that the crystalline bridges occur randomly, the probability of linking any number of crystal blocks can be defined by a single parameter p, which defines the probability of a crystalline sequence linking two adjacent lamellae. The distribution of crystal lengths can then be predicted by a simple theory which is exactly analogous to that which describes a stepwise condensation reaction. It has been shown that the observed crystal length distributions are in good agreement with this model, although we should note the reservations expressed by Thomas et al. [116]. Clements and Ward [118] found that the effect of varying draw temperature in tensile drawing was to change the overall scale of the morphology. The long period and the average crystal length increased together with increasing draw temperature, so that the modulus/draw ratio relationship which (as discussed in detail below) depends only on their ratio, remained invariant.

The high deformation ratios imposed on polyethylene also lead to a high degree of non-crystalline orientation in solid phase deformed material, as indicated by the orientation of the non-crystalline component in the broad line NMR [119], and the

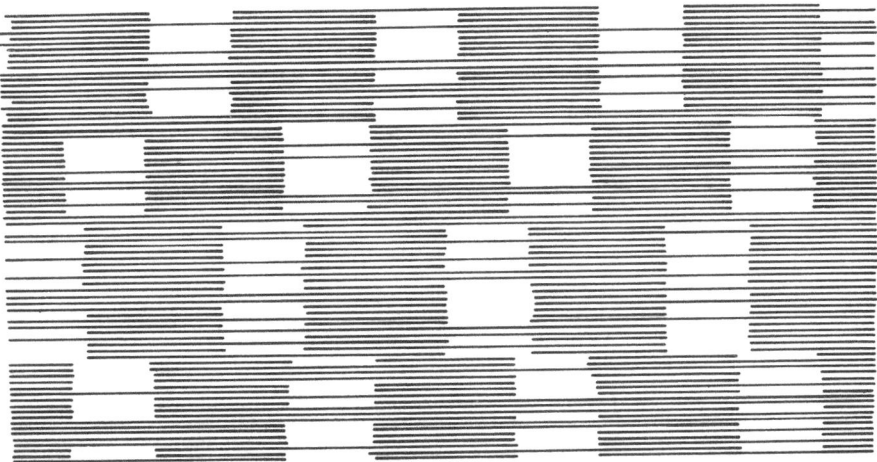

Fig. 29. A schematic representation of the structure of the crystalline phase in highly oriented LPE (constructed for p = 0.4)

very large optical birefringence [120]. The non-crystalline orientation is important with regard to superheating effects and barrier properties, to be discussed later.

Structural studies on solution spun and gel spun and drawn polyethylene fibres have so far not lead to a clear structural model. A very high degree of crystalline orientation has been observed, and the micro fibrillar nature of the materials emphasised [121]. Grubb [122] has suggested that the presence of long fibrous crystals is common to both solution fibres gel spun and drawn fibres and solid phase deformed LPE. It has then to be assumed that defects within these long crystals are responsible for their mean lengths in the dark field microscopy appearing comparatively small (~ 350 Å). It is only possible to conclude tentatively that for these materials the stiffening elements at a molecular level are tie-molecules rather than crystalline bridges.

In POM a careful examination of WAXS and SAXS measurements on a wide range of drawn samples, showed that the average crystal length, even for the highest modulus material was only comparable with the long period [123]. Again it must be concluded that the crystalline bridge model is not applicable.

5 Mechanical Properties

5.1 Dynamic Mechanical Behaviour

The extensional dynamic storage modulus E' and the loss factor $\tan \delta_E$ for a series of linear polyethylene tapes of different draw ratios are shown in Fig. 30(a) and (b). There are two features worthy of particular note. First, the modulus at low temperatures is about 160 GPa, which is about one half of the theoretical modulus and the maximum value obtained from neutron diffraction and other measurements. Secondly, the α and γ relaxations are both clearly visible even in the highest draw ratio material, although the magnitude of $\tan \delta_E$ for the γ relaxation reduces with increasing draw ratio.

The early attempts to interpret the dynamic mechanical behaviour in structural terms include that of Smith et al. [109] where the plateau modulus was correlated with the fraction of non-crystalline material f_a determined by NMR. Plots of the plateau compliances at -60 °C and -160 °C as a function of f_a suggested a modified Takayanagi series model, with a constant amount of non-crystalline material in parallel with the simple series model. The model showed good internal consistency, with values for the compliances of the non-crystalline regions which were acceptable in physical terms.

An entirely different model for explaining the mechanical stiffness was proposed by Barham and Arridge [124] who considered that high modulus polyethylene consists of needle-like crystals embedded in a matrix of remaining material. A constant fraction of such crystals was assumed (~ 0.8) and the increase in modulus with increasing draw ratio attributed to the increasing aspect ratio of the crystals, which were assumed to deform affinely. Again, a good fit to the model was obtained.

The debate moved forward a further stage [110] with the observation that the average crystal length (sometimes termed the longitudinal crystal thickness), determined

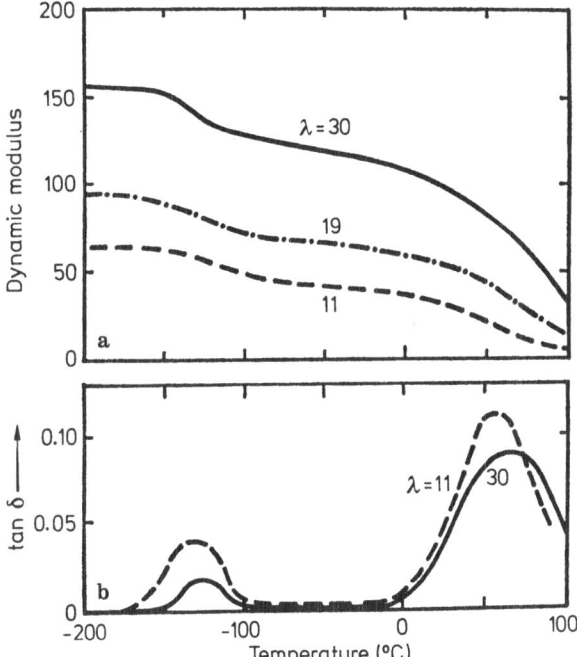

Fig. 30a. Dynamic modulus and **(b)** tan δ plotted against temperature for drawn LPE at indicated draw ratio (λ)

from the integral breadth of the 002 reflection in the WAXS pattern, increases with draw ratio to values of 500–600 Å. This compares with a long period in the range 150–250 Å, determined from the observed two-point SAXS pattern. The long period remains approximately constant with increasing draw ratio, although the intensity of the SAXS pattern diminishes. These observations led Gibson, Davies and Ward [117] to propose a structure for ultra high modulus polyethylene in which the crystal blocks of the Peterlin model are linked by crystalline bridges. With regard to mechanical properties the crystalline bridges provide a degree of crystal continuity, which acts in parallel coupling with the remaining crystal blocks and non-crystalline material. In structural terms, the crystalline bridges are equivalent to Peterlin's taut tie molecules. The crystalline bridge model was used to provide a quantitative correlation between the Young's modulus and the average crystal length, primarily determined by WAXS, but supported by other structural measurements including dark field electron microscopy and nitric acid etching followed by gel permeation chromatography. The parameter p defines the probability of a crystalline sequence traversing the intercrystalline region to link two adjacent lamellae, and is given by

$$p = (\bar{L}_{002} - L)/(\bar{L}_{002} + L) \tag{7}$$

where $\bar{L}_{002}$ is the average crystal thickness and L is the long period spacing.

A Takayanagi-type model (Fig. 31) can be used to quantify the relationship with the storage modulus at $-50\,^\circ$C (in the plateau region between the α and γ-relaxations). It can be shown that

$$E'_{-50\,^\circ C} = E_c \chi p(2 - p) \tag{8}$$

where χ is the volume fraction of crystallinity and E_c is the crystal modulus of LPE in the chain direction.

The quantity $\chi p(2 - p)$ is simply the volume fraction of material incorporated in the crystalline bridges. It has been shown [118] that the effect of increasing draw temperature is to increase both the long period, as known from the work of Peterlin and coworkers [110] for low draw material, and the average longitudinal crystal thickness, in very nearly the same proportion. The relationship expressed by Eq. (8) is therefore maintained, as shown in Fig. 32. This model in its simplest form of Eq. (8) was also shown [125] to explain the remarkable strain hardening results reported by Barham, Arridge and Keller [126]. It was found that the modulus fell on heating highly drawn materials to temperatures about 125 °C, followed by rapid quenching, but that the initial high modulus could be restored to substantially its original value if the samples were left at room temperature for a day or so. It was shown that the initial heat treatment led to a substantial fall in the longitudinal crystal thickness, followed by a rise in the latter, accompanied by the corresponding rise in modulus, after the samples had been maintained at constant length for the appropriate time [125].

It was also recognized that an alternative representation, equivalent to the Takayanagi model of Fig. 31, is to regard the high modulus LPE as a short fibre composite, in which the crystalline bridge sequences play the role of the short fibres, and the remaining crystal blocks and non-crystalline material act as the matrix. Although this has some formal mathematical relationship with the Barham and Arridge model, there is a completely different physical interpretation. In the model proposed by Gibson et al., drawing increases the *proportion* and length of the fibre phase, which is envisaged

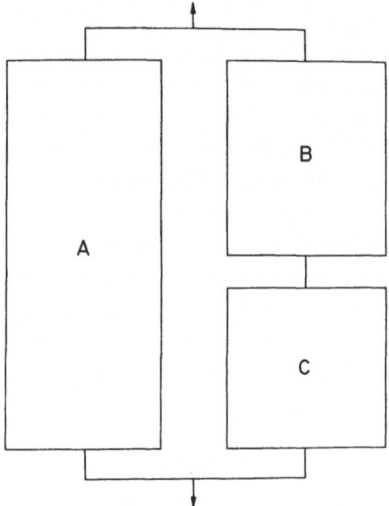

Fig. 31. The assumed mechanical connectivity of the fibre-phase (A) lamellar phase (B) and amorphous phase (C) of highly oriented LPE

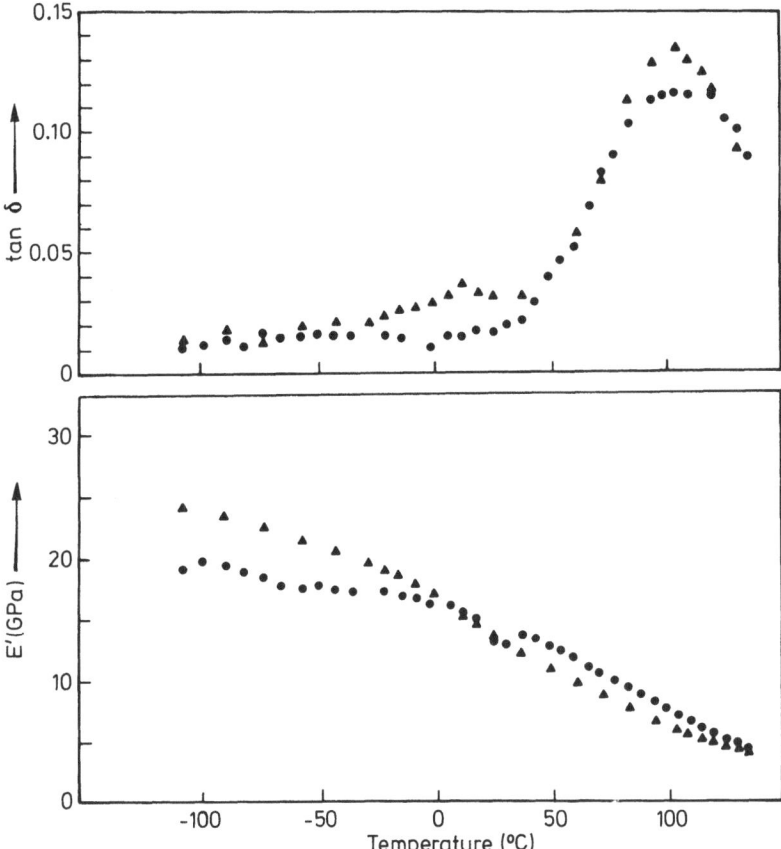

Fig. 32. Storage modulus, E′, and mechanical loss factor, tan δ, for GM61 polypropylene (M_w = 400,000). Frequency 5 Hz; λ = 15. ● As-drawn, ▲ after 60 min annealing at 135 °C

as clumps of crystalline regions of diameter ∼100 Å rather than the submacroscopic needle-like crystals, which were identified with fibrillar like entities. It was shown by Gibson et al. [127] that this fibre composite model could be used to embrace both shear and tensile relaxation behaviour in oriented LPE. In addition to a satisfactory prediction of the correct magnitudes of the ∼50 °C plateau moduli, the model also predicted the fall in the tensile storage modulus with temperature and the fall in tan δ_E with increasing crystal continuity. The essence of the model in this respect was to use the measured change in the isotropic shear modulus with temperature to predict the change in the shear lag factor and hence the reduced effectiveness of the crystalline bridges with increasing temperature due to the onset of the α relaxation.

In the case of the γ relaxation, it was concluded that there are two mechanisms for the change in modulus and hence for tan δ_E. The first is an increase in the efficiency of stress transfer (i.e. the shear lag factor) with falling temperature due to the quenching of molecular motions. These are predominantly if not entirely in the non-crystalline regions. Secondly, the quenching of these molecular motions also gives rise to an increase in the stiffness of the non-crystalline regions, which then contribute directly to the overall stiffness of the sample.

In a recent investigation, the five complex stiffnesses of higly oriented LPE produced by hydrostatic extrusion and die drawing have been determined by ultrasonic measurements [128]. Although the very high stiffnesses obtained at the highest draw ratios were attributed to increasing crystal continuity, the low temperature ultrasonic behaviour could be predicted to a good approximation by the reorienting unit aggregate model. This surprising result suggests that the overall orientation may still be the key parameter at low temperatures and high frequencies where there is no molecular mobility in the structure.

The dynamic mechanical behaviour of ultra high modulus polypropylene [129,130] is shown in Fig. 32. As in LPE, the modulus is temperature dependent, rising to a value of 25 GPa at −140 °C, which is rather more than half the value of 42 GPa obtained from crystal measurements. Although the α and γ relaxations of the isotropic polymer can be seen in the highly drawn material, the β-relaxation is undetectable. On annealing, the modulus at high temperatures is markedly reduced, and a β-relaxa-

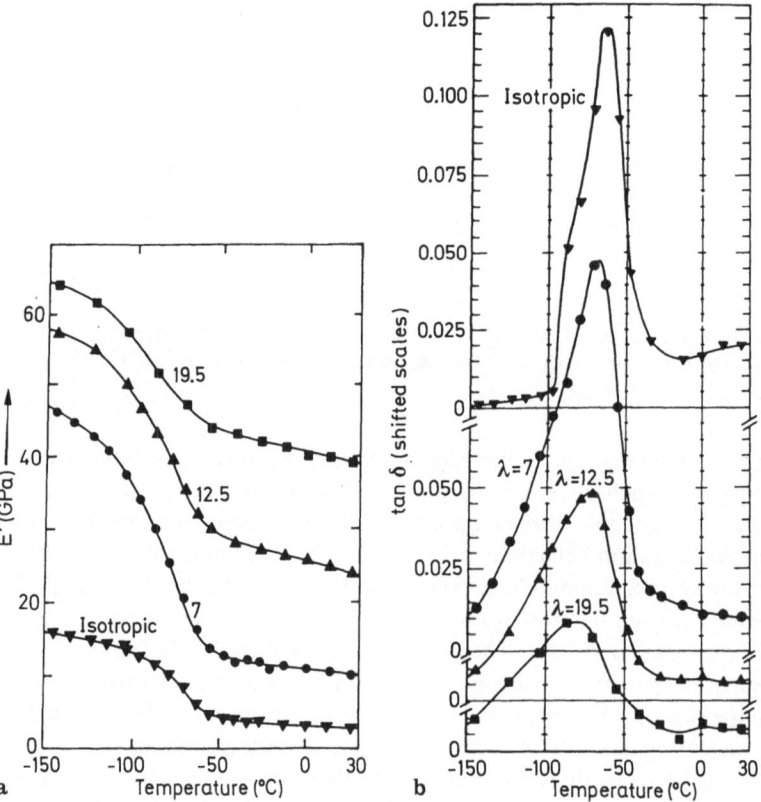

Fig. 33a. Storage modulus, E′ (at 5 Hz), as a function of temperature for drawn and isotropic POM Delrin 500) samples. Numbers on curves refer to deformation ratio; **b.** The mechanical loss factor, tan δ, corresponding to the data of **(a)**

tion is observed. This suggests that the high draw reduces the mobility of the non-crystalline regions, probably by the production of taut-tie molecules, which can be relaxed by annealing.

Dynamic mechanical results [131] for isotropic and drawn POM are shown in Fig. 33. There is again a large temperature dependence, and the modulus in the draw ratio 19.5 sample rises from 40 GPa to 65 GPa with decreasing temperature, which can be compared with a theoretical estimate of 106 GPa. Drawing reduces the magnitude of the γ-relaxation very appreciably, primarily due to the increased orientation of the non-crystalline regions.

5.2 Creep, Recovery and Stress Relaxation Behaviour

Because of the potential applications of high modulus polyethylene for reinforcement of brittle matrices, it was soon recognized that the creep and recovery behaviour should be studied in some detail.

In the first of a series of publications, Wilding and Ward [132] compared the creep and recovery behaviour of comparatively low molecular weight drawn monofilaments (Rigidex 50 Grade, $\overline{M}_n = 6,180$; $\overline{M}_w = 101,450$) of draw ratios 10, 20 and 30 with a higher molecular weight monofilament (Rigidex H020 Grade, $\overline{M}_n = 33,000$; $\overline{M}_w = 312,000$) of draw ratio 20. As shown in Fig. 34, the high draw reduces the overall creep compliance, and also makes for a more nearly linear response. A more instructive way of examining the data is, however, to follow the example of Sherby and Dorn,

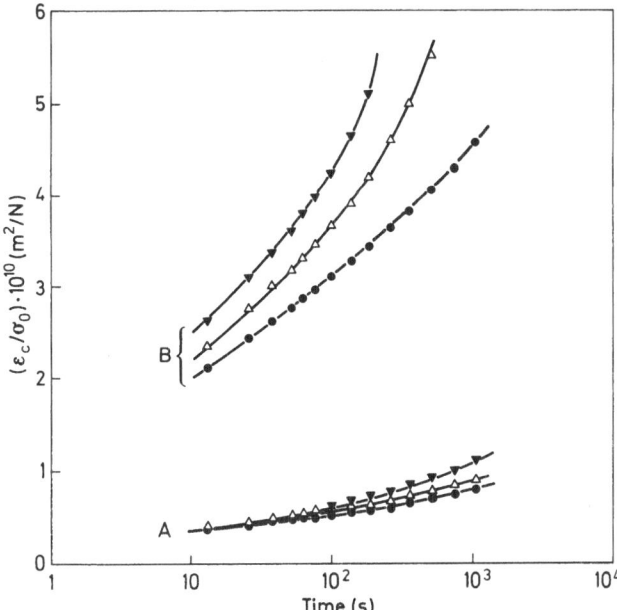

Fig. 34. Creep compliance (ε_c/σ_0) of drawn Rigidex 50 at 0.1 (●), 0.15 (△) and 0.2 GPa (▼) applied stress (σ_0) as a function of time; (———), are least squares fits to the mechanical model. A, $\lambda = 30$; B, $\lambda = 10$

and plot the creep strain rate $\dot{\varepsilon}_c$ (on a logarithmic scale) as a function of total creep strain ε_c.

It was found that increasing draw ratio in the low molecular weight sample gives a reduction in the creep rate at any given strain, and at all draw ratios the strain rate eventually reaches a constant creep rate. The observation of a final constant creep rate is extremely important, because it suggests that the samples effectively achieve a constant structure where creep is controlled by a defined flow process. From a practical standpoint the existence of a constant creep process is very unsatisfactory because there is no reason to suppose that failure will not eventually occur.

Wilding and Ward showed that the creep and recovery behaviour of the low molecular weight samples could be represented to a good approximation by the model representation shown in Fig. 35(b), which consists of a Maxwell and Voigt element in series, on the basis that the parameters E_m, E_v, η_m and η_v are dependent on the stress level. Data for the creep response ε_c of the samples under discussion at a constant applied stress σ_0 were therefore fitted to the equation

$$\varepsilon_c = \frac{\sigma_0}{E_m} + \frac{\sigma_0}{E_v}(1 - e^{-t/\tau}) + \frac{\sigma_0 t}{\eta_m} \tag{9}$$

where E_m, E_v, η_m and η_v correspond to the springs and dashpots of the elements in Fig. 35(b), and τ is the retardation time of the Voigt element, given by $\tau = \eta_v/E_v$. The full lines in Fig. 36 are the creep curves calculated from a least squares procedure to

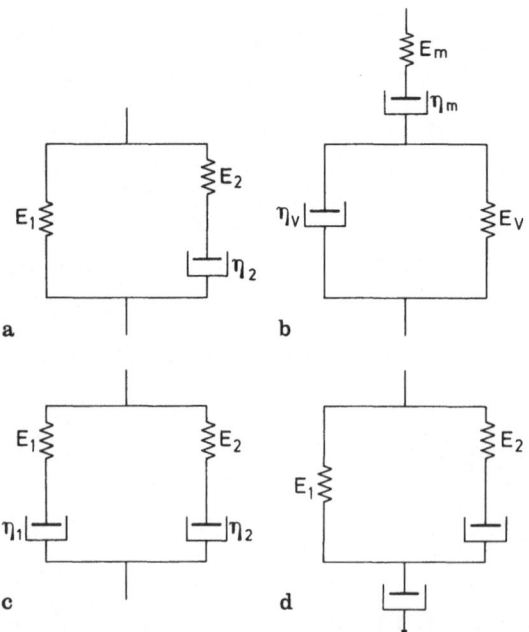

Fig. 35a–c. Mechanical models of creep and recovery. (**d**) Modified model for viscoelastic creep

obtain optimum values for the parameters E_m, E_v, η_m and η_v, and the dotted lines are the predicted recovery curves. Although the latter are not identical in shape to the experimentally observed recovery curves (as might be anticipated, a spectrum of retardation times would be required) the overall levels of recovery strain at long times are predicted very well. It is of particular importance that this representation enables a separation between the recoverable creep (the Voigt element) and non-recoverable creep (the Maxwell dashpot). Moreover, there are two further important facts which emerge from the detailed modelling. First, the values of E_m and E_v increase with draw ratio, but are comparatively insensitive to stress level. This is consistent with the fact that the short term response (e.g. the 10 sec isochronal modulus) depends only on draw ratio. Secondly, η_v is much less dependent on stress level and is very much smaller in magnitude than η_m. Moreover the stress dependence of η_m was shown to correspond to that expected for a thermally activated process as proposed by Eyring. This latter observation forms the basis for a more profound observation, namely that the plateau creep behaviour corresponds closely to a creep rate $\dot{\varepsilon}_p$ represented by the Eyring equation

$$\dot{\varepsilon}_p = \dot{\varepsilon}_0 \exp \frac{-\Delta H}{kT} \sinh \frac{\sigma v}{kT} \qquad (10)$$

where ΔH represents the activation energy and
v the activation volume for the thermally activated process.

The broader implications of this result will be discussed later. At this stage, it is appropriate to consider the appropriate development of the modelling which enables us to represent the continuous fall in creep rate observed at low stress levels in higher molecular weight samples [132]. In this case, it appears that there is no plateau creep rate, which is of considerable importance in terms of practical applications because the creep rate falls to zero and the strain is totally recovered when the load is removed. For such materials an apparent "critical stress" can be defined, below which the samples will creep to a limiting strain of $\sim 1\%$. In subsequent investigations [133-135], Wilding and Ward showed that this type of behaviour was observed for ethylene-hexene-1 copolymer and for drawn samples prepared by γ-irradiation of isotropic low molecular weight polymer prior to drawing, as well as for the high molecular weight H020 drawn polymer discussed above. A satisfactory representation of this behaviour is given by the mechanical model shown in Fig. 35(c), where both η_1 and η_2 represent thermally activated processes. The behaviour of the dashpot η_2 is similar to the dashpot η_m in the earlier Fig. 35(b), i.e. this governs the response at high stresses. A key assumption is that there is a very marked difference between the nature of the two activated processes expressed by η_1 and η_2. It is assumed that the dashpot η_1 has a larger activation volume and a smaller pre-exponential factor $\dot{\varepsilon}_0$ than the corresponding quantities for the dashpot η_2. The behaviour given by this representation can then be shown schematically in Fig. 37(a), where the contributions of the two processes are seen together with the total response. In Fig. 37(a) the stress is plotted as a function of log strain rate, whereas in Fig. 37(b) we show log plateau creep rate $\dot{\varepsilon}_p$ as a function of applied stress. These figures emphasise the kinship between creep and yield. The proposed representation is familiar in the context of yield behaviour, having been proposed by Roetling [137], Bauwens [138] and others. The large activation volume for

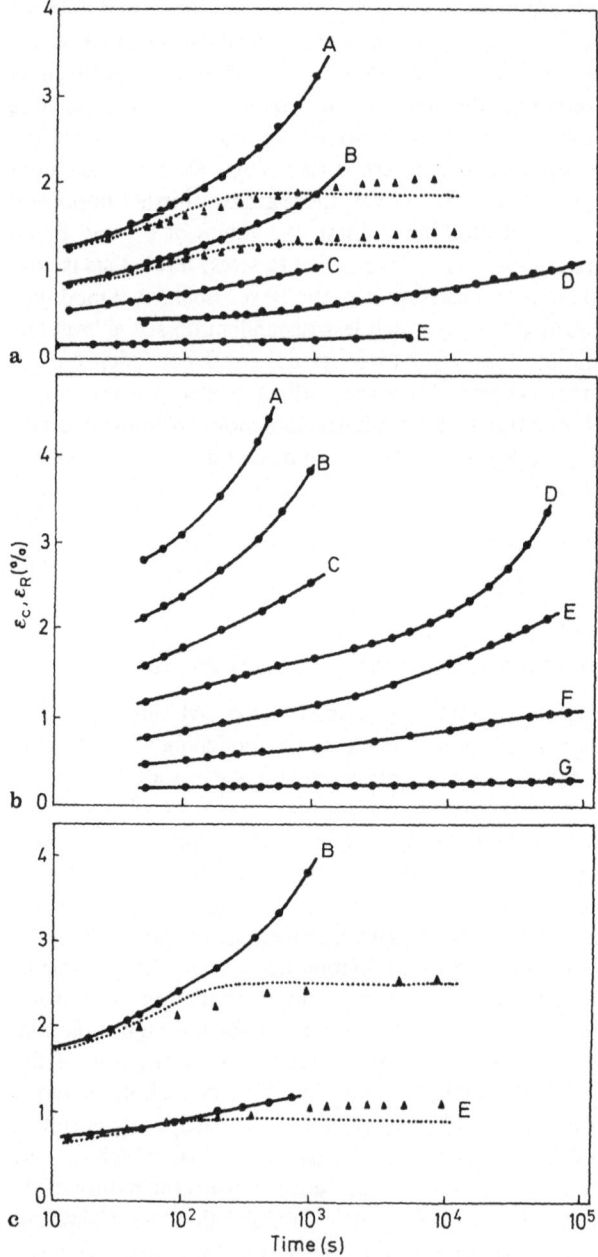

Fig. 36a–c. ε_c (●) and recovery strain ε_R (▲) vs. time for (a) Rigidex 50, $\lambda = 20$ A, 0.2; B, 0.15; C, 0.1; D, 0.05; E, 0.25 GPa and (b) and (c) H020-548, $\lambda = 20$. A, 0.5; B, 0.4; C, 0.3; D, 02; E, 0.15; F, 0.1; G, 0.05 GPa; (————), in (a) and (c) are least square fits to the mechanical model, and the dotted lines are the respective recovery curves; (————), in (b) are visual fits to the data

the dashpot η_1, means that σv_1, is generally large compared with kT. We can therefore write

$$\sigma = \frac{2.3 \text{ kT}}{v_1}\left[\log \dot{\varepsilon}_p - \log \frac{[\dot{\varepsilon}_0]_1}{2}\right] + \frac{\Delta H_1}{kT} + \frac{kT}{v_2}\sinh^{-1}\left[\frac{\dot{\varepsilon}_p}{[\dot{\varepsilon}_0]_2}\exp\frac{\Delta H_2}{kT}\right]$$

(11)

where the subscripts identify the quantities similar to those of Eq. (10) referring to the two processes.

The 'critical stress' can now be seen to be essentially an experimental limitation. The smallest strain reading on our present creep apparatus is $\sim 5 \times 10^{-5}$. The anticipated plateau strain rate of 10^{-12} s^{-1} for a typical sample with a 'critical stress' of ~ 0.2 GPa would therefore only produce a measurable creep response after 500 days, which is on the limit of the time scale of present creep tests at Leeds University.

Fig. 38 shows a set of curves fitted to plateau creep data on the basis of the two-process model, and the corresponding activation parameters are shown in Table 1. The activation volume for the η_2 dashpot is in the range of 100 Å^3, and, as remarked by Wilding and Ward, this comparatively small activation volume could be consistent with a slip process in the crystalline regions. Support for this hypothesis comes from examination of the temperature dependence of this process. Results for selected samples show that there is a good linear relationship between $\log \dot{\varepsilon}_p$ and $1/T$. At high stress levels the behaviour corresponds to that of an apparent single activated process, with

$$\Delta H_{\text{eff}} = \frac{v_2\Delta H_1}{v_1 + v_2} + \frac{v_1\Delta H_2}{v_1 + v_2}$$

(12)

Since $v_2 \ll v_1$, $\Delta H_{\text{eff}} \approx \Delta H_2$.

Values for $\Delta H_{\text{eff}} \sim 30$ kcal/mol were obtained, which are in the range for the α-relaxation process in polyethylene, when studied by dynamic mechanical measurements.

The identification of process 2 with the crystalline regions is consistent with the intercrystalline bridge model for the low-strain mechanical behaviour. In this model the crystalline continuity is provided by crystalline bridges, and at high temperatures the effectiveness of these bridges is reduced by the α-relaxation process, which permits shear of chains past each other. The actual movement of the chains through the crystalline regions must involve the movement of a defect, and the Reneker defect is one possibility. There is a distinct tendency for the activation volume v_2 to fall with increasing draw ratio, suggesting that the activated event becomes more localised as the structure reaches a higher degree of alignment and perfection.

It has already been noted that Process 1 has a much larger activation volume than Process 2, and the value of ~ 500 Å^3 is close to values obtained for amorphous polymers by yield stress measurements. It is therefore attractive to speculate that this process might relate to the molecular network acting in parallel with the crystalline bridges and crystalline material. This general proposition is supported in a direct intuitive way by the observation that the contribution of Process 2 increases with polymer molecular weight and with cross-linking by γ-irradiation prior to drawing,

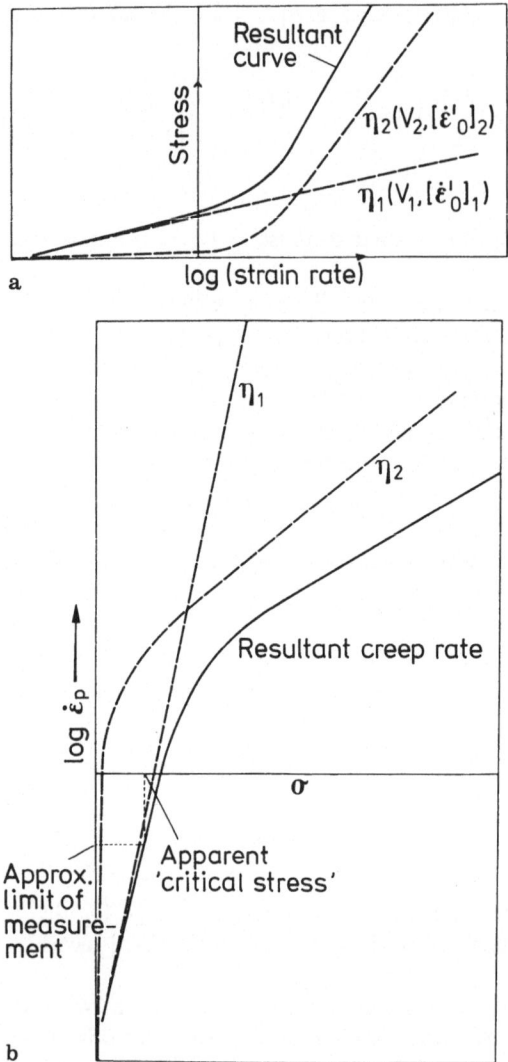

Fig. 37a and b. Schematic representation of plastic flow for the 2-process model; (**a**) yield stress vs. applied strain rate; (**b**) creep strain rate $\dot{\varepsilon}_p$ vs. applied stress

both of which would increase the number of network junction points. There is also the possibility that there is a small entropic component of 10–20 MPa to be added to the flow stress, corresponding to the shrinkage force which also plays an important role in the thermal expansion behaviour (see Sect. 6.4). The exact nature of Process 1 and the modelling in this respect must therefore be regarded as open.

Although there are no data for the creep behaviour of solution-spun or gel spun fibres of a comparable nature to those for the melt spun and drawn materials, a few results can be presented.

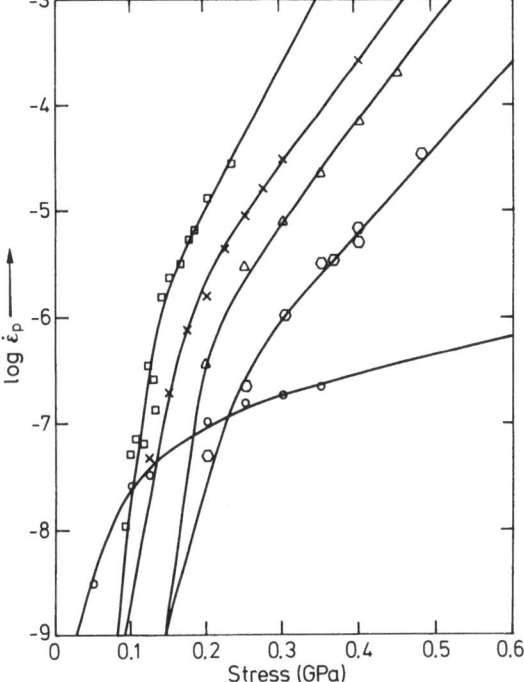

Fig. 38. Curves fitted to plateau creep data on the basis of the two-process model with no constraint on parameters: (□) Rigidex 50, λ = 20; (X) 006.60, λ = 20; (△) γ-irradiated Rigidex 50, λ = 20; (○) H020, λ = 20; (o) Hostalen GUR, solution-spun fibre

Table 3.

Sample	V_1 (Å^3)	$[\dot{\varepsilon}_0']_1$ (s^{-1})	V_2 (Å^3)	$[\dot{\varepsilon}_0']_2$, (s^{-1})
Rigidex 50, λ = 20	460	2.2×10^{-8}	77	1×10^{-6}
006-60, λ = 10	580	1.2×10^{-13}	370	2.8×10^{-6}
006-60, λ = 30	370	2.4×10^{-13}	123	1.5×10^{-6}
006-60, λ = 30	510	2.4×10^{-17}	78	6×10^{-7}
H020-54P, λ = 10	...	...		
H020-54P, λ = 20	260	1.1×10^{-13}	106	3.1×10^{-7}
γ-irradiated Rigidex 50, λ = 20	490	1.8×10^{-17}	110	9×10^{-7}
002-55, λ = 20	580	5.2×10^{-21}	86	3.8×10^{-7}
Hostalen GUR solution-spun fiber	307	1.5×10^{-10}	15	1×10^{-7}

Figure 38 and Table 3 include a set of results for a solution spun fibre, prepared by spinning GUR Hostalen grade polymer from n-decane (Prof. A. J. Pennings, State University of Groningen, The Netherlands). It can be seen that, although the creep rates at high stress are much lower than those for melt spun and drawn fibres, there is a much lower critical stress. It also appears that the activation volume for the high stress process (~ 15 Å^3) is very much smaller than those for the other fibres. There is clearly a major difference in behaviour due to the major differences in structure between the two types of material.

The improvements in creep behaviour produced by crosslinking with γ-irradiation prior to drawing have already been discussed. The effects of γ-irradiation on the creep and recovery behaviour of drawn monofilaments of Rigidex 50 were also studied by Wilding and Ward [134]. It was found that there was a substantial increase in the creep rate for a given stress level, although the recovery behaviour at long times was almost complete, as would be expected for the production of a permanent molecular network by crosslinking. These results were extremely disappointing, and in view of the importance of achieving a more positive result, this area has been examined again recently by Woods, Busfield and Ward [139].

It has now been shown that it is possible to obtain dramatic improvements in the creep behaviour of polyethylene fibres either by electron irradiation in vacuum to a dose of 50 MRad or by irradiation in acetylene to a smaller dose of 20 MRad, both of which give a substantial degree of crosslinking. Figure 39 summarises the key effects on the stress-strain curves at a range of strain rates and temperature. It can be seen that the strain rate dependence is much reduced and that a much more nearly linear stress strain curve is seen in all cases i.e. the creep behaviour is quenched, so that the material passes through a ductile-brittle transition. It is particularly important to note the improved retention of properties at higher temperatures, a result which is brought out very clearly by the comparison of modulus and strength between the irradiated and control samples shown in Fig. 40. Finally, a comparison of creep and yield behaviour combined in the manner of Fig. 37(a) and (b) above, also makes the same point, and suggests a 'critical stress' of ~ 0.25 GPa for the crosslinked fibres (Fig. 41).

The equivalence of creep and constant strain rate tests has already been discussed and shown to hold to a good degree of approximation. Recent fundamental studies have examined the validity of the two process-model of Fig. 35(c) for describing stress relaxation behaviour [136]. The problem of numerical evaluation of this model was circumvented by noting that at low stresses this model can be modified to that shown in Fig. 35(d). Although it is not identical in numerical terms with regard to the dashpots, the springs do have the same significance as in the two process-model. Hence it was possible to parameterise the data using the simpler model of Fig. 35(d) to find E_1 and E_2, and then to use the two process model with the dashpot parameters already determined (Table 3 above) to predict stress-relaxation data. Figure 42 shows the excellent fits to the creep data using the model of Fig. 35(d), and Fig. 43 shows the predicted stress relaxation curve (full line) obtained in this way. Bearing in mind the assumptions inherent in the calculation of the predicted curve, and the difficulty of obtaining reliable values for the activation parameters of the low stress Process 1, the agreement is good. Although it is not clear from Fig. 42, at very long times the predicted curve would display a second relaxation during which the stress falls to

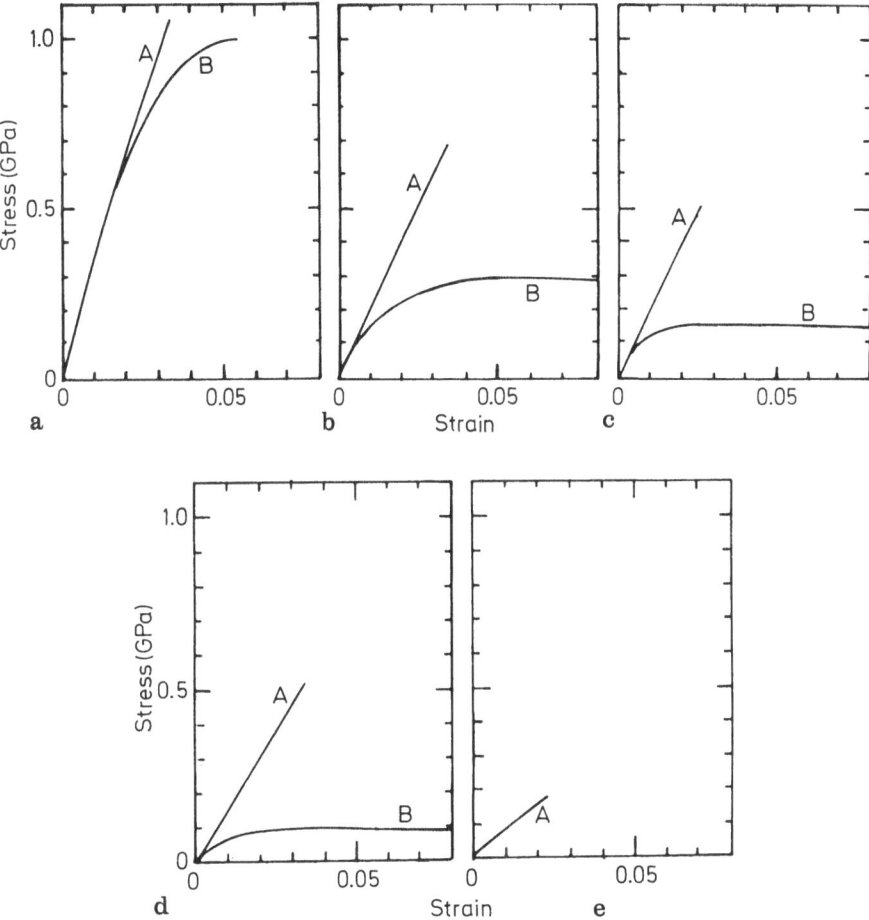

Fig. 39a–e. Stress-strain curves for (A) electron irradiated fibres (20 MRad in acetylene) (B) unirradiated fibres. Strain rate (s^{-1}): 8.3×10^{-3} (**a**); 8.3×10^{-5} (**b**), (**d**), (**e**); 2.1×10^{-6} (**c**). Temperature (°C): 23 (**a**), (**b**), (**c**); 70 (**d**), 130 (**e**)

zero. That this does not occur within the experimental time scale is consistent with the prediction of the two process model.

Recent work [140] has shown that the creep and recovery behaviour of ultra high modulus polypropylenes is very similar to that of LPE. Again the Sherby-Dorn plots form a good entry to the detailed examination of the creep response. Plateau creep behaviour similar to that of LPE has been observed, and the high stress process correlates well with the α-relaxation process in terms of its activation energy.

5.3 Tensile Strength

The major feature of the tensile strength data for highly drawn polyethylene is the markedly larger values of strengths recorded for the fibres prepared from dilute solu-

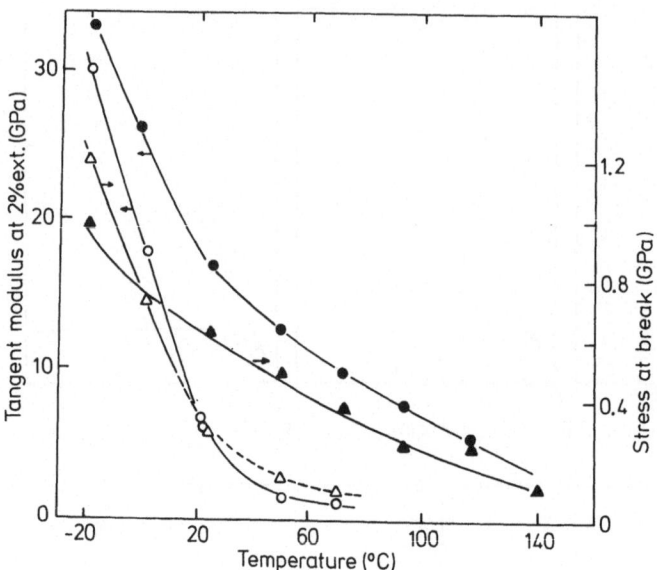

Fig. 40. Tangent modulus ○ ● and failure stress △ ▲ as a function of temperature, open symbols untreated, full symbols irradiated: 20 MRad in acetylene. ------- yield ——— fracture. Strain rate 8.3×10^{-5} s^{-1}

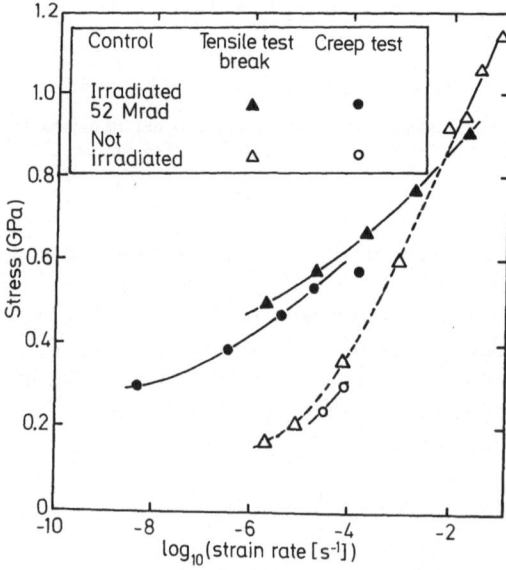

Fig. 41. Stress as a function of log (strain rate). Constant strain rate tests △ ▲ (------- yield, ——— fracture), Creep tests ○ ●. Open symbols untreated, full symbols 50 MRad in vacuum

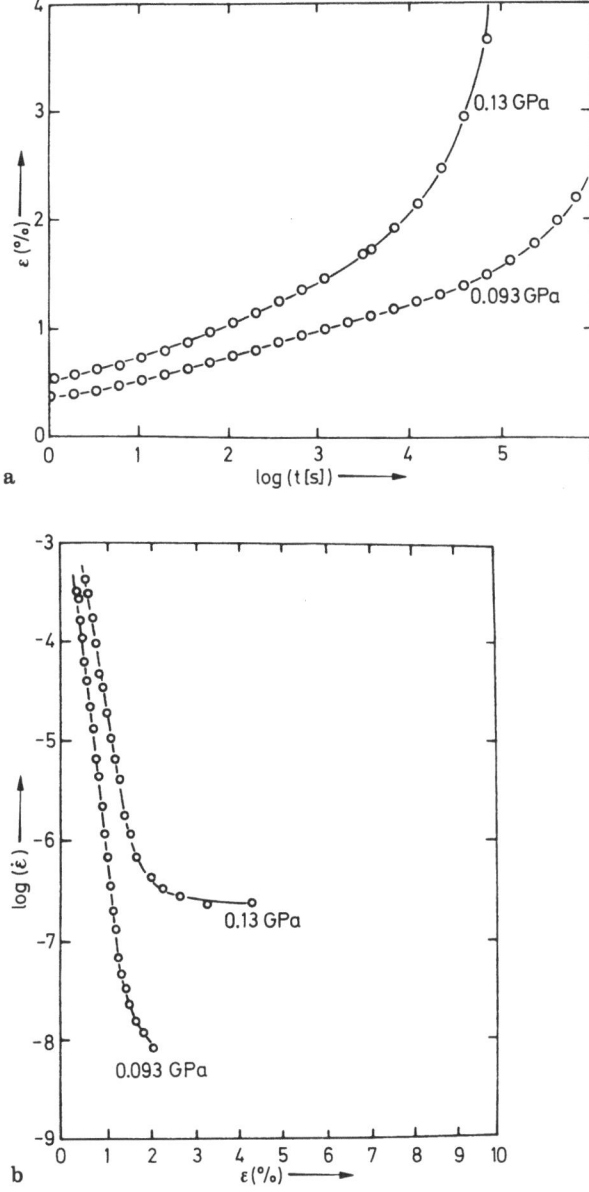

Fig. 42a and b. Creep against log time curves for Rigidex 50, λ = 20 at 20 °C. **(b)** Log $\dot{\varepsilon}$ against $\dot{\varepsilon}$ plots for Rigidex 50, λ = 20 at 20 °C

tion or gel spinning and drawing, compared with those prepared by melt spinning and drawing. In Fig. 2 results were shown for the dilute solution fibres where tensile strengths as high as 3.5 GPa have been obtained. It is considered that key reason for the differences between fibres from the different types of preparation routes lies in the polymer molecular weight differences. This conclusion follows from the expe-

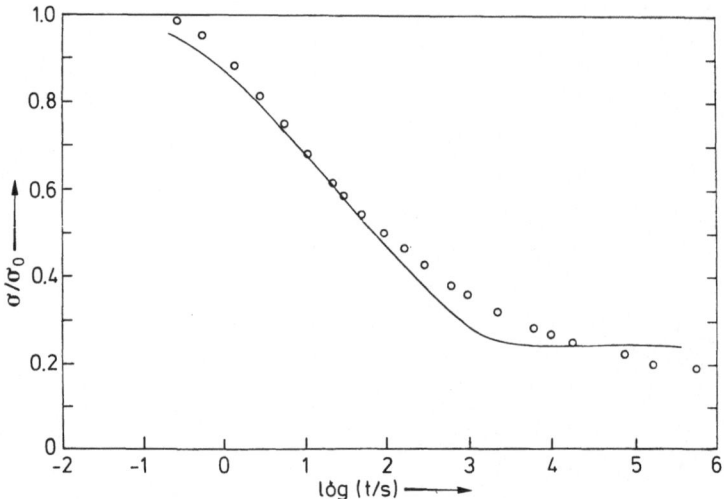

Fig. 43. Stress-relaxation data for Rigidex 50, $\lambda = 20$ at 20 °C

rience of fibre technology in other polymers such as polyesters and nylon, including Flory's early prediction [141] of the dependence of tensile strength on number average molecular weight.

Smith and Lemstra [142] have shown a correlation between tensile strengths σ_T for a given level of Young's modulus E and polymer molecular weight for a series of gel spun and hot drawn fibres. The results are shown in Fig. 44 together with solid lines calculated on the basis of a linear dependence between $\log \sigma_T$ and $\log E$, which can be written as

$$\sigma_T = mE^n \tag{13}$$

Although Smith et al. [143] and Wu and Black [144] have given evidence for a similar dependence of strength on molecular weight for melt spun and drawn fibres, the results would not appear to be so clear cut. For example Smith et al. show tensile strengths approaching 1 GPa at a Young's modulus of 50 GPa for melt spun and drawn filaments of $\overline{M}_w = 280,000$; $\overline{M}_n = 18,000$. Wu and Black in their study of the effect of molecular weight and molecular weight distribution on strength quote a value of tensile strength of 1.15 GPa for a modulus of 49 GPa for polymer with $\overline{M}_w = 115,000$; $\overline{M}_n = 28,000$ and 1.4 GPa for a modulus of 48 GPa for polymer with $\overline{M}_w = 84,000$; $\overline{M}_n = 25,000$.

As pointed out by Cansfield et al. [145] the tensile behaviour of ultra high modulus polyethylene is very dependent on strain rate. A tensile strength of ~ 1.5 GPa can therefore be obtained at high strain rates (similar to those adopted by Wu and Black) even for very low molecular weight polymers.

With these reservations in mind regarding the exact absolute magnitude of tensile strength figures from different workers, it is nevertheless appropriate to attempt to summarise their main findings. For melt spun and drawn fibres, Wu and Black [144] concluded that the number average molecular weight was the predominant factor in determining tensile strength, and could find no significantly positive trends with $\overline{M}_w$ or polydispersity. Smith, Lemstra and Pijpers [143], on the other hand, concluded that a reduction in polydispersity from 8 to 1.1 gave an increase in strength by a factor of about two for a fibre with $\overline{M}_w \sim 10^5$. In retrospect, this result could equally well have been interpreted as arising from an increase in $\overline{M}_n$. This alternative explanation is consistent with the results of Smith et al. [143], which show that two fibre samples with equivalent Young's moduli of 50 GPa have closely similar tensile strengths for molecular weight characteristics of $\overline{M}_n = 120{,}000$, $\overline{M}_w = 800{,}000$ and $\overline{M}_n = 110{,}000$ and $\overline{M}_w = 120{,}000$ respectively.

Smith et al. [143] made a comparison between solid state extruded materials, melt spun and drawn fibres, gel spun and drawn fibres and surface growth fibres. It was concluded that the method of preparation was immaterial with regard to final mechanical properties. They adopted the approach of examining the effect of process and molecular weight on the ratio between the tensile strength and the modulus. It was clear that at constant molecular weight this ratio was not affected by such process variables as draw temperature and draw rate. They did however suggest that the lower comparative strengths of solid state extruded material compared with melt spun and drawn fibres might be due to the effect of specimen size on strength, as would occur if the latter is flaw-dominated. Significantly, however, Smith et al. observed no differences in strengths between melt spun and drawn fibres and gel-spun and drawn fibres, where both sets of fibres were produced from polymer with $\overline{M}_n = 18{,}000$, $\overline{M}_w = 280{,}000$. Their most striking results were obtained for melt spun and drawn fibres (Fig. 44) and for a wide range of molecular weight polymers (Fig. 45). These results suggest strong molecular weight effects with regard to both $\overline{M}_n$ and $\overline{M}_w$. It is, however, necessary to inject a note of caution. It is by no means certain that the process was

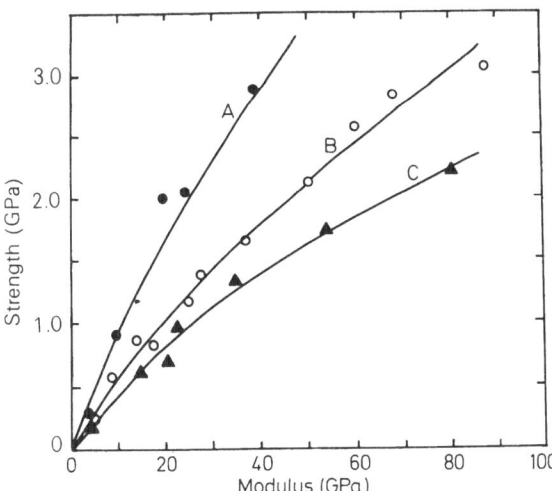

Fig. 44. Tensile strength vs. Young's modulus of solution spun/drawn polyethylene filaments having various molecular weights: $A - \overline{M}_w = 4 \times 10^6$, $B - \overline{M}_w = 1.5 \times 10^6$, and $C - \overline{M}_w = 8 \times 10^5$ kg per kmol. Solid lines calculated according to Eq. (13). With permission of the publishers John Wiley & Sons. Inc. (C)

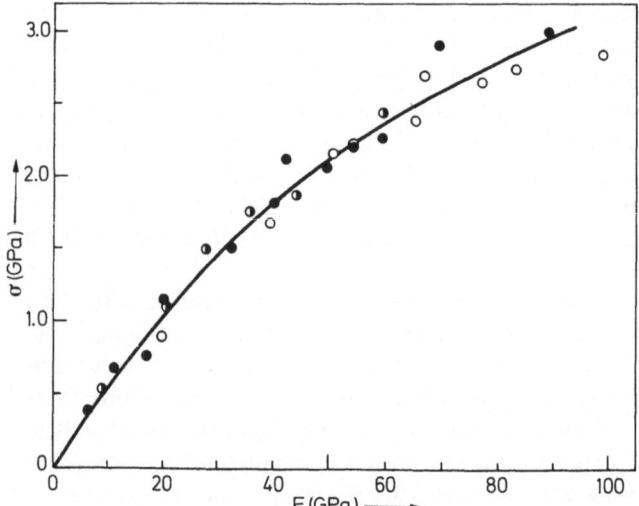

Fig. 45. Tensile strength vs. Young's modulus of high molecular weight polyethylene sample 9 ($\overline{M}_n$ = 200×10^3; $\overline{M}_w$ = $1,500 \times 10^3$). ○ — surface grown (Zwijnenburg and Pennings); ● — solution-spun/drawn wet; ○ — solution-spun/drawn dried. With permission of the publishers John Wiley & Sons. Inc. (C)

optimized for each polymer examined. In this regard, it is somewhat disconcerting to note that Cansfield et al. report tensile strengths well in excess of 1 GPa for Alathon 7050 polymer ($\overline{M}_n$ = 22,000, $\overline{M}_w$ = 60,000) drawn to a modulus of only 40 GPa. This result is clearly at variance with the data shown in Fig. 45 and 46.

Another result which may conflict with the results of Smith et al. [143], is the observation of a clear diameter effect in gel spun and drawn fibres by Pennings and his

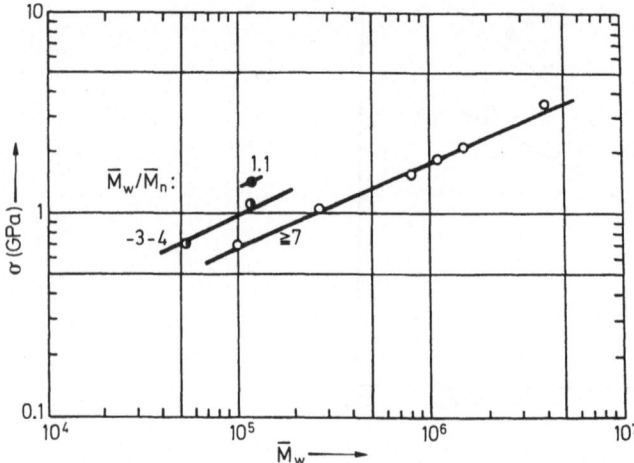

Fig. 46. Tensile strength of highly oriented polyethylene at Young's modulus = 50 GPa versus weight average molecular weight for various polydispersities (indicated in graph). With permission of the publishers John Wiley & Sons. Inc. (C)

collaborators [146]. As can be seen in Fig. 47, the tensile strength increases from 2 GPa to 4.5 GPa as the fibre diameter is reduced from 0.12 mm to 0.02 mm (20 μ). Following the Griffith relationship, the fibre strength σ_B is given by

$$\sigma_B = \left[\frac{G_c E}{Y^2 c}\right]^{1/2} \qquad (14)$$

where G_c is the strain energy release rate
　　　E　Young's modulus
　　　c　surface crack length
and 　 Y^2 a geometrical factor
The results of Fig. 46 suggest that

$$\sigma_B \alpha \frac{1}{d^{1/2}} \qquad (15)$$

as shown in Fig. 48, where $1/\sigma_B$ is plotted versus $d^{1/2}$.

If the straight line fit is extrapolated to zero diameter a maximum value of 26 GPa is obtained for the fibre strength. Pennings et al. consider that this is the strength in the absence of surface flaws, and point out that it compares well with a theoretical estimate of 25 GPa, obtained from Morse potential calculations of the C—C bond strength.

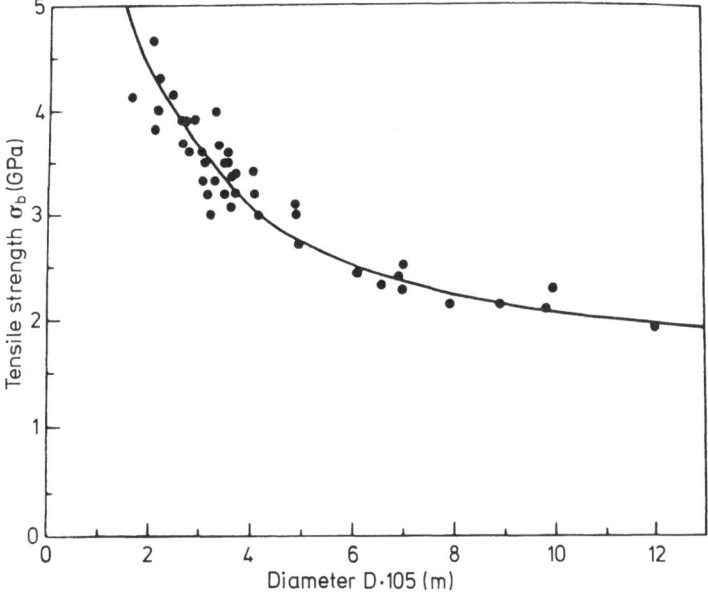

Fig. 47. Tensile strength at break σ_b of fully oriented polyethylene filaments as a function of fibre diameter. With permission of the publishers Pergamon Press Inc. (C)

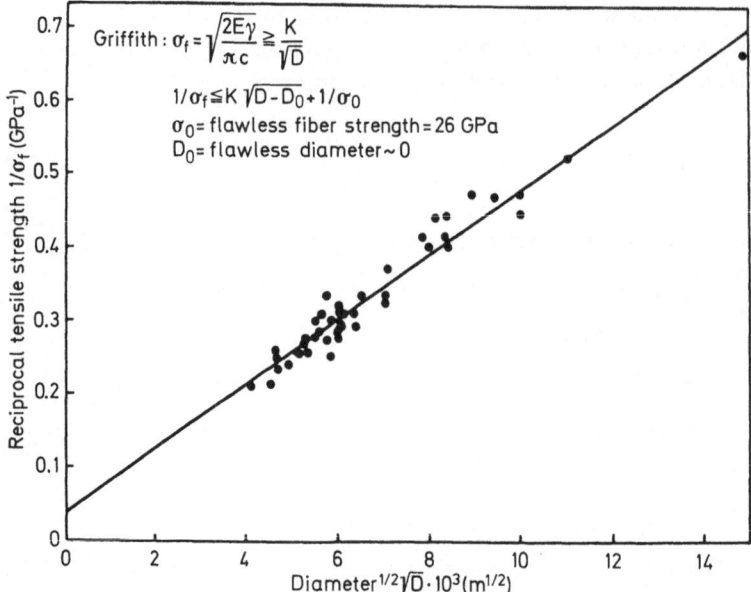

Fig. 48. Linear strength-diameter relationship as observed for ultraoriented high molecular weight polyethylene fibres. With permission of the publishers Pergamon Press Inc. (C)

5.4 Tearing Behaviour

The tearing behaviour of oriented polyethylene films is of especial interest because of the potential application of such materials due to their very good gas barrier properties and chemical resistance. Clements and Ward [147] carried out trouser leg tearing tests on a wide range of polyethylene sheets, to examine the influence of draw ratio, draw temperature, molecular weight and copolymerisation. In all cases the values of the fracture surface energy G_c were in the range 10^2 to 10^4 Jm^{-2}. Large differences were observed between low and high molecular weight grades. At high molecular weight the energy for crack propagation parallel to the direction of orientation fell by a factor of approximately two as the draw ratio was increased from 10 to 20. At low and medium molecular weights, increasing the draw ratio above a value of 10 had no significant effect on the fracture surface energy. An interesting observation was that the values of G_c measured at low temperatures were very similar for all the oriented materials at a comparable draw ratio.

Observation of the fracture surfaces by scanning electron microscopy suggested a correlation between the nature of the fracture surface and the value of G_c. It appeared that the changes in G_c correlated with the *lateral* morphology. At medium and low molecular weights the fibrillar nature is complete at comparatively low draw ratios, and further drawing only produces changes within fibrils, which do not affect the tearing energy. High molecular weight material on the other hand only shows fibrillation at high draw ratios.

It was also shown that the draw temperature affects G_c, and the best prospects for producing oriented PE sheets with adequate tearing resistance are obtained by drawing high molecular weight grades at high draw temperatures. It was concluded that it is important to retain a molecular network which embraces the whole structure and prevents fibrillation. It appears that at low temperatures the reduction in mobility is sufficient to ensure that no molecular pull-out occurs between fibrils even in low molecular weight material.

6 Thermal Properties

6.1 Thermal Conductivity

The thermal conductivity of hydrostatically extruded LPE samples has been measured parallel $(K_{||})$ and perpendicular $(K_{\perp})$ to the extrusion direction over a wide temperature range [148]. Figure 49 illustrates the effect of extrusion ratio on $K_{||}$ and $K_{\perp}$ at 100 K. There is a marked increase in $K_{||}$ with increasing draw ratio. In very simple terms this could be regarded as partly due to an increase in the phonon velocity due to the increase in modulus, and partly due to an increase in the mean free path due to higher crystal lengths. In fact, it has been shown that $K_{||}$ at 100 K can be related to the degree of crystal continuity, which on a simple Takayanagi model is quantified in terms of $\chi p(2 - p)$, as shown in the discussion of dynamic mechanical behaviour. Gibson

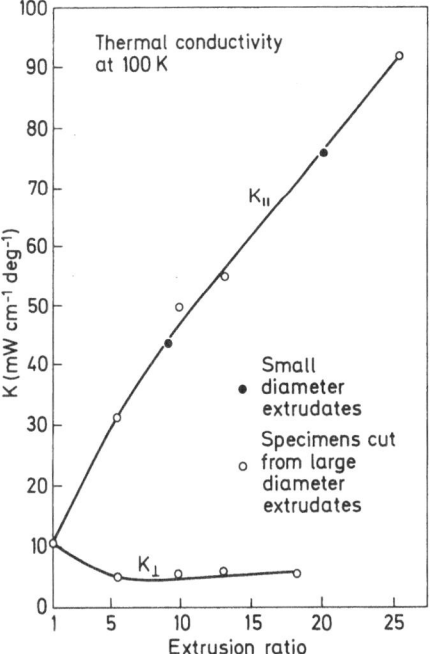

Fig. 49. Variation of thermal conductivity at 100 °K for various values of extrusion ratio λ; (●) small-diameter extrudates (○) specimens cut from large-diameter extrudates

et al. [149] suggested a direct proportionality between $K_{||}/K_{c||} = E/E_c$ and $\chi p(2-p)$ where $K_{c||}$ and E_c are the conductivity and modulus of the crystalline regions parallel to the extrusion direction, and the other symbols have been defined above; indeed there is an exactly linear correlation between $K_{||}$ at $-173\,°C$ (100 K) and the plateau modulus at $-50\,°C$.

$K_\perp$ is very little affected by draw ratio so that $K_{||}/K_\perp$ at 100 K reaches ~ 20 at the highest deformation ratio, where the value of $K_{||} \sim 90$ mW cm^{-1}deg^{-1} approximates to that of stainless steel. Figure 50 shows the variation of conductivity over a wide temperature range and again illustrates the significant increase in $K_{||}$ with deformation ratio. Unlike stainless steel, where K increases to ~ 150 mW cm^{-1}deg^{-1} at 300 K, the value of $K_{||}$ for LPE materials remains virtually constant between 100–300 K. In contrast, $K_\perp$ steadily decreases over the same temperature range.

6.2 Shrinkage and Shrinkage Force

Measurements of irreversible shrinkage and shrinkage force have been undertaken on oriented polyethylene and polypropylene over a wide range of temperatures [150]. In the case of polyethylene, the behaviour was studied as a function of draw ratio and molecular weight for several linear polymers and several copolymers. A remarkable result is that complete retraction was observed on heating above the polymer melting point, which suggests that at least part of the molecular network present in the isotropic state retains its identity during drawing. At lower temperatures the shrinkage is much decreased in high draw material compared with low draw material. The

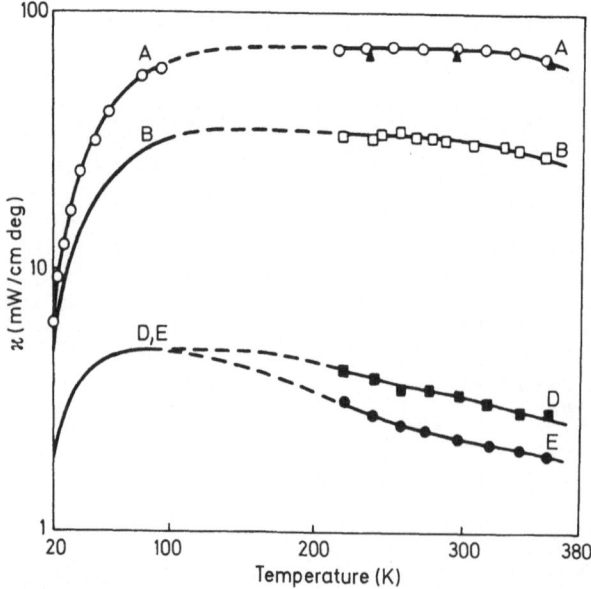

Fig. 50. Temperature dependence of thermal conductivity. The various curves and points represent: A, $K_{||}$ for $\lambda = 18$; B, $K_{||}$ for $\lambda = 5.4$; D, $K_\perp$ for $\lambda = 5.4$; E, $K_\perp$ for $\lambda = 18$

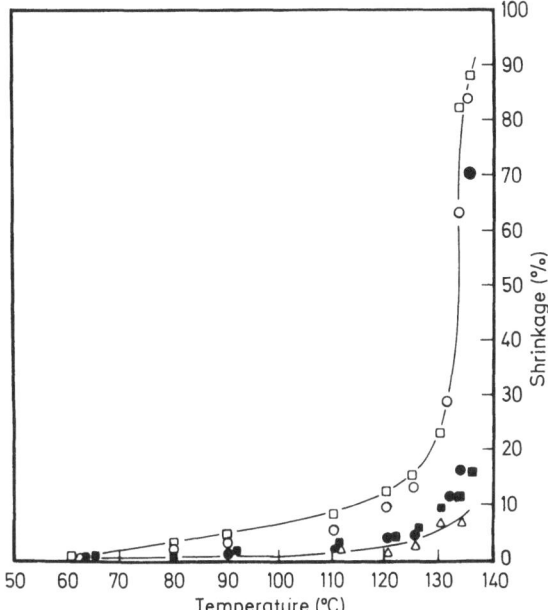

Fig. 51. Shrinkage results as a function of temperature R50: ($\bigcirc$) $\lambda = 11$, ($\bullet$) $\lambda = 20$, ($\triangle$) $\lambda = 25$ to 30; H020-54P: ($\square$) $\lambda = 11$, ($\blacksquare$) $\lambda = 20$. The draw temperature was 75 °C

results shown in Fig. 51 indicate that shrinkage relates to draw ratio, and not to molecular weight. The increasing stability with increasing draw ratio is attributed to the increasing degree of crystal continuity i.e. the crystalline bridges, and it has been proposed that shrinkage requires mobility in the crystalline phase. The shrinkage forces are very high (Fig. 52) and show time dependence reminiscent of amorphous PET [150] and PMMA [84], with a rapid build up and decay of stress at high temperatures, and a stable situation at lower temperatures. The magnitude of the shrinkage force depends somewhat on molecular weight, increasing with increasing molecular weight. The effect of copolymerisation was found to be explicable essentially in terms of a reduction in the melting temperature, and results for different molecular weight polymers can also be reduced to a single curve by plotting shrinkage versus the temperature difference between the peak melting temperature and the shrinkage temperature.

Similar results have been obtained for polypropylene [152], where again the high modulus products show a greatly reduced shrinkage, and comparatively high shrinkage forces.

6.3 Melting Behaviour and Superheating

The melting behaviour of the highly oriented polyethylenes [153,154] was found to show a dependence on heating rate, an effect generally known as superheating. The magnitude of this effect, as well as the maximum melting temperature, increased markedly at high draw ratio and high molecular weight. It was clear that the superheating behaviour of unconstrained samples was not identical to that of pressure-

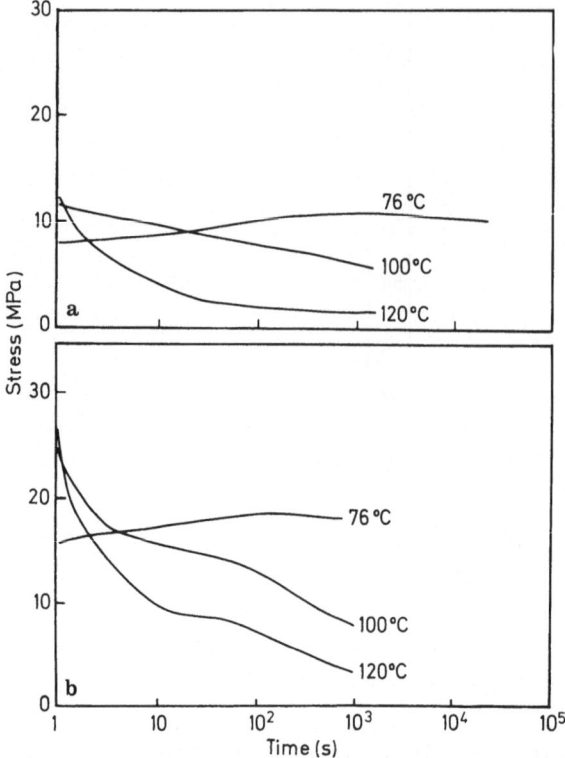

Fig. 52. Variation of shrinkage force with time and temperature for R50 (a) drawn to $\lambda = 11$ (b) drawn to $\lambda = 20\text{–}25$

crystallised polymer. It was therefore concluded that the high modulus LPE materials did not contain chain-extended material. This result is consistent with the other structural measurements of crystal length by WAXS, dark field electron microscopy, etc. In restrained samples, the superheating effects are very large indeed [154], as illustrated in Fig. 53, and this is attributed to the presence of an extended molecular network. The entropic restrictions on molecules which connect two or more crystalline regions, which lead to superheating have been discussed quantitatively by Zachmann [155].

6.4 Thermal Expansion

There is a very large anisotropy in the linear thermal expansivities of these oriented polymers. Early studies [74, 156] showed that in polyethylene the expansivity parallel to the orientation direction $\alpha_{\shortparallel}$ was negative and apparently very close to the value $(-12 \times 10^{-6} \text{ K}^{-1})$ for the c-axis expansion of the crystalline regions obtained from X-ray measurements. This result was attributed to the high degree of crystal continuity, and did not appear to be controversial. More recent work [157], however, has

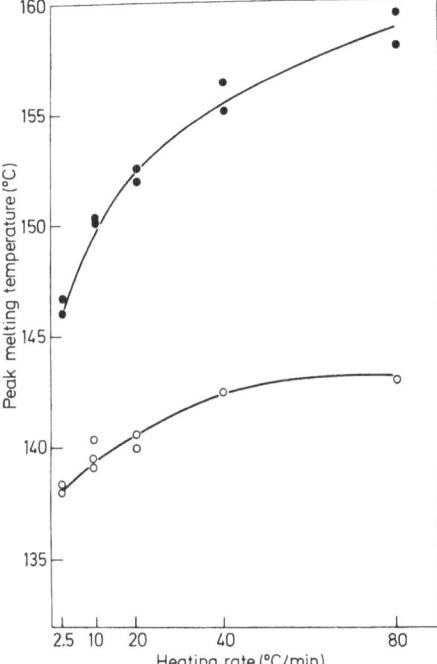

Fig. 53. Effect of heating rate on the peak melting temperature of sample 2: ○, unrestrained; ●, restrained

shown that in many polymers $\alpha_{||}$ can have a large negative value at high temperatures and this is also true for high draw polyethylenes. The linear thermal expansivity perpendicular to the orientation direction is positive and has a value $\sim 10^4 \, K^{-1}$, which is similar to that for isotropic material.

The large negative value for $\alpha_{||}$ was attributed by Choy et al. [158] to the contraction force exerted on the structure by tie molecules. Orchard et al. [157] have pointed out that the Choy model predicts that $\alpha_{||} - \alpha_0$, where α_0 is the linear thermal expansivity for the crystalline regions, will become closer to zero as temperature increases, which is contrary to observation. Instead, Orchard et al. proposed an alternative model which is in the spirit of ideas proposed by Struick [159] on the effects of frozen-in internal stresses in polymers. It is imagined that the structure consists of two components acting mechanically in parallel. The first component is responsible for the stiffness of the polymer and may consist of lamellae, crystalline bridges, taut tie molecules, etc. The second component, which gives rise to an internal stress, is considered to be a large scale molecular network which intermingles with the first component. This network is considered to contribute negligbly to the modulus, but provides the retraction stress measured in a macroscopic shrinkage force measurement.

It can be shown that this model gives the general result

$$\alpha_{||} - \alpha_0 = -\frac{B}{E}\left[1 - \frac{T}{E}\frac{dE}{dT}\right] \tag{16}$$

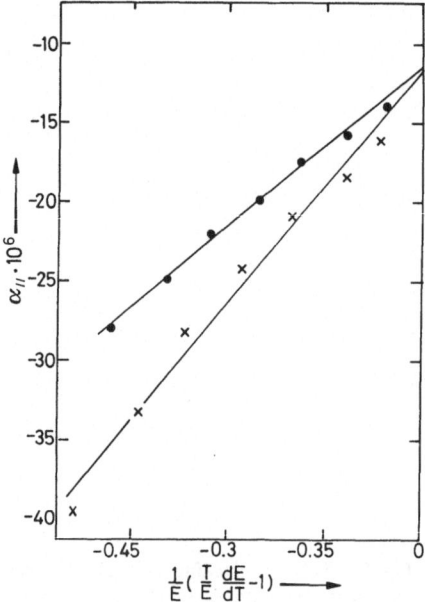

Fig. 54. Film thermal expansion coefficient v modulus function from equation (16). ● 006-60 homopolymer $\lambda = 11$, × 002-55 copolymer $\lambda = 11$

where E is the axial modulus and T is absolute temperature. For simplicity it is assumed that the retractive force $\sigma = -BT$, i.e. an ideal rubber network. (Numerical calculations have shown that this simplification, which makes the understanding of the theory more accessible, is fully justified). Figure 54 shows the excellent correlation between $\alpha_{||}$ and

$$\frac{1}{E}\left[\frac{T}{E}\frac{dE}{dT} - 1\right]$$

for a homopolymer and a copolymer of polyethylene. It can be seen that the straight line fits extrapolate to a value of -11×10^6 K^{-1} for α_0. The slopes of the line gives values for the internal shrinkage stress which correspond quite well with the measured values of shrinkage force i.e. in the range 10–30 MPa at 76 °C.

This theoretical treatment shows that the major reason for the large negative expansion coefficient with increasing temperature is the fall in sample stiffness with temperature, so that the internal stress (which actually increases somewhat with increasing temperature) has an increasing effect. With falling temperature the values of $\alpha_{||}$ for all high draw samples converge to the c-axis expansion level of -12×10^6 K^{-1}. At high temperatures there is a wide range of values, which can be attributed partly to the different stiffnesses and dependences of stiffness on temperature, and partly to the different residual stresses introduced during processing.

7 Barrier Properties and Chemical Resistance

7.1 Gas Barrier Properties

Measurements of the permeability and diffusivity have been undertaken for helium and oxygen for a range of highly oriented polyethylene films [160]. The deduced solubilities for both gases are proportional to the amorphous volume fraction showing that the non-crystalline regions are the transport medium in all instances. The solubility obtained for oxygen is about ten times that for helium. This result is consistent with the proposal of Michaels and Bixler [161] which relates the amorphous phase solubility $\overset{*}{S}$ to the Lennard-Jones force constant $\varepsilon/\bar{k}$ of the diffusant.

We have

$$\ln \overset{*}{S} = 0.022\varepsilon/\bar{k} - 5.07 \tag{17}$$

Results for drawn polyethylene give a value of $\varepsilon/\bar{k} = 124$ K which agrees well with the value of 118 K calculated by Michaels and Bixler from the second virial coefficient.

The changes in the diffusion coefficient with draw ratio are more complex than those for the solubility, because the former depends both on the size of the diffusant molecule and on the structure of the polymer. The results for helium are shown in Fig. 55. It can be seen that there is no change in the diffusion coefficient up to draw ratio 9 followed by a rapid drop between draw ratios 10 and 12 and then a more gradual fall at higher draw ratios. No differences can be detected between either homopolymers of different molecular weight or between these and the hexene-1 copolymers.

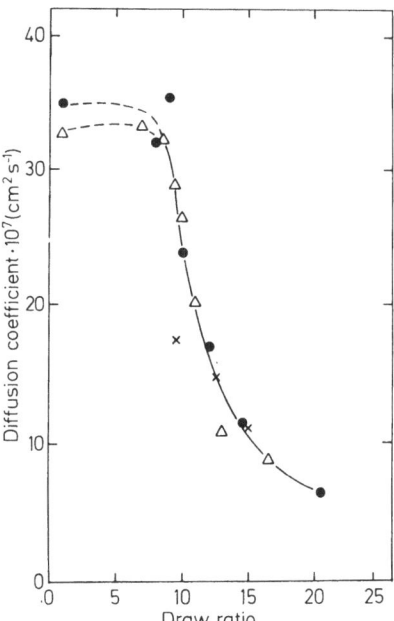

Fig. 55. Variation of oxygen diffusion coefficient with draw ratio. Key as for Fig. 56

These results are consistent with current views of a morphological changes occurring during drawing proposed by Peterlin, and extended more recently to include high draw materials [108]. The initial microspherulitic morphology is converted to a micro-fibrillar structure at about draw ratio 12, and this gives a dramatic reduction in the diffusion coefficient due to increased tortuosity of the diffusion paths. The further slow decline above draw ratio 12 can be attributed to further orientation of the non-crystalline regions, and possibly the production of crystalline bridges.

The diffusion results for oxygen are shown in Fig. 56. There are several features worthy of note. First, the reduction in the diffusion coefficient with draw ratio is much more marked than for helium (a maximum factor of 50 compared with 6). This reduction could be of some technological importance, because the highly drawn PE films are then similar to oriented polyester films. Secondly, the changes in diffusion coefficient for oxygen are evident at the lowest draw ratios and show a continuous fall with draw ratio. It appears that the larger oxygen molecule is not able to penetrate the deformed microspherulitic structure during the early stages of drawing. Finally, it is very important to note that oxygen diffusion discriminates differences in structure between the different grades of polyethylene and also differences due to different drawing conditions. It is of particular interest that the diffusion coefficient is lower for the Rigidex 002-55 copolymer at the corresponding draw ratio. There are also differences between samples drawn at different temperatures. Hizex 7000F drawn at 100 °C shows lower values than that drawn at 115 °C while the values for Rigidex 006-60 drawn at 75 °C are lower still. This is consistent with a tighter structure being formed at the lower draw temperatures.

Measurements of helium diffusion coefficients over the temperature range 25–55 °C showed no significant changes in activation energies throughout the drawing range.

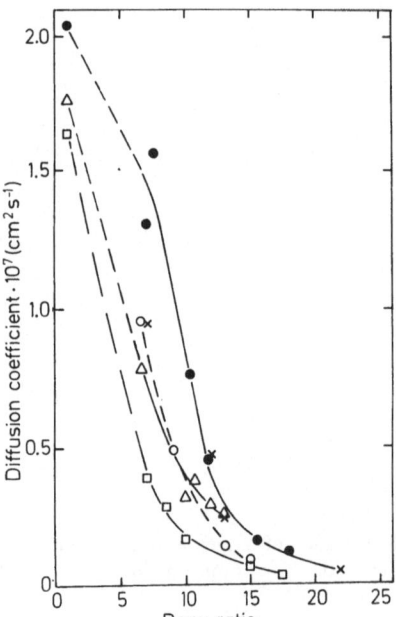

Fig. 56. Variation of helium diffusion coefficient with draw ratio. ● Hizex 7000F 2 mm feedstock T_D 115 °C, ○ Hizex 7000F 2 mm Feedstock T_D 100 °C, × Hizex 7000F 0.5 mm feedstock T_D 115 °C △ Rigidex 002-55 0.5 mm feedstock T_D 115 °C, □ 006-60 0.5 mm feedstock T_D 75 °C

It was therefore concluded that the changes in diffusion which occur on drawing result from shorter jumps between activation sites as a result of greater blockage of available paths.

7.2 Sorption and diffusion of solvents

The equilibrium sorption and diffusion coefficient were determined for toluene and tetrahydrofuran in oriented linear polyethylene pipe produced by hydrostatic extrusion [162]. The results, which are summarised in Table 4, show that the sorption is about a factor of ten less than that for isotropic material, and the diffusion coefficients are a factor of a hundred less. It is clear from these results, taken in conjunction with those for gases discussed above, that the improvements in barrier properties increase with molecular size. This can be attributed to the increased difficulty of penetrating the structure as the molecular size increases.

Table 4.

	Equilibrium sorption (weight %)	Diffusion coefficient (cm²s⁻¹)
Isotropic pipe Toluene	12	1.3×10^{-7}
THF	10	1.1×10^{-7}
Oriented pipe Toluene	1.5	1.6×10^{-9}
THF	1.0	1.9×10^{-9}

7.3 Chemical Resistance

Even in its isotropic form polyethylene possesses good resistance to acids and alkalis, although it is affected by some hydrocarbons and solvents e.g. hot xylene. The highly oriented products show a very much increased resistance. It has been shown that the swelling in organic solvent such as xylene is very much reduced, and tests have shown that weathering behaviour and resistance to sunlight is considerably improved. Detailed studies of nitric acid etching were undertaken for the structural investigation described above. The very high drawn samples showed exceptional resistance, as reported by Capaccio and Ward [114,115] and other workers [163]. For $\lambda = 30$ no measureable weight loss occurred even after 10 days etching in fuming nitric acid at 60 °C.

8 Conclusions

The initial stimulus for the preparation of high modulus polyethylenes came from the recognition of the gap between the available stiffness and the theoretical stiffness. As indicated in this review the gap has now been largely bridged, which is emphasised still further by consideration of the total mechanical anisotropy as shown in Table 5.

Table 5.

Axial Young's Modulus[a]	−175 °C	20 °C	Theoretical[f, g]
λ = 30–35 20 Hz	160	100	260–320
Transverse Young's Modulus[b]			
λ = 25 1–2 mins loading	—	1.35	5–10
Shear Modulus[c]			
λ = 25 1 Hz	1.9	1.3	1—3
Poisson's Ratio[d, e]			
λ = 20–24 1 min loading	—	0.4–0.6	0.01–0.6

[a]   J. B. Smith, G. R. Davies G. Capaccio & I. M. Ward, J. Polym. Sci., Polym. Phys. Edn. *13*, 2331 (1975);
[b]   S. A. Jawad & I. M. Ward, J. Mater. Sci., *13*, 1381 (1978);
[c]   A. G. Gibson, S. A. Jaward, G. R. Davies & I. M. Ward, Polymer, *23*, 349 (1982);
[d, e]   A. M. Zihlif, R. A. Duckett & I. M. Ward, J. Mater. Sci., *13*, 1837 (1978); *17*, 1125 (1982);
[f]   A. Odajima & M. Maeda, J. Polym. Sci., *C15*, 55, (1966);
[g]   K. Tashiro, M. Kobayashi & H. Tadokoro, Macromolecules, *11*, 914 (1978)

The measured mechanical behaviour is extremely anisotropic and comparatively close to that expected on theoretical grounds. Any further progress in this respect will be marginal.

Similar achievements have been made for polypropylene and polyoxymethylene. In PP the measured low temperature dynamic mechanical moduli of 25 GPa have been reported [129] compared with a theoretical value of 46 GPa. The corresponding values for POM are 65 GPa and 106 GPa respectively [131].

Table 6.

Fibre	Property						
	Tensile Modulus	Tensile Strength	Elong- ation at Break	Density ϱ	Specific Modulus	Specific Strength	Maximum Working Tempera- ture
	(GPa)	(GPa)	%	g/cm³	GPa/ϱ	GPa/ϱ	°C
Carbon	230	1.5	1.5	1.8	128	0.8	>1500
Glass	75	2.0	2.5	2.5	30	0.8	250
Kevlar 49	125	3.0	3.0	1.45	86	2.1	≈180
Polyethylene	50–100	1.0–4.0	4–18	0.96	52–104	1–4.2	130

From a technological viewpoint, it is interesting to compare the stiffness and strength of the polyethylene fibres with those of carbon, glass and Kevlar. As can be seen from Table 6, the values for the specific modulus and specific strength are very favourable to polyethylene. Taken in conjunction with the comparatively large extension to break which leads to high energy absorption capabilities and comparative ease of processing, these results emphasise the potential of the high modulus polyethylene fibres for reinforced composites and similar applications.

For applications in the form of solid sections, the enhancement of stiffness and strength is also important, but other properties such as the reduced permeability to gases and solvents, and the improved thermal stability can also play an important role. It is now a considerable challenge to capitalise on an extremely innovative area of polymer science and technology.

9 References

1. Meyer, K. H. and Lotmar, W.: Helv. Chim. Acta., *19*, 68 (1936)
2. Capaccio, G. and Ward, I. M.: Nature Physical Science *243*, 143 (1973) Brit. Patent Appl. 10 746/73 (filed 6 March 1973)
3. Gibson, A. G. et al.: J. Mater. Sci. *9*, 1193 (1974)
4. Gibson, A. G. and Ward, I. M.: Patent Appl. 30823/73 (filled June 1973)
5. Weeks, N. E. and Porter, R. S.: J. Polym. Sci., Polym. Phys. Edn., *12*, 635 (1974)
6. Zwijnenberg and Pennings, A. J.: J. Polym. Sci., Polym. Letters Edn., *14*, 339 (1976)
7. Smith, P. and Lemstra, P. J.: J. Mater. Sci. *15*, 505 (1980)
8. Barham, P. J. and Keller, A.: J. Polym. Sci., Polym. Letters Edn., *17*, 591 (1979)
9. Wobser, G. and Blasenberg, S.: Kolloid Z. Z. Polym. *241*, 985 (1970)
10. Ultra-High Modulus Polymers (ed. A. Ciferri & I. M. Ward), Applied Science Publishers, London 1979, Chapters 1, 2 and 3
11. Parsons, B., and Ward, I. M.: Plastics Rubber Proc. Appl. *2*, 215 (1982)
12. Ward, I. M.: Makromol Chem. *109/110*, 25 (1982)
13. Pennings, A. J., and Kiel, A. M.: Kolloid Z., *205*, 160 (1965)
14. Pennings, A. J.: 'Crystal Growth', Proc. Int. Conf. on Crystal Growth, Peiser, H. S. (ed.), Boston (1966)
15. Pennings, A. J., Mark, J. M. A. A., and Kiel, A. M.: Kolloid Z. *237*, 336 (1970)
16. Frank, F. C.: Proc. Roy. Soc. (London) *A319*, 127 (1970)
17. Frank, F. C., Keller, A. and Mackley, M. R.: Polymer *12*, 467 (1976)
18. Mackley, M. R., and Keller, A.: Phil Trans. Roy. Soc. (London) *278*, 29 (1975)
19. Mackley, M. R., and Keller, A.: Polymer *14*, 16 (1973)
20. Frank, F. C. and Mackley, M. R.: J. Polm. Sci., Polym. Phys. Edn., *14*, 1121 (1976)
21. Berry, M. V., and Mackley, M. R.: Phil. Trans. Roy. Soc. (London) *287*, 1 (1977)
22. Keller, A.: J. Polym. Sci., Polym. Symposia *C58*, 395 (1977)
23. Mackley, M. R.: Phys. Technol. *9*, 13 (1978)
24. Keller, A.: 'Ultra-High Modulus Polymers', Ciferri, A., Ward, I. M., (eds.) London, Applied Science Publishers, 1979
25. Zwijnenberg, A., and Pennings, A. J.: Colloid Polym. Sci., *253*, 452 (1975)
26. Pennings, A. J., and Meihuizen, K. E.: Ultra-High Modulus Polymers, Ciferri, A., Ward, I. M. (eds.), London, Applied Science Publishers, 1979
27. Coombes, A., and Keller, A.: J. Polym. Sci., Polym. Phys. Ed., *17*, 1637 (1979)
28. Kalb, B., and Pennings, A. J.: Polymer *21*, 3 (1980)
29. Pennings, A. J. et al.: Pure Appl. Chem. *55*, 777 (1983)
30. Smook, J., and Pennings, A. J.: J. Mater. Sci. *19*, 31 (1984)
31. Smook, J., Ph. D. thesis, Groningen, 1984.
32. Smook, J., and Pennings, A. J.: J. Appl. Polym. Sci., *27*, 2209 (1982)

33. Matsuo, M., and Manley, R. St. J.: Macromolecules, *15*, 985 (1982)
34. Matsuo, M., and Manley, R. St. J.: Macromolecules, *16*, 1500 (1983)
35. Matsuo, M., Tsuji, M., and Manley, R. St. J.: Macromolecules, *16*, 1505 (1983)
36. Peguy, A., and Manley, R. St. J.: Polym. Commun. *25*, 39 (1984)
37. Smith, P., Lemstra, P. J., and Booij, H. C.: J. Polym. Sci., Polym. Phys. Edn., *19*, 877 (1981)
38. van der Vegt, A. K., and Smith, P. P. A.: Advances in Polymer Science, Monograph 26, London Soc. Chem. Ind. p. 313, 1967
39. Southern, J. H., and Porter, R. S.: J. Appl. Polym. Sci. *14*, 2305 (1970)
40. Southern, J. H., and Porter, R. S.: J. Macromol. Sci., Phys. *B4*, 541 (1970)
41. Crystal, R. G., and Southern, J. H.: J. Polym. Sci., Polym. Phys. Edn., *A2*, *9*, 1641 (1971)
42. Perkins, W. G., Capiati, N. J., and Porter, R. S.: Polym. Eng. Sci., *16*, 200 (1976)
43. Odell, J. A., Grubb, D. T., and Keller, A.: Polymer *19*, 617 (1978)
44. Andrews, J. M., and Ward, I. M.: J. Mater. Sci., *5*, 411 (1970)
45. Capaccio, G., and Ward, I. M.: Polymer, *15*, 233 (1974)
46. Capaccio, G., and Ward, I. M.: Polym. Eng. Sci., *15*, 219 (1975)
47. Capaccio, G., and Ward, I. M.: Polymer, *16*, 239 (1975)
48. Capaccio, G., Crompton, T. A., and Ward, I. M.: J. Polym. Sci., Polym. Phys. Edn. *14*, 1641 (1976)
49. Capaccio, G., Crompton, T. A., and Ward, I. M.: ACS Polym. Preprints *18*, (21, 343 (1977)); Polym. Eng. Sci., *18*, 533 (1978)
50. Capaccio, G., Chapman, T. J., and Ward, I. M.: Polymer *16*, 469 (1975)
51. Capaccio, G., Crompton, T. A., and Ward, I. M.: Polymer *17*, 644 (1976)
52. Capaccio, G., Crompton, T. A. and Ward, I. M.: J. Polym. Sci., Polym. Phys. Edn., *18*, 301 (1980)
53. Fatou, J. G. and Mandelkern, L.: J. Phys. Chem. *60*, 21, 417 (1965)
54. Way, J. L., Atkinson, J. R. and Nutting, J.: J. Mater. Sci., *9* (1974) 293
55. Barham, P. J. and Keller, A.: J. Mater. Sci., *11*, 27 (1976)
56. Capaccio, G., Ward, I. M. and Smith, F. S.: Brit. Pat. Appln. 9795/74, filed 5 March 1974.
57. Cansfield, D. L. M., Capaccio, G., and Ward, I. M.: Polym. Eng. Sci., *16*, 721 (1976)
58. Brew, B., and Ward, I. M.: Polymer, *19*, 1338 (1978)
59. Taylor, W. N. Jr., and Clark, E. S.: Polym. Eng. Sci., *18*, 518 (1978)
60. Capaccio, G., and Ward, I. M.: Brit. Pat. Appln. 52644/74 filed 3 Oct 1973
61. Clark, E. S., and Scott, L. S.: Polym. Eng. Sci. *14*, 682 (1974)
62. Capaccio, G. et al.: J. Macromol. Sci. Phys., *B15*, 381 (1978)
63. Smith, J. B., Manuel, A. J., and Ward, I. M.: Polymer, *16*, 57 (1975)
64. Jarecki, L. and Meier, D. J.: Polymer *20*, 1078 (1979)
65. Jarecki, L. and Meier, D. J.: J. Polym. Sci., Polym. Phys. Edn., *16*, 2015 (1978)
66. Capaccio, G., and Ward, I. M.: J. Polym. Sci., Polym. Phys. Edn., *22*, 475 (1984)
67. Clements, J., et al.: Polymer (in press)
68. Imada, K., et al.: J. Mater. Sci., *6*, 537 (1971)
69. Nakamura, K., Imada, K., and Takayanagi, M.: Int. J. Polym. Mater. *2*, 71 (1972)
70. Imada, K. and Takayanagi, M.: Int. J. Polym. Mater. *2*, 89 (1972)
71. Maruyama, S., Imada, K., and Takayanagi, M.: Int. J. Polym. Mater. *2*, 105 (1973)
72. Maruyama, S., Imada, K., and Takayanagi, M.: Int. J. Polym. Mater. *2*, 125 (1973)
73. Capiati, N. J. et al., J. Mater. Sci., *13*, 334 (1977)
74. Zachariades, A. E., Mead, W. T., and Porter, R. S.: Ultra-High Modulus Polymers (ed. A. Ciferri and I. M. Ward), Applied Science Publishers, London 1979, Chapter 2.
75. Farrell, C. J., and Keller, A.: J. Mater. Sci., *12*, 966 (1977)
76. Nakayama, K., et al.: Eng. Edn., *3*, 1489 (1974)
77. Gupta, R., and McCormick, P. G.: J. Mater. Sci., *15*, 619 (1980)
78. Griswold, P. D., Zachariades, A. E., and Porter, R. S.: Polym. Eng. Sci., *18*, 861 (1978)
79. Buckley, A., and Long, H. A.: Polymer Eng. Sci., *91*, 115 (1969)
80. Alexander, J. M., and Wormell, P. J.: Ann. C.I.R.P. *19*, 28 (1971)
81. Williams, T.: J. Mater. Sci., *8*, 59 (1973)
82. Gibson, A. G., and Ward, I. M.: J. Polym. Sci., Polym. Phys. Edn., *16*, 2015 (1978)
83. Coates, P. D., and Ward, I. M.: J. Polym. Sci., *16*, 2031 (1978)
84. Kahar, N., Duckett, R. A., and Ward, I. M.: Polymer, *19*, 316 (1978)

85. Hope, P. S., Ward, I. M., and Gibson, A. G.: J. Mater. Sci., 15, 2207 (1980)
86. Hope, P. S., Gibson, A. G., and Ward, I. M.: J. Polym. Sci., Polym. Phys. Edn., 18, 1243 (1980)
87. Hope, P. S., and Parsons, B.: Polym. Eng. Sci., 20, 589 (1980)
88. Hope, P. S. et al.: Polym. Eng. Sci., 20, 540 (1980)
89. Coates, P. D., Gibson, A. G., and Ward, I. M.: J. Mater. Sci., 15, 359 (1980)
90. Hope, P. S., and Ward, I. M.: J. Mater. Sci., 16, 1511 (1981)
19. Hope, P. S. et al.: J. Mech. Work, Tech. 5, 223 (1981)
92. Hoffman, O., and Sachs, G.: "Introduction to the Theory of Plasticity" (McGraw-Hill, New York and London 1953)
93. Avitzur B.: Metal forming: Process and Analysis, New York and London, (McGraw-Hill, 1968)
94. Coates, P. D., and Ward, I. M.: J. Mater. Sci., 13, 1957 (1978)
95. Hope, P. S., Richardson, A., and Ward, I. M.: Polym. Eng. Sci., 22, 307 (1982)
96. Curtis, A. C., Hope, P. S., and Ward, I. M.: Polym. Composites 3, 138 (1982)
97. Ward, I. M.: Proc. Phys. Soc., 80, 1176 (1962)
98. Brody, H., and Ward, I. M.: Polym. Eng. Sci., 11, 139 (1971)
99. Sahari, J. B., Parsons, B., and Ward, I. M.: J. Mater. Sci., in press.
100. Bassett, D. C., and Turner, B.: Phil. Mag. 29, 925 (1974)
101. Coates, P. D., and Ward, I. M.: Polymer, 20, 1553 (1979)
102. Gibson, A. G., and Ward, I. M.: J. Mater. Sci., 15, 979 (1980)
103. Gibson, A. G., and Ward, I. M.: Polym. Eng. Sci., 20, 1229 (1980)
104. Hope, P. S., Richardson and Ward, I. M.: J. Appl. Polym. Sci., 26, 2879 (1981)
105. Richardson, A., Hope, P. S., and Ward, I. M.: J. Polym. Sci., Polym. Phys. Edn., 21, 2525 (1983)
106. Richardson, A., Ania F., Rueda D. R. Ward, I. M. and Balta Calleja F. J.: Polym. Eng. Sci., (in press)
107. Richardson, A. et al.: (to be published)
108. Peterlin, A., Ultra-High Modulus Polymers, Ciferri, A., Ward, I. M. (eds.), London, Applied Science Publishers, 1979, Chapter 10
109. Smith, J. B. et al.: J. Polym. Sci., Polym. Phys. Edn., 13, 2331 (1975)
110. Clements, J., Jakeways, R., and Ward, I. M.: Polymer, 19, 639 (1978)
111. Peterlin, A., and Corneliussen, R.: J. Polym. Sci., A2, 6, 1273 (1968)
112. Frye, C. J., et al.: Polymer 20, 1310 (1979)
113. Frye, C. J., et al.: J. Polym. Sci., Polym. Phys. Edn., 20, 1677 (1982)
114. Capaccio, G., and Ward, I. M.: J. Polym. Sci., Polym. Phys. Edn., 19, 667 (1981)
115. Capaccio, G., and Ward, I. M.: J. Polym. Sci., Polym. Phys. Edn., 20, 667 (1981)
116. Sherman, E. S., Porter, R. S., and Thomas, E. L.: Polymer, 23, 1069 (1982)
117. Gibson, A. G., Davies, G. R., and Ward, I. M.: Polymer, 19, 683 (1978)
118. Clements, J. and Ward, I. M.: Polymer, 24, 27 (1983)
119. Smith, J. S., Manuel, A. J., and Ward, I. M.: Polymer, 16, 57 (1975)
120. Capaccio, G., Gibson, A. G., and Ward, I. M.: Ultra-High Modulus Polymers, Cifferi, A., Ward I. M., London, Applied Science Publishers, 1979, Chapter 1.
121. Smith, P. et al.: Colloid Polym. Sci., 259, 1070 (1981)
122. Grubb, D. T.: J. Polym. Sci., Polym. Phys. Edn., 21, 165 (1983)
123. Jungnitz, S., Ph. D. thesis Leeds University 1983
124. Barham, P. J., and Arridge, R. G. C.: J. Polym. Sci., Polym. Phys. Edn., 15, 1177 (1977)
125. Clements, J. et al.: Polymer 20, 295 (1979)
126. Arridge, R. G. C., Barham, P. J., and Keller, A.: J. Polym. Sci., Polym. Phys. Edn., 15, 389 (1977)
127. Gibson, A. G. et al.: Polymer, 23, 349 (1982)
128. Leung, W. P. et al.: Polymer, 25, 447 (1984)
129. Wills, A. J., Capaccio, G., and Ward, I. M.: J. Polym. Sci., Polym. Phys. Edn. 18, 493 (1980)
130. Jawad, S. M. et al.: (to be published)
131. Brew, B. et al.: J. Polym. Sci., Polym. Phys. Edn., 17, 351, (1979)
132. Wilding, M. A., and Ward, I. M.: Polymer, 19, 969 (1978)
133. Wilding, M. A., and Ward, I. M.: Polymer, 22, 870 (1981)
134. Wilding, M. A., and Ward, I. M.: Plastics Rubber Proc. Appl. 1, 167 (1981)
135. Ward, I. M., and Wilding, M. A.: J. Polym. Sci., Polym. Phys. Edn., 22, 561 (1984)

136. Wilding, M. A., and Ward, I. M.: J. Mater. Sci., *19*, 629 (1984)
137. Roetling, J. A.: Polymer *6*, 311 (1965)
138. Bauwens-Crowet, C., Bauwens, J. C., and Homes: J. Polym. Sci., A2, *7*, 735 (1969)
139. Woods, D. W., Busfield, W. K., and Ward, I. M.: Polym. Commun. (in press)
140. Duxbury, J., and Ward, I. M.: (to be published)
141. Flory, P. J.: J. Amer. Chem. Soc., *67*, 2048 (1945)
142. Smith, P., and Lemstra, P. J., J. Polym. Sci., Polym. Phys. Edn., *19*, 1007 (1981)
143. Smith, P., Lemstra, P. J., and Pijpers J. P. L.: J. Polym. Sci., Polym. Phys. Edn., *20*, 2229 (1982)
144. Wu, W., and Black, W. B.: Polym. Eng. Sci., *19*, 1163 (1979)
145. Cansfield, D. L. M., et al., Polym. Commun. *24*, 130 (1983)
146. Smook, J., Hamersma, W., and Pennings, A. J.: J. Mater. Sci., *19*, 1359 (1984)
147. Clements, J., and Ward, I. M.: J. Mater. Sci., *18*, 2484 (1983)
148. Gibson, A. G. et al.: J. Polym. Sci., Polym. Lett. Edn., *15*, 183 (1977)
149. Gibson, A. G., Greig, D., and Ward, I. M.: J. Polym. Sci., Polym. Phys. Edn., *18*, 1481 (1980)
150. Capaccio, G., and Ward, I. M.: Colloid Polym. Sci., *260*, 46 (1982)
151. Pinnock, P. R., and Ward, I. M.: Trans. Farad. Soc., *62*, 1308 (1966)
152. Orchard, G. A. J., and Ward, I. M.: (to be published)
153. Clements, J., Capaccio, G., and Ward, I. M.: J. Polym. Sci., Polym. Phys. Edn., *17*, 693 (1979)
154. Clements, J. and Ward, I. M.: Polymer, *23*, 935 (1982)
155. Zachmann, H. G.: Kolloid Z. Z. Polym. *206*, 25 (1965); *216–217*, 180 (1967)
156. Gibson, A. G., and Ward, I. M.: J. Mater. Sci., *14*, 1838 (1979)
157. Orchard, G. A. J., Davies, G. R., and Ward, I. M.: Polymer *25*, 1203 (1984)
158. Choy, C. L., Chen, F. C. and Young, K.: J. Polym. Sci., Polym. Phys. Edn., *19* 335 (1981)
159. Struick, L. C. E.: Polym. Eng. Sci., *18*, 799 (1978)
160. Holden, P., Orchard, G. A. J., and Ward, I. M.: J. Polym. Sci., Polym. Phys. Edn., (in press)
161. Michaels, A. S., and Bixler, H. J.: J. Polym. Sci., *50*, 393 (1961)
162. Marshall, J. M., Hope, P. S., and Ward, I. M.: Polymer, *23*, 142 (1982)
163. Weeks, N. E., Mori, S., and Porter, R. S.: J. Polym. Sci., Polym. Phys. Edn., *13*, 2031 (1975)

K. Dušek (Editor)
Received December 7, 1984

Synthetic Ion-Exchange Resins

Mukul Biswas*
Department of Chemistry, Indian Institute of Technology, Kharagpur 721302, India

S. Packirisamy**
Department of Polymer Science, Tokyo Institute of Technology, 0-Okayama,
Meguro-ku, Tokyo 152, Japan

This review highlights the developments in the field of cation-exchange, anion-exchange, composite and thermally regenerable resins. The main focus is on the literature on synthesis and physico-chemical properties. Cation-exchange and anion-exchange resins have been discussed on the basis of the nature of monomers involved in the synthesis of polymeric matrices of the resins. Composite ion-exchange resins which incorporate inert materials and magnetic particles are discussed. The polymerization techniques adopted to synthesize thermally regenerable ion-exchange resins possessing the maximum possible thermally regenerable capacity are dealt with.

It appears that enough scope exists in framing novel synthetis and modification procedures based on the resin systems discussed in this review so as to render them more useful and economically viable.

* Present address: Institute of Polymer Science, The University of Akron, Akron, Ohio 44325, U.S.A. (March '84–May '85)
** Present address: Research & Development, Berger Paints India Ltd., 14 & 15 Swarnamoyee Road, P.O. Botanic Garden, Howrah-3, India

Abbreviations

MA	Methyl acrylate
EA	Ethyl acrylate
BA	Butylacrylate
AAm	Acrylamide
MAAm	Methacrylamide
DAA	Diallylamine
MDAA	Methyldiallylamine
TAA	Triallylamine
TAT	1,3,5-Triacryloylhexahydrotriazine
NVC	N-Vinylcarbazole
DVB	Divinylbenzene
HEXA	1,6-bis(N,N-diallylamino)hexane
EGDMA	Ethylene glycol dimethacrylate
ABME	Allyl benzoin methyl ether
AIBN	Azo-bis-isobutyronitrile
PVC	Poly(Vinyl chloride)
SDVB	Styrene-divinylbenzene copolymer
PNVC	Poly(N-Vinylcarbazole)
NVCF	N-Vinylcarbazole-furfural copolycondensate
PNVCF	Poly(N-Vinylcarbazole)—furfural copolycondensate
NVCPA	N-Vinylcarbazole—phthalic anhydride condensate
NVCDVB	N-Vinylcarbazole—divinylbenzene copolymer
NVCFO	N-Vinylcarbazole—formaldehyde copolycondensate
PNVCS	Sulfonic acid resin from PNVC
NVCFS	Sulfonic acid resin from NVCF
PNVCFS	Sulfonic acid resin from PNVCF
NVCDVBS	Sulfonic acid resin from NVCDVB
NVCFOS	Sulfonic acid resin from NVCFO
SFS	Sulfonic acid resin from styrene—furfural copolycondensate
PFS	Sulfonic acid resin from α-pinene—furfural copolycondensate
KU—1	Commercial sulfonic acid resin from phenol—formaldehyde resin
KU—2	Commercial sulfonic acid resin from SDVB copolymer
PNVCP	Phosphonic acid resin from PNVC
NVCFP	Phosphonic acid resin from NVCF
PNVCFP	Phosphonic acid resin from PNVCF
SFP	Phosphonic acid resin from styrene—furfural copolycondensate
SFAP	Phosphonic acid resin from acetylated styrene—furfural copolycondensate
PFP	Phosphonic acid resin from α-pinene—furfural copolycondensate
PFAP	Phosphonic acid resin from acetylated α-pinene—furfural copolycondensate
FAT	Condensation product of furfural with tetramethyl diuram disulfide and polyethylenepolyamine
AN-2F	Condensation product of methylol phenol derivatives and polyethylenepolyamine

PP-type resin   Porous SDVB resin obtained by using a nonsolvating diluent (pre-
                    cipitant) in the polymerizing mixture
PPS-type resin   Porous SDVB resin obtained by using a nonsolvating diluent along
                    with a solvating diluent in the polymerizing mixture
PM-type resin   Porous SDVB resin obtained by incorporating a preformed linear
                    polymer in the polymerizing mixture
PMS-type resin   Porous SDVB resin obtained by incorporating a linear preformed
                    polymer and a solvating diluent in the polymerizing mixture
PMP-type resin   Porous SDVB resin obtained by incorporating a preformed linear
                    polymer and a nonsolvating diluent in the polymerizing mixture
IPNs   Interpenetrating polymeric networks
meq/g   Milliequivalents per gram
F_M Volume fraction of monomer (M)
D_i Apparent co-efficient of interdiffusion in $cm^2\ sec^{-1}$
K_H^A Relative affinity co-efficient for the exchange $\bar{H}^+ + A^+ \rightleftharpoons \bar{A}^+ + H^+$

1 Introduction

One of the most abundant uses of chemically modified polymers is in the form of synthetic ion-exchange resins. The first synthetic industrial ion exchanger was prepared in 1903 by Harm and Rümpler. However, a spectacular development began only in 1935 with the discovery by Adams and Holmes, that crushed phonographic records exhibit ion-exchange properties. Since then, rapid developments in the science of polymerization techniques, the availability of a vast range of organic starting materials and the need to select the proper resin and techniques to be adopted for each application spurred the growth of ion-exchange technology.

There are different natural and synthetic products which show ion-exchange properties. However, as far as the practical applications are concerned the organic resins are by far the most important ion exchangers. Their main advantages are high chemical and mechanical stability, high ion-exchange capacity, and high exchange rate. An additional advantage is the possibility of selecting the fixed ionogenic groups and the degree of crosslinking of the matrix according to the intended applications. For these obvious reasons, by now a mammoth literature exists on the synthesis, characterization, and applications of organic ion-exchange resins.

This review attempts to highlight the major advances in the science of ion-exchange resins made over the last 15 years. The developments in the field of cation-exchange and anion-exchange resins have been discussed on the basis of the nature of monomers involved in the synthesis of polymeric matrices of the resins. In addition, two well-recognized novel types of resins, namely composite and thermally regenerable ion-exchange resins, have been discussed. Emphasis has been given mainly on the synthesis and physico-chemical properties, avoiding detailed technological applications except of cases of composite and thermally regenerable resins where relevant reference to the technological data has been made while explaining principle and function vis-a-vis the basis of the synthesis. In view of the voluminous literature dealt with to focus the above points, it has not been possible to give proportional emphasis on the application of these resins. However, a few selective cases have been discussed to highlight the utility of special-purpose resins.

2 Cation-Exchange Resins

2.1 Styrene-based Resins

(a) *Resins from SDVB Copolymers.* Most of the commercial cation-exchange resins are synthesized from SDVB copolymers. In addition to the cheap and abundant availability of these monomers, the SDVB copolymers have excellent physical strength and are not easily subject to degradation by oxidation, hydrolysis, or elevated temperatures. Moreover, one active group can be introduced for each aromatic ring in the copolymer giving rise to high exchange capacities. The literature on SDVB copolymer systems has been frequently reviewed. As such the present article will highlight the significant advances made in the modification of conventional SDVB resins.

The cation-exchange resins obtained from SDVB copolymers prepared by conventional pearl polymerization technique mostly have the gel structure. The conventional SDVB copolymer has been modified by using a solvent or a linear polymer in the monomer mixture during the polymerization to make porous SDVB copolymers.

Resins from Solvent-Modified SDVB Copolymers. Millar et al. [1-4] have made a systematic study of the modification of SDVB copolymer networks by carrying out the polymerization in presence of diluents. From such polymers ion-exchange materials have been prepared whose mechanical properties, exchange kinetics, and equilibria differ from those of corresponding materials prepared from SDVB copolymers obtained by conventional method. The diluent may be a solvent, nonsolvent or a poor solvent for the copolymer, and, depending on the nature of the diluent used, the properties of the products differ.

When a solvating diluent (e.g., toluene) is used for a given DVB content, both the degree of entanglement and the extensibility of the final network are determined by the dilution of the polymerizing system when the primary network is formed, i.e., at the time of gelation. Thereafter, the polymer network continues to be formed within the ambit of the primary network with a little or no shrinkage during polymerization and with the polymer chains fully solvated. In consequence, the nuclei are less entangled and with increasing dilution the growing chains reach greater lengths before mutual termination occurs. When the polymerization is completed and the solvent is removed, the expanded network collapses but reversibly so that on the addition of a good solvating agent such as toluene, it re-expands to its earlier state [1].

The toluene-modified copolymers containing >27% DVB quite rapidly with evolution of air bubbles-adsorb liquids such as cyclohexane and heptane which are taken up by other copolymers to a negligible extent only. At higher DVB contents the amount of organic solvent taken up is approximately equal to the amount of inert diluent originally used. In contrast to the conventional hydrocarbon copolymers, these materials can even accommodate appreciable quantities of water without observable change in volume. These phenomena can be ascribed to the existence of macropores in the copolymers containing more than 27% DVB.

In the toluene-modified network the heterogeneity of the structure can be increased in two independent ways by a decrease in volume fraction of monomers (F_M) or by an increase in DVB content of the monomer solution. Thus, a toluene-modified network prepared with 15% DVB content (15% tol resin) has an expanded network structure and the network obtained by using 27% DVB content (27% tol resin) has a structure which is intermediate between an expanded and a macroporous structure. The network prepared with 34% DVB content (34% tol resin) and 55% DVB content (55% tol resin) contain an appreciable portion of their porosity in the form of macropores [2].

When the diluent is a nonsolvent, there is no comparable swollen state and the final structure is one in which large entangled nuclei are connected by a relatively small number of coiled and crumpled internuclear chains and the polymer and diluent phases are segregated. As with the solvent-modified materials, collapse of the system of interconnected nuclei occurs as the diluent is removed, but the large size of the nuclei will lead to the appearance of macroporosity at considerably lower DVB contents and dilutions. In fact, the macroporosity appears at lower DVB contents and dilutions (i.e., higher F_M values) in copolymers made with heptane (nonsolvent) as diluent rather than with toluene [3].

The characteristic structure of the diluent-modified SDVB copolymers is retained in their sulfonated derivatives. A comparison [1] of the solvent regains of macroporous sulfonic acid resin, conventional sulfonic acid resin, and their parent hydrocarbon matrices is made in Table 1. The sulfonated macroporous product like its parent hydrocarbon copolymer also takes up normally incompatible solvents and the close similarity of their regains in the sulfonated material to the water regain of the hydrocarbon matrices, expressed in compatible units, shows that the macropores have largely survived the sulfonation [1].

The kinetics and the equilibrium characteristics of ion-exchange of solvent-modified resins differ from those of conventional resins. In both the conventional resin (7% DVB) and 15% tol resin D_i (apparent coefficient of interdiffusion in cm^2 sec^{-1}) decreases rapidly with increasing ionic diameter of the exchanging ion, consistent with the purely expanded structure postulated for the 15% tol resin. For 34% tol and 55% tol resins the D_i values undergo a small change with ionic diameter, showing the ease with which ions can move through the macropores [2]. The macroporous resins differ from the other resins in their selectivity towards inorganic ions [4], the 34% tol resin shows higher values of K_H^A (relative affinity coefficient) at low values of exchange ($\bar{X}$) compared with a conventional resin of the same apparent crosslinking.

The ion-exchange properties of toluene-modified and heptane-modified resins are compared in Table 2. Although the macroporous toluene-modified resin has a higher DVB content at lower F_M value, both the specific water regains and the rates of exchange of the smaller ion (Na$^+$) are substantially the same. However, the heptane-modified resin has a lower rate of exchange for the large organic ion (NEt$_4^+$), reflecting the lower proportion of exchange sites in the macropores of this material [3].

Matrices having different pore structures and pore-size distributions compared to the ones obtained by the use of one diluent have been prepared by Sederel and De Jong [5] by using two diluents (one solvating and one nonsolvating) at the same time

Table 1. Comparison of solvent regains of a macroporous resin with a conventional resin [1]

Solvent	Solvent regain of macroporous resin (ml/g)		Solvent regain of conventional resin (ml/g)	
	Hydrocarbon copolymer	Sulfonated copolymer[a]	Hydrocarbon copolymer	Sulfonated copolymer[a]
Toluene	2.21	1.34	0.79	nil
Cyclohexane	2.10	0.92	nil	nil
n-Heptane	2.09	0.96	nil	nil
Nitromethane	2.01	2.09	0.18	0.26
Water	1.10	2.24	nil	1.86
Water on Li$^+$ form		2.18		1.82
Water on Na$^+$ form		2.19		1.64
Water on K$^+$ form		2.23		1.37

[a] Measured on the dry H$^+$-form unless otherwise stated

Table 2. Comparison of properties of toluene-modified and heptane-modified resins [3]

Characteristics	Heptane-modified resin	Toluene-modified resin
DVB ($\%$ v/v)	21.00	27.00
Volume fraction of monomers	0.62	0.51
Weight capacity (meq/g)	4.75	4.36
Specific water regain (g/meq)	0.20	0.22
$Na^+ - H^+$ exchange: Q^a	1.00	1.00
$Na^+ - H^+$ exchange: $10^6 D_i^b$	2.80	3.00
$NEt_4^+ - H^+$ exchange: Q^a	0.80	0.87
$NEt_4^+ - H^+$ exchange: $10^6 D_i^b$	0.13	0.23

[a] Q is the limiting extent of echange in meq/meq of total capacity;
[b] D_i is the apparent co-efficient of interdiffusion in $cm^2 sec^{-1}$

in the polymerizing mixture. Kun and Kunin [6] suggest that the pore-structure formation can be divided into three stages: (i) building of microspheres, (ii) agglomeration of microspheres, and (iii) fixation of agglomerates within the bead. In presence of a nonsolvent in the polymerization system, the first stage is the formation of nuclei and their agglomeration. During the second stage of pore structure formation the entanglement continues. The agglomerations of the nuclei which form the microspheres are expected to be compact only when the nonsolvent is used as a diluent (porous by precipitant; PP type). However, when a solvent is present along with a nonsolvent in the reaction mixture (porous by precipitant and solvent; PPS type), polymer chains in the nuclei are less entangled as a consequence of the increased solvating character of the system. Also, during the second stage of pore-structure formation the agglomerations of the nuclei are more solvated compared to those formed in the absence of solvent. The microspheres are now less compact and will contain more pores of smaller size. In addition to these, in presence of added solvating diluent the phase separation has a less drastic character and hence there will be a regular formation of microspheres within the bead. As compared to the PP-type resin, the pores and the pore volumes of the PPS resin are smaller.

Resins from Linear Polymer-Modified SDVB Copolymers. PM-type resins (porous by macromolecular material) with improved rate, capacity, and high regeneration efficiency have been prepared [7-12] by dissolving or incorporating a preformed linear polymer in a mixture of styrene and DVB, polymerizing the mixture, extracting the polymer with a solvent from the cured product, and then sulfonating the SDVB copolymer. Polystyrene, poly(vinyl toluenes), poly(vinyl xylenes), polyacrylics, and copolymers of vinylacetate and dimethacrylate have been used for the modification of SDVB copolymers. The added linear polymer prevents the resulting network from having a complete gel structure. The pore-structure of the polymeric network is influenced by the molecular weight and the amount of linear polymer used [5].

Sederal and De Jong [5] propose a further novel modification in controlling the porosity of SDVB matrices by using a solvating or a nonsolvating diluent along with a linear polymer such as polystyrene. The modified polymeric materials PMS (porous by macromolecular material and solvent) and PMP (porous by macromolecular material and precipitant) differ in their pore structure from the polymeric matrix obtained in presence of a solvent or a nonsolvent alone. The effect of variation of solvent and nonsolvent in pore structure formation of PMS- and PMP-type matrices are summarized below.

Resin   Variation of pore structure

PMS type Increase in solvent results in increase in pore volume within certain limits without change in pore-size distribution.

PMP type Increase in nonsolvent increases the pore volume and also the pore-size distribution.

Hilgen et al. [13] concluded that in a macromolecular network, pores can collapse as a consequence of cohesion when the solvated agglomerations of the polymer chains are approaching each other by loss of solvent. This process can be counteracted by:

1) A decreased interaction between polymer chains and the fluid which can be attained by: (a) substituting the solvent for a less solvating one and (b) a high concentration of DVB in the network — which will result in a restriction of the mobility of the polymer chains.

2) The presence of large pores —·which prevents the agglomerations from contacting each other. Large pores can be formed by the addition of high concentration of nonsolvent (PP type) or linear polymer (PM type) to the polymerizing mixture or by a combination of both (PMP type).

Also, with ion-exchange resins the same principles ruling pore stability are encountered as were found in their parent matrices. Ion-exchange resins prepared from matrices having sufficient DVB content and relatively wide pores, mostly connected with a high pore volume, retain their volume during removal of solvent better than types having lower pore volumes and smaller pores.

Resins from Interpenetrating Polymeric Networks (IPNs). IPNs have been prepared by Millar et al. [14] by causing conventional DVB-crosslinked polystyrene to imbibe a calculated amount of a suitable mixture of styrene and DVB solution containing an initiator and then polymerizing the imbibed monomers within the original polymeric network. In some instances the imbibition and subsequent polymerization have been carried out on the secondary intermeshed copolymer so produced giving the corresponding copolymer containing three IPNs. Sulfonation of the secondary and tertiary intermesh hydrocarbon copolymers yield corresponding intermeshed cation-exchange materials.

The ion-exchange resins synthesized from SDVB copolymers obtained by two-stage copolymerization, described as above, have higher porosity, real density, and exchange capacity as compared to those prepared from SDVB copolymers obtained by one-stage polymerization of styrene and DVB [15]. The extent of improvement depends on the composition of the first stage. The best properties are obtained when the first stage contains 10 % DVB units.

Haeupke et al. [16] found that gel and half-porous copolymers may become porous after being modified with a second monomeric phase containing an inert diluent. Such modified copolymers are called 'double porous copolymers' (Haeupke et al. [16]) or 'hybrid copolymers' (Barrett and Clemens [17]).

A hybrid copolymer has been prepared [17] by the modification of SDVB macroreticular copolymer with methyl acrylate and DVB. Under the conditions of intermesh copolymerization, about 14 % of methyl acrylate-DVB copolymer could be introduced. After saponification of ester groups followed by sulfonation, a strongly acidic cation-exchange resin has been obtained.

A study [18,19] on the variation of the degree of crosslinking of SDVB copolymer used for the preparation of an intermeshed cation-exchange resin from methacrylic acid and DVB suggests the following: (i) a double-porous carboxylic acid cation-exchange resin can be obtained by the modification of porous SDVB copolymers with methacrylic acid and DVB in the inert diluent, n-heptane, if the crosslinking degree of the starting copolymer does not exceed 10–15%, (ii) during the modification of starting porous copolymers having a high degree of crosslinking, filling of pores with methacrylic acid-DVB copolymer occurs, whereas with copolymers having a low degree of crosslinking intermesh copolymerization occurs, and (iii) modification of a gel copolymer (with degree of crosslinking 3%) under identical conditions does not result in the production of a porous cation-exchange resin.

Resins fron SDVB Copolymers of Uniformly Large Particle Size. SDVB copolymers of uniformly large particle size [20,21] can be prepared by the copolymerization of styrene and DVB to obtain copolymer particles of smaller size (first stage), swelling the particles to a larger size in styrene monomer, and then fixing the size by subsequent polymerization. Agglomeration of polymer particles which usually takes place can be avoided by adopting a procedure in which during the course of polymerization (in the first stage) the suspension stabilizer is temporarily eliminated by the addition of hydrochloric acid before the addition of fresh monomer and initiator, and after the addition of fresh monomer and initiator the suspension stabilizer is restored by the addition of fresh alkali [22]. The polymer particles obtained by this procedure can be converted to cation-exchange resins by conventional means.

(b) *Resins from Polystyrene Crosslinked by Friedel-Crafts Reaction.* Conventional crosslinked polystyrene gels and ion-exchange resins produced by copolymerization of styrene and DVB are characterized by inhomogeneous structure which is allegedly responsible for their low osmotic ability and their insufficient permeability to large organic ions [23]. The inhomogeneity is the result of different reaction rates of the comonomers. Such inhomogeneity can be avoided if the polymers are crosslinked after the polymerization process [24,25]. It is reasonable to assume that prior to the crosslinking process the crosslinking agent is evenly distributed throughout the volume of the polymer solution. As the concentration of polymer is high relative to the crosslinking agent, the crosslink bridges formed as a result of the reaction should have the same statistical distribution in the gel as they have in the initial polymer solution. These materials are thus termed 'isoporous' to distinguish them from resins that are formed in an inhomogeneous way.

The isoporous macrocrosslinked networks are obtained by the introduction of crosslinking bridges on polystyrene [26–28] or slightly crosslinked polystyrene [23] through Friedel-Crafts reaction. Bis-chloromethylbenzene [24], bis-chlormethylbiphenyl [23,24,29], p-xylenedichloride [23,29], 2,3,5,6-tetramethylbenzylchloride [30], and various other monofunctional and bifunctional agents [31–41] have been used as crosslinking agents. In this particular method of synthesis it is difficult to control the degree of crosslinking of the polymer. In addition, small deviations from the conditions of synthesis can result in change of the crosslinking density of the final network [42].

Negere et al. [43] reported that the condition of isoporosity is dependent on the concentration of polystyrene in the initial solution. Isoporous resins are obtained only for dilutions of 1 g/cc and the materials have narrow pore-size distributions centered around 25 Å.

For the same degree of crosslinking, isoporous polystyrene resins swell more than SDVB copolymers [29,44]. Both SDVB copolymers and isoporous polystyrene resins show a decrease in swelling with increasing degrees of crosslinking. However, when crosslinking is carried out beyond 0.15 mol of crosslinks per mole of repeating unit for isoporous materials, the reverse phenomenon is true; increasing the number of crosslinks leads to an increase in its volume. As the number of crosslinks increases, the crosslink bridges keep the polymer chains better separated from one another and hence higher swelling results. Another unusual feature that distinguishes the isoporous polystyrene resins from SDVB copolymers is their ability to swell in non-solvents, e.g., water [45]. Isoporous polystyrene resins also show an internal surface area of up to 30 % greater than that of SDVB copolymers.

Tsyurupa et al. [24] reported the synthesis of sulfonic acid resins from isoporous polystyrene networks obtained through the crosslinking of polystyrene with bis-chloromethylated benzene or biphenyl. The sulfonated resins possess a high exchange capacity which decreases only insignificantly with the growing degree of crosslinking (Table 3). The exchange capacity amounts to 4.2 meq/g for the isoporous resins even at crosslinking degrees of 40–100 %. On the contrary, for the sulfonated SDVB co-polymers a sharp decrease of exchange capacity is observed when the DVB content is higher than 10 %. This shows that the isoporous resins have high accessibility for chemical reagents.

The structure of the isoporous sulfonic acid resins is as fully accessible [24] for the tetrabutylammonium ions as are the standard resins with 1–2 % DVB (Table 4). The isoporous resins irrespective of the type and amount of crosslinking can be completely saturated with the tetrabutylammonium ions. On the contrary, it is well known that the tetrabutylammonium ion uptake of standard SDVB resins fall sharply with the increasing degree of crosslinking.

In general, the ion-exchange resins synthesized from the isoporous polymers are characterized by high exchange capacity, high permeability for large organic ions, and low volume fluctuations.

Table 3. Comparison of ion-exchange capacity of sulfonated isoporous and standard resins [24]

Degree of crosslinking (%)	Ion-exchange capacity of isoporous resins (meq/g)		Ion-exchange capacity of standard SDVB resins (meq/g)
	I[a]	II[b]	
4.17	5.29	5.25	5.20
11.10	5.25	5.28	4.59
25.00	5.02	5.12	
43.00	5.00	4.25	
66.50	4.50	4.40	
100.00	4.26	4.16	

[a] Crosslinking agent: bis-chloromethylated biphenyl;
[b] Crosslinking agent: bis-chloromethylated benzene

Table 4. Comparison of ion-exchange capacity of sulfonated isoporous and standard resins towards $(C_4H_9)_4N^+$ and Na^+ ions [24]

Degree of crosslinking (%)	Ratio of ion-exchange capacities of isoporous resins towards $(C_4H_9)_4N^+$ and Na^+ (%)		Ratio of ion-exchange capacities of standard SDVB resins towards $(C_4H_9)_4N^+$ and Na^+ (%)
	I[a]	II[b]	
1.4			98.0
1.7			110.0
2.3			110.0
11.1	98.0	99.0	51.5
43.0	91.0	88.0	
100.0	107.0	102.0	

[a] Crosslinking agent: bis-chloromethylated biphenyl;
[b] Crosslinking agent: bis-chloromethylated benzene

More recently, Biswas and Chatterjee [46-50] have prepared polystyrene electrophilically substituted with phthalic anhydride [46], pyromellitic dianhydride [47], trimellitic anhydride [48], and 1,2,3,4-tetrahydrophthalic anhydride [49] by Friedel-Crafts reaction. These anhydride-modified polystyrenes have further been converted into sulfonic acid resins [47,50] by conventional sulfonation reaction. The thermal stabilities of these modified polymers are comparatively better than that of unmodified polystyrene. Ion-exchange capacities also compare favourably with the polystyrene-based commercial resins.

2.2 N-Vinylcarbazole-based Resins

Polymers of N-vinylcarbazole (NVC) — the most reactive monomer in the N-vinyl monomer series — are well known for their superior thermal stability due mainly to the presence of the carbazole moiety in the structure. It is expected that appropriate chemical modification of poly(N-vinylcarbazole) (PNVC) through sulfonation or phosphorylation would yield ion-exchange resins of improved thermal stability. Yet another novel way of achieving this objective involves the chemical modification of NVC and PNVC through copolymerization or copolycondensation followed by introduction of $-SO_3H$ or $-PO_3H_2$ groups. To this end, some specific contribution has been made by Pielichowski and Morawiec [51] and by Biswas et al. [52-62] who modified PNVC [51,52], NVC-furfural (NVCF) [53-55], PNVC-furfural (PNVCF) [53-55], NVC-phthalic anhydride (NVCPA) [56], NVC-formaldehyde (NVCFO) [57] copolycondensates, and NVC-DVB copolymer [58,59] to cation-exchange resins. The characteristic features of these resins are summarized in Table 5.

Exhaustive studies on the NVC/PNVC-based resins lead to the following conclusions:

i   In general, the isothermal stability of cation-exchange resins synthesized from chemically modified NVC/PNVC is higher than that of the resins synthesized from unmodified PNVC.

Table 5. Comparison of NVC/PNVC-based cation-exchange resins

Resin	Ionogenic groups	Total capacity (meq/g)	Salt-splitting capacity (meq/g)	Loss in capacity[a] on heating with water (%)	Ref.
PNVCS	$-SO_3H$	4.50	3.90	25.50	52)
NVCFS	$-SO_3H$ $-COOH$	4.62	4.02	15.00	53)
PNVCFS	$-SO_3H$ $-COOH$	4.01	3.01	5.23	53)
NVCPAS	$-SO_3H$	4.73	4.38	1.27	54)
NVCDVBS	$-SO_3H$	4.90	4.40	14.30	59)
NVCFOS	$-SO_3H$ $-COOH$	5.20	4.60	1.80	57)
PNVCP	$-PO_3H_2$	3.17	data not available	data not available	51)
NVCFP	$-PO_3H_2$ PO_2H	4.64	0.74	no loss	54, 55, 60)
PNVCFP	$-PO_3H_2$ PO_2H	4.84	0.75	no loss	54, 55, 60)

[a] Resins were heated with deionized water in sealed glass ampoules for 24 h at 100 °C

ii Capacity-wise these resins are comparable to those of styrene-based commercial resins.

iii No generalization may be made so far as the comparative thermal stability of the resins are concerned. However, the addition copolymerization of NVC with DVB is found to produce polymers of higher thermal stability than those obtained from the condensation of phthalic anhydride or furfural with NVC. Sulfonation enhances[62] the thermal stability of NVCPA, but reduces the same in the case of NVCF and NVCDVB.

iv A remarkable feature of NVC/PNVC-based phosphonic acid resins is the appreciable enhancement of thermal stability of the copolycondensates on phosphorylation.

v A study on the thermal stability of NVCFP and PNVCFP resins in 5N nitric acid (at 100 °C for 24 h) reveals the introduction of nitro groups into the polymeric matrices and the formation of carboxylic acid groups, probably by the oxidation of furfurylic bridges and/or any other aliphatic bridges[55].

vi Surprisingly, NVCFP and PNVCFP resins show a very high increase in capacity on heating in air at 200 °C (Fig. 1), which is rationalized from the observation of Armitage and Lyle[63] that the thermal decomposition of ion-exchange resins yields carboxyl groups on the aromatic nucleus, formed by oxidative cleavage of the aliphatic backbone of the polymer, especially at points where crosslinking occurs. PNVC after crosslinking with furfural will have furfurylic crosslinks and also ethylenic backbones. Moreover, during phosphorylation the solvent, e.g., 1,2-dichloroethane, used may also introduce additional aliphatic ethylenic bridges. These aliphatic backbones/crosslinks and furfurylic bridges probably undergo oxidative thermal degradation forming carboxylic acid groups[55].

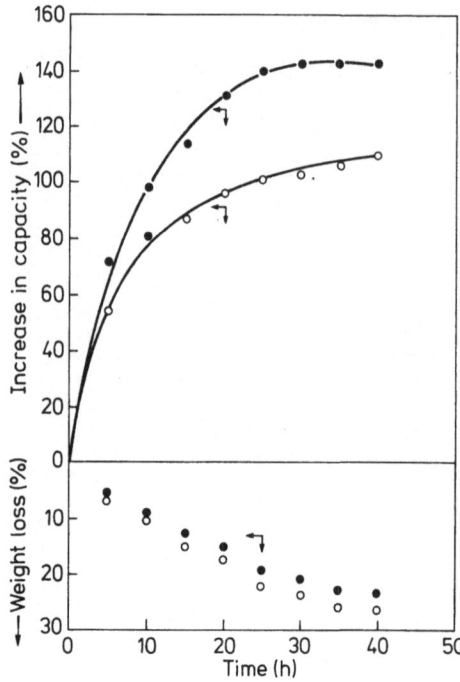

Fig. 1. Isothermal oxidative degradation studies of NVCFP and PNVCFP. (○) increase in capacity of NVCFP; (●) increase in capacity of PNVCFP; (○) weight loss of NVCFP; (●) weight loss of PNVCFP [55]

2.3 Poly(vinyl chloride)-based Resins

Dima et al. [64–68] studied in detail the synthesis and properties of polyene sulfonic acid cation-exchange resins obtained by the sulfonation of poly(vinyl chloride) (PVC) using chlorosulfonic acid in 1,2-dichloroethane solvent. Depending upon the synthesis conditions, resins having sulfonic exchange capacity of 2–4.3 meq/g and carboxylic exchange capacity of 0.3–1.5 meq/g have been obtained. The resins synthesized from suspension-polymerized PVC show greater sulfonic and carboxylic exchange capacities than those obtained from an emulsion-polymerized PVC [65].

During the sulfonation process, extensive gas (HCl, SO_2) formation takes place [64]. The amounts of hydrogen chloride and sulfur dioxide gases, as well as the rate of their release, increase and the reaction time decreases when the amount of chlorosulfonic acid used in the reaction is increased incrementally [66]. During the sulfonation process using chlorosulfonic acid, other reactions such as dehydrochlorination, oxidation, quaternary degradation, and quaternary regrouping also take place.

Owing to the conjugated double bonds contained in the polyene sulfonic acid resin, a part of the carboxylic acid groups possesses ion-exchange properties like those of sulfonic acid groups [64]. For a polyene sulfonic acid cation-exchange resin obtained by sulfonation of PVC with chlorosulfonic acid in 1,2-dichloroethane at a ratio of 1:5:2, respectively [67], the capacity decreases on treatment with 5N sulfuric acid or 5N sodium hydroxide at 100 °C. When treated with water at 180 °C for 6 h, the capacity decreases from 4.56 to 1.43 meq/g. When heated in air at 180 °C up to 30 h, the resin loses 80 % of its exchange capacity within 6 h and regains 20 %

of its exchange capacity in subsequent 24 h, indicating an oxidation yielding —COOH groups.

Dehydrohalogenated PVC on sulfonation with sulfuric acid gives bifunctional cation-exchange resins containing —SO₃H and —COOH groups [69,70]. The amount of sulfonic acid groups present in the resin is nearly independent of the temperature of treatment, whereas the amount of introduced carboxylic acid groups increases with increase in temperature.

Arylation of PVC with benzene at 0 °C followed by sulfonation with chlorosulfonic acid in 1,2-dichloroethane at 20 °C and 90 °C gives cation-exchange resins [71]. At 20 °C, the sulfonic acid group is fixed only on the benzene ring, but at 90 °C the sulfonic acid capacity increases by one unit because of the simultaneous sulfonation of the aromatic rings and of the dehydrochlorinated segments of PVC chains. The carboxylic capacity is only 0.25 meq/g when the sulfonation is carried out at 20 °C and it increases to 1.5 meq/g when the temperature is raised to 90 °C, as a result of the greater number of carboxylic acid groups being formed due to oxidative degradation reaction.

2.4 Poly(vinyl alcohol)-based Resins

The synthesis of sulfonic acid resin [72] from the condensation product of benzaldehyde with poly(vinyl alcohol) and the modification of poly(vinyl alcohol) to phosphonic acid resins [73-76] have been reported.

Arai and Ogiwara [77] synthesized a poly(vinyl alcohol)-grafted ion-exchange resin from modified Amberlite 120 B (SDVB sulfonic acid resin). Amberlite 120 B has been isopropylated, and photoinduced grafting of vinyl acetate on the modified resin has been carried out using UV radiation. The poly(vinyl acetate)-grafted resin has been saponified to give poly(vinyl alcohol)-grafted resin. However, the unmodified Amberlite 120 B does not undergo grafting with vinyl acetate due to the possibility that the α-hydrogen of the styrene-sulfonic acid repeating unit is not labile enough to allow grafting. After isopropylation, the grafting is possible, because the isopropyl group would be relatively labile towards UV radiation. The Fe(III) salt form of the isopropylated Amberlite 120 B gives larger graft ratios than the sodium salt form due to the photosensitizing effect of Fe(III). The catalytic activity of poly(vinyl alcohol)-grafted resin on the hydrolysis of amylose increases with increasing graft ratio (in the range of 0.1 to 0.13), but the activity is lower than that of the parent resin.

2.5 Acrylic Monomer-based Resins

The copolymerization of acrylic acid or methacrylic acid [78-80] and a crosslinking agent such as DVB [78] or ethylene dimethacrylate [80] gives weak acid cation-exchange resins. The pearl polymerization technique can be used if esters instead of water soluble acids are copolymerized [81,82].

Porous methacrylic acid-DVB resin having excellent sorptive-desorptive properties and suitable for the pharmaceutical isolation of polymyxin E can be prepared by the suspension copolymerization method [83,84].

The suspension copolymerization of acrylic esters, acrylonitrile, and DVB in presence of a free radical catalyst followed by the hydrolysis of the bead copolymer with alkali gives a carboxylic acid cation-exchange resin [85] with exchange capacity of 4.2–4.6 meq/ml.

A carboxylic acid cation-exchange resin has been synthesized by polymerizing a mixture of methacrylic acid, methyl methacrylate, DVB, and ethylvinylbenzene and then hydrolyzing the resulting copolymer. The increase in degree of crosslinking makes the resin to be weaker with a higher apparent pK value [86].

Carboxylic acid cation-exchange resins having high exchange capacity and swelling have been prepared [87] by copolymerizing methacrylic acid with 4,4'-diisopropyl diphenyl oxide at 90 °C for 8–9 h using benzoyl peroxide as the initiator. Interstingly, the copolymer has a capacity of 6.7–9.9 meq/g and swelling of 1.2–3.1 ml/g which exceeds that observed for a cation-exchange resin from methacrylic acid and DVB. The presence of 4,4'-diisopropyl diphenyl oxide units leads to a looser structure and therefore improved swelling.

Resins with good hydrodynamic properties and sorption capacities for proteins and enzymes have been prepared [88–90] by the polymerization of acrylic acid or methacrylic acid and a crosslinking monomer such as 1,3,5-triacryloylhexahydrotriazine (TAT) in aqueous acetic acid using a radical initiator. Methacrylic acid-TAT copolymer prepared by using a redox catalyst [91] is useful as a cation-exchange resin for selective and reversible sorption of proteolytic enzymes such as terrilytin.

Polyelectrolytes useful as ion-exchange resins have been prepared [92] by heating a mixture of acrylonitrile, acetone, formaldehyde, and sodium bisulfite in a strongly alkaline aqueous solution and then saponifying the resulting polymer.

A carboxylic acid resin having an exchange capacity of 5.5–6.5 meq/g has been prepared [93] by polymerizing furylacrylic acid with 1–9% benzene sulphonic acid in a sealed ampoule at 100 °C for 7.5 h and then treating the polymer with sodium hydroxide.

2.6 Aldehyde-based Resins

(a) *Resins from Formaldehyde*. Nowadays, the conventional phenolformaldehyde sulfonic acid cation-exchange resins are not used. In place of phenol, other reactants are used in order to obtain resins possessing imroved properties.

A novolac-type of resin has been prepared [94] from the condensation product of anisole with formaldehyde. Anisole has been condensed with an equimolecular amount of formaldehyde in the presence of sulfuric acid and the product sulfonated at ≤ 120 °C with sulfuric acid. The sulfonated product has been made insoluble by treatment with formaldehyde or by heating at 140°–145 °C, and the resin prepared in this way has a superior resistance to the action of oxidizing agents such as 1N nitric acid and 0.1N potassium permanganate.

The condensation of resorcinol, chloroacetic acid, and formaldehyde in alkaline medium gives a bifunctional cation-exchange resin containing —COOH and —OH groups [95]. The condensation of salicylic acid with formaldehyde in acid medium followed by treatment with phenol in alkaline medium also gives a bifunctional resin which is suitable for separation and purification of proteins [96]. A bifunctional cation-exchange resin containing $-SO_3H$ and $-OH$ groups has been prepared by heating phenol with acetone disulfonic acid and a formaldehyde-paraformaldehyde mixture [97].

Phenoxyalkylphosphonic acids $PhO(CH_2)_nPO(OH)_2$ (n = 1, 2, 3, or 6) prepared by treating triethoxy phosphine with phenoxyalkyl bromides followed by acid hydrolysis of the diethylphosphonates have been condensed with formaldehyde [98] in presence of 1% hydrochloric acid to give ion-exchange resins containing $-PO(OH)_2$ groups at the end of the aliphatic side chains of various lengths and having exchange capacities of 6.6–8.9 meq/g. In general, the polycondensation of formaldehyde with phenoxy ethers $PhOCH_2X$ or $PhOCH_2CH_2X$, where X = COOH, SO_3H, or PO_3H_2, gives resins containing active groups in the aliphatic side chains bound with the aromatic nucleus through oxygen [99].

A resin, non-swelling, insoluble in organic acids, and with a decreased degree of crosslinking has been prepared from 9-carbazole acetic acid, butyraldehyde, and formaldehyde in a two-stage polymerization [100]. Reaction of 9-carbazole acetic acid and butyraldehyde gives a linear polymer, which in the second stage has been reacted with formaldehyde to give a crosslinked network with static exchange capacity of 2.80 meq/g and swelling capacity of 1.4 (in 0.5N sodium hydroxide solution).

(b) *Resins from Furfural.* Furfural has widely been used in place of formaldehyde for the synthesis of cation-exchange resins, as the former gives polymers of improved thermal stability due mainly to the presence of the heterocylic ring. Furfural has been condensed with phenols, ethers, organic acids, styrene, α-pinene, cyclohexane, shellac, and lignin to give polymeric end products which are used as matrices for the synthesis of ion-exchange resins.

Phenols. Phenol has been condensed with furfural [101] in presence of hydroxylamine hydrochloride and the condensate has been sulfonated in petroleum oil at 70–80 °C for 7–8 h and 3% hydrolyzed lignin has been added to give KFFUR-1,2 resin with a total capacity of 2.5–2.9 meq/g at 1:1 mole ratio and 3.5 meq/g at 2:1 mole ratio of phenol to furfural. In a similar way, p-phenolsulfonic acid can be polymerized with furfural in the mole ratios 1:2 and 1:3 to give KFFUR-1,2,3 with a static exchange capacity of 3 meq/g for calcium chloride and 3.5–4.0 meq/g for sodium hydroxide. The mechanical properties and exchange capacity of this resin can be improved by conducting the polycondensation in a magnetic field [102].

The phosphorylation of phenol-furfural condensate [103] affords a cation-exchange resin with improved thermal and chemical resistance. The heat treatment [104] of phosphonic acid resins from furfural resins, phenol-resorcinol-formaldehyde resin, and polystyrene at 100–180 °C for 10–48 h shows that the furfural-based phosphonic acid resins possess higher thermal stability than those from the other two polymers.

A polyfunctional cation-exchange resin [105] prepared by heating a mixture of furfural and p-phenolsulphonic acid with a prepolymer from furylacrylic acid at 60 °C for 5–6 h is stable up to 100 °C, swells 135% in water (in H^+ form), and possesses static exchange capacity of 4.9–5.0, 2.3, 2.7, and 3.5 meq/g for 0.1N NaOH, 0.1N $CaCl_2$, $-COOH$ groups, and 0.1N salsolidine, respectively.

The condensation of furfural with resorcinol [106] in presence of hydrochloric acid at 34–35 °C followed by sulfonation of the condensate affords a polyfunctional cation-exchange resin containing $-SO_3H$, $-COOH$, and $-OH$ (phenolic) groups. The carboxylic activity of this resin has been attributed to the oxidation of furfurylic entities. The resin possesses the total exchange capacity of 7.9 meq/g and the salt-splitting capacity of 2.1 meq/g.

Organic Acids. The polycondensation of furfural with *p*-toluenesulfonic acid or β-naphthalenesulfonic acid affords heat-resistant cation-exchange resins [107,108]. Carboxylic acid cation-exchange resins having high mechanical and chemical stability and high selectivity towards Cu^{2+} ions have been synthesized by condensing furfural with hydroxybenzoic acids such as *p*-hydroxybenzoic acid and salicylic acid [109]. Carboxylic acid resins have also been synthesized through the polycondensation of furfural with β-resorcylic acid or 3-hydroxy-2-napthoic acid [110].

Furfural-salicylic acid and furfural-β-naphthalenesulfonic acid resins possess thermal stability [111,112] exceeding that of KU-1 (sulfonated phenol-formaldehyde resin) and approaching that of KU-2 (sulfonated SDVB copolymer). The heat resistance of toluenesulfonic acid-furfural resin is quite similar to that of KU-2 resin. The improved thermal stability of furfural-based cation-exchange resins is attributed to the presence of heterocyclic and aromatic rings in these resins.

Cation-exchange resins suitable for the purification of sewage and industrial wastes containing monovalent and bivalent metal ions have been synthesized by treating furfural with humic acids (prepared by natural or artificial oxidation of coal) in presence of 4–5% sulfuric acid at 100 °C for 3–3.5 h followed by thermal processing [113]. The polycondensation of peat humic acids with furfural in an acid medium in presence of saw-dust gives a cation-exchange resin having improved physico-chemical properties and homogeneity [114].

Ethers. The polycondensation of diphenyl ether and furfural (1:1.5 mole ratio) in presence of sulfuric acid yields a resin which has been sulfonated by sulfuric acid at 50 °C for 10 h to give a cation-exchange resin [115]. The swelling and the exchange capacity of the product increase with the amount of sulfuric acid present during the polycondensation and are higher for a resin obtained by polycondensation in carbon tetrachloride than the one prepared in absence of solvent. The thermal stability [116] of sulfonated diphenyl ether-furfural resin is more than that of KU-1 and comparable to that of KU-2. The sulfonic acid resin derived from the copolymer obtained by the polymerization of diphenyl ether and furfural at a mole ratio of 1:2 in carbon tetrachloride has optimum adsorption and catalytic activity [117].

Poly(iso-butyl vinyl ether)-furfural condensate [118] obtained through the condensation of furfural with poly(iso-butyl vinyl ether) in benzene medium in presence of hydrochloric acid at 50 °C on treatment with sulfuric acid yields a carboxylic acid cation-exchange resin with a capacity of 5.8 meq/g.

Styrene. The condensation of furfural with styrene in presence of zinc chloride followed by sulfonation with sulfuric acid yields a bifunctional cation-exchange resin (SFS) containing $-SO_3H$ and $-COOH$ groups [119]. The resin has a total capacity of 4.2 meq/g and a sal-splitting capacity of 3.1 meq/g and can be used up to 100 °C without any appreciable fall in capacity [120]. In column operation the bed volume of the resin has been observed to be constant in a series of cycles carried out, indicating the stability of the resin in different regenerant and influent solutions.

Styrene-furfural and acetylated styrene-furfural copolymers on phosphorylation afford [121] phosphonic acid (SFP) and hydroxyphosphonic acid (SFAP) cation-exchange resins, respectively. The equilibrium studies [122] reveal that both resins have a high affinity for hydrogen and divalent ions. The values of the separation factor

Table 6. Values of separation factor $\alpha_{Na^+}^{M^{n+}}$ for different metal ions on phosphorus-containing ion-exchange resins, in equinormal solutions of sodium chloride, and metal ion [121]

Resin	$\alpha_{Na^+}^{Pb^{2+}}$	$\alpha_{Na^+}^{Zn^{2+}}$	$\alpha_{Na^+}^{Cd^{2+}}$	$\alpha_{Na^+}^{Ca^{2+}}$	$\alpha_{Na^+}^{Mg^{2+}}$	$\alpha_{Na^+}^{Th^{2+}}$
SFP	70.43	5.03	4.16	3.76	2.85	1.27
SFAP	70.43	6.94	5.25	4.55	4.00	2.91
Duolite ES-63	70.43	7.07	5.67	4.37	3.76	—

for different metal ions on SFP, SFAP, and Duolite ES-63 (commercial SDVB-based phosphonic acid resin) is shown in Table 6. Obviously, the separation factor for lead is very high, indicating high preference of the phosphorus-containing resins for lead ions. The values of the separation factor for other ions are much lower but still exhibited a considerable difference for lead ions. Table 6 suggests that the affinity of the phosphorus-containing resins for different ions lies in the order, $H^+ > Pb^{2+} > Zn^{2+} > Cd^{2+} > Ca^{2+} > Mg^{2+} > Th^{4+}$. From the separation factor values, it is clear that these resins show wide differences in their affinities for different ions and this property has been used for the chromatographic separation of these metal ions. Due to the higher affinity of these phosphorus-containing resins for divalent and mutivalent ions compared to univalent ions, these resins can be used for the isolation of traces of divalent ions, Cu^{2+}, Pb^{2+}, and Zn^{2+}, from solutions containing a high concentration of monovalent ions.

α-*pinene*. The condensation of furfural with α-pinene in presence of anhydrous zinc chloride or aluminium chloride gives a crosslinked polymer [123,124]. As in the case of styrene-furfural copolymer, the sulfonation of α-pinene-furfural copolymer also gives a bifunctional cation-exchange resin (PFS) containing $-SO_3H$ and $-COOH$ groups [124]. The phosphorylation of α-pinene-furfural and acetylated α-pinene-furfural copolymers affords phosphonic acid (PFP) and hydroxyphosphonic acid (PFAP) resins. The equilibrium studies [125] indicate that in the acid medium the affinity of these resins towards Na^+ and Ca^{2+} follow the sequence, PFAP < PFP < PFS and towards Fe^{2+} the sequence is, PFP < PFAP < PFS. For the Na^+ form of these resins, in the neutral medium the selectivity trend follows the sequence, PFP > PFAP > PFS for Ca^{2+} and PFAP > PFP > PFS for Mg^{2+} exchange. For the $2 Na^+ \rightleftharpoons Ca^{2+}$ system, the equilibrium constants are found to be 1.48, 4.48, and 9.8 for PFS, PFAP, and PFP, respectively. For the $2 Na^+ \rightleftharpoons Mg^{2+}$ system, the equilibrium constants are 1.15, 2.66, and 3.27 for PFS, PFP, and PFAP, respectively. From the equilibrium constants it is clear that both PFP and PFAP prefer Ca^{2+} to Mg^{2+}. Hence, it is possible that these resins in Na^+ form can be advantageously used for the separation of Ca^{2+} and Mg^{2+} in a mixture. In acid medium, PFP and PFAP resins show a higher affinity for Fe^{2+} than for Na^+ and Ca^{2+}. This property can be exploited for the removal of ferrous iron impurities.

Miscellaneous. The condensation products of furfural with cyclohexane [126–128], shellac [129], lignin [130,131], or tetrahydrofuran [132] have also been modified to cation-exchange resins.

The sulfonation or phosphorylation of cyclohexane-furfural copolymer [126–128] affords ion-exchange resins which are stable in air up to 150 °C. The phosphonic acid cation-exchange resin shows selective sorption for Cu^{2+} from a solution containing copper sulfate and cobalt sulfate.

The sulfonation of shellac-furfural [129] and lignin-furfural [130, 131] copolymers gives bifunctional cation-exchange resins containing $-SO_3H$ and $-COOH$ groups. The breakthrough capacity of the cation-exchange resin from lignin-furfural copolymer depends on the sulfonation conditions, and addition of silver sulfate or manganese dioxide as a catalyst during the sulfonation increases its value [131]. Furfural-tetrahydrofuran condensate on treatment with sulfuric acid gives a weak acid resin [132] with a total capacity of 16.2 meq/g.

3 Anion-Exchange Resins

3.1 Styrene-based Resins

The preparation and properties of SDVB copolymers have been described in the preceding section. SDVB copolymers can be converted to anion-exchange resins through chloromethylation with monochlorodimethyl ether using a Friedel-Crafts catalyst followed by treatment with different amines [133-147]. Anderson [133] studied in detail the influence of various Friedel-Crafts catalysts for the chloromethylation of SDVB copolymers. A chlorosulfonic acid — methylal mixture has also been used for the chloromethylation of SDVB copolymers [148-152].

Anion-exchange resins have also been prepared through the nitration of SDVB copolymers with sulfuric acid-nitric acid mixture followed by the reduction of nitro groups with sodium sulfide [153-155]. Such resins containing $-NH_2$ groups have been further modified with ethylenimine [154] or hydroxylamine [155]. Aminated polystyrene can be crosslinked with dichloroethane to give an anion-exchange resin [156].

Belfer et al. [157-161] reported that irreversible structural transformations take place during the chloromethylation of isoporous polystyrene resins. A styrene — 4% DVB copolymer undergoes irreversible structural reorganization during alkylation with 5-chloromethyl-8-hydroxyquinoline in the dry impregnated state in presence of anhydrous aluminium chloride [157]. Similarly, styrene — 1% TAT copolymer undergoes striking morphological transformations during chloromethylation with chlorosulfonic acid and methylal [158]. A reasonably homogeneous resin is obtained when the ratio of chlorosulfonic acid to methylal is 1:1 and in excess of chlorosulfonic acid the resin obtained has a heterogeneous substructure with uneven chlorine distribution [162]. The solvent used for the chloromethylation also affects the morphology. When chlorinated hydrocarbon solvents are used as swelling solvents, two different textures exist in the same grain of the resin and the shell consists of numerous small holes, whereas the inner space consists of voids and large cavities. When a nonswelling, nonpolar solvent is used, the resin obtained has a highly developed inner structure with well-distributed channels connected by inner voids [162].

3.2 Vinyl Pyridine-based Resins

Anionic gels and macroporous resins can be prepared by the copolymerization of 2-vinylpyridine and DVB using a radical initiator [163]. Macroporous resins have been prepared by using n-octane as diluent and the resins are characterized by low bulk density and high solvent absorption with swelling. Increasing the diluent concentration produces high resin absorptivity with decreased swelling. With increased macroporosity, high exchange capacity is obtained at any level of crosslinking.

Granular anion-exchange resins have been prepared from N-containing C-vinyl heterocyclics, polyvinyl compounds and one or more aliphatic vinyl compounds in an aqueous phase in presence of water-insoluble suspension stabilizers [164]. Thus, a mixture of 4-vinylpyridine, ethylacrylate, and p-DVB on polymerization using hydroxylapatite as suspension stabilizer and AIBN as the initiator gives an anion-exchange resin.

The suspension copolymerization of 2,5-divinylpyridine with 2-[β-(dimethylamino) ethyl]-5-vinyl pyridine at 70 °C using gelatin as dispersant and AIBN as initiator gives an anion-exchange resin [165] having the capacities of 7.5, 7.7, 5.9, and 14.5 meq/g for HCl, PhOH, $CuSO_4$, and $Th(NO_3)_4$. 2,5-divinylpyridine has also been copolymerized with 1,3-bis(dimethylamine)-2-(5-vinyl-2-pyridyl) propane in bulk, using AIBN as initiator to give an anion-exchange resin having capacities of 7.8, 8.6, and 7.4 meq/g for HCl, PhOH, and $CuSO_4$.

Anion-exchange resins which are heat resistant and possessing high mechanical strength have been prepared through the copolymerization of 2-methyl-5-vinylpyridine with vinylbenzylbromide followed by saponification [166]. Anion-exchange resin obtained from the copolymer of vinylbenzylchloride-4-vinylpyridine possesses good sorption capacity [167]. Anion-exchange resins have also been prepared by copolymerization of 2-methyl-5-vinylpyridine with hexamethylene-dimethacrylamide, diaminodiphenylmethane dimethacrylamide. or m-phenylene dimethacrylamide followed by chloroalkylation with benzyl chloride. These resins possess good thermal and chemical stability [168].

Macroporous anion-exchange resin, useful for the sorption of uranyl nitrate, has been prepared by suspension polymerization of 4-methyl-2-vinylpyridine with trimethylene glycol dimethacrylate [169].

3.3 Acrylic Monomer-based Resins

The synthesis of anion-exchange resins from acrylonitrile — SDVB copolymer has been reported [170]. The copolymer beads are converted to imino ester form by passing dry hydrogen chloride gas into a slurry of the beads in anhydrous methanol at 5 °C. These beads, after washing with anhydrous methanol, are refluxed with a mixture of ethylene diamine and anhydrous methanol. The resin thus synthesized has a capacity of 5 meq/g in Cl^- form. Other diamines used for the synthesis of such type of resins are aminoethanolamine, neopentyldiamine, and trimethylenediamine [170]. Other monomers which are used for the synthesis of polymer matrices are methylacrylonitrile, vinylidene cyanide, cyanoalkyacrylates, and methacrylates, o-, m-, and p-cyanostyrene, 1-cyanobutadiene, 2-cyanobutadiene, methyl, ethyl, propyl,

and butyl α-cyanocrylates, 1,3,5-trivinylbenzene, 1,2,4-trivinylbenzene, trivinyl-
propane, the isomeric divinylxylenes, divinyltoluenes, and divinylnaphthalenes, N,N'-
methylenediacrylamide, N,N'-methylenedimethacrylamide, and N,N'-divinylethylene
urea. These resins are stable to osmotic shock and resist attrition during acid-base
treatment [170].

A high-capacity, weak-base anion-exchange resin has been prepared by suspension
polymerization of methacrylic acid with DVB followed by the amination of the beads
with an alkylamine. The resin has good anion-exchange properties and dimentional
stability, and can be used in deionization of water as a substitute for silica [171].

Anion-exchange resins especially useful for sorption and chromatographic separation of biopolymers have been prepared [172] by treating ethylenedimethacrylate-2-
hydroxyethylmethacrylate or 2-hydroxyethylacrylate-methylenebisacrylamide copolymer with 1-chloro-2-diethylaminoethane hydrochloride or with epichlorohydrin
and triethanolamine or with triethylglycidylammonium chloride.

Strong and weak-base anion-exchange resins can be synthesized through the
copolymerization of N-(N,N-dimethylaminoethyl)methacrylamide or N-(N,N,N-
trimethylaminoethyl)methacrylamide with N,N'-hexamethylenedimethacrylamide [173].
Acrylamide-N,N'-bis(acrylamidoethyl)ethylene urea copolymer or acrylamide-N,N'-
bis(methylacrylamidomethyl)triethylene urea copolymer on treatment with para-
formaldehyde and diethylamine afford anion-exchange resins which are useful for
protein sorption [174].

Anion-exchange resin useful for the extraction of copper from esters has been obtained by the amination of methylmethacrylate — DVB copolymer with hydrazine [175]
or guanidine [176]. Acryloyl chloride undergoes spontaneous polymerization with
polyethylenepolyamine at room temperature to give an anion-exchange resin with
high capacity and selectivity for bivalent metal ions [177].

3.4 Epichlorohydrin-based Resins

Polymerization of ammonia with epichlorohydrin at 90–100 °C gives an ammonia-
epichlorohydrin copolymer having an exchange capacity of 4.8–5.6 meq/g, of which
~5% is strongly basic [178]. Ammonia-epichlorohydrin copolymer has been crosslinked with aqueous ammonia, aliphatic polyamines, iso-propanol, epichlorohydrin,
or ethylenediamine — epichlorohydrin copolymer to give anion-exchange resins
having high resistance to poisoning by organic materials [179,180]. Anion-exchangers
with selectivity for heavy metals have been prepared by hardening ammonia-epichlorohydrin oligomer with alkali metal sulfides or ammonium sulfide [181]. Ammonia-
epichlorohydrin copolymer has also been treated with diamino isopropanol or
p-phenylenediamine to give anion-exchange resin useful for water purification [182].

Anion-exchange resins prepared from polyalkylenimine, epichlorohydrin, and
third components capable of producing crosslinking have larger life times than is
possible with polyalkylenimine and epichlorohydrin alone [183]. Such useful third
components include·phenol, resorcinol, and aldehydes.

Diamine and polyamine crosslinked polyepichlorohydrin anion-exchange resins have been prepared by reacting polyepichlorohydrin with amine in the mole ratio 1:2 or 1:3 at 80 °C in a mixture of dimethylformamide and acetone for 6 h using 1% aluminosilicate as a catalyst. Depending on the amine structure, the capacity of these resins varies with respect to 0.1N hydrochloric acid from 2.9 to 9.1 meq/g [184].

Melamine-formaldehyde-epichlorohydrin resin, useful for the separation of molybdenum and tungsten, has been prepared by treating melamine with formaldehyde at 60–65 °C and then condensing the reaction product with epichlorohydrin [185]. Polyethylenepolyamine has been polymerized with formaldehyde and phenol-epichlorohydrin or ammonia-epichlorohydrin reaction products to give anion-exchange resins [186]. Pretreatment of polyamine with semicarbazide or its salts increases the exchange capacity.

3.5 Furfural-based Resins

The condensation of furfural with tetramethyldiuram disulfide and polyethylenepolyamine affords an anion-exchange resin (FAT) having high selectivity towards molybdenum [187,188]. Increasing the amount of furfural decreases its exchange capacity towards molybdenum. Resins having selectivity towards molybdenum have also been prepared by the condensation of butanol solution of phenol disulfide and polyethylenepolyamine with furfural [189]. FAT resin is more heat resistant [190] than AN-2F resin (condensation product of methylol phenol derivatives and polyethylenepolyamine). AN-2F resin undergoes 25% loss in capacity on heating with water at 100 °C for 24 h, whereas FAT resin undergoes 12.5% loss. The functional-group cleavage begins at 100 °C and $\geq$150 °C for AN-2F and FAT resins, respectively.

Anion-exchange resins with good heat and chemical resistance and high sorption rates have been prepared by polycondensation of furfural with polyethylenepolyamine and PVC. An increase in PVC content decreases the sorption properties of the resins [191,192].

The polycondensation of furfural with m-phenylene diamine, p-phenylene diamine, or hexamethylene diamine in dimethylformamide at 80–85 °C using zinc chloride or carbon tetrachloride as a catalyst affords anion-exchange resins [193–198] having high resistance to heat, chemicals, and irradiation. The maximum ion-exchange capacity of these resins towards 0.1N hydrochloric acid are 5.7, 5.4, and 6.5 meq/g, respectively.

The styrene-furfural condensation product on chloromethylation followed by amination with dimethylamine affords a weak-base resin [199] having a capacity of 1.4–1.5 meq/g. The relative affinities of this resin for anions follow the sequence, $CO_3^{2-} > NO_3^- > SO_4^{2-} > HCO_3^- > OAc^-$ for the Cl^- form of the resin. The operational efficiency of this resin increases with lowering the pH of the solution to be treated.

4 Composite Ion-Exchange Resins

Blending of two independent ion-exchange resins would result in the development of a composite ion-exchange resin. Ion-exchange resin which incorporates an inert material in its matrix is also classified as a composite resin. Usually, the inert material is incorporated in the crosslinked network of the ion-exchange resin during the course of polymerization.

The incorporation of a weak-acid and a weak-base resin into a crosslinked inert matrix or adjustment of the polymerization procedure of a mixture of weak-acid and weak-base monomers (or their derivatives) so that discrete acidic and basic domains would be formed has led to the development of a new class of ion-exchange resins known as thermally regenerable resins which will be discussed in the next section.

Blending of inert materials with ion-exchange resins to an extent that the desirable properties such as thermal stability and exchange capacity are not affected can partially replace the polymer content thereby reducing the production cost. Vasudevan et al. [200–202] reported the preparation of phenol-formaldehyde composite ion-exchange resins by mixing calculated quantities of sulfonated phenol, formaldehyde, and sulfonated products of coal, saw-dust, lignite, or coconut charcoal and polymerizing them together in presence of sulfuric acid. The p-nitrophenol adsorption studies show that the resins are macroporous in nature. On being heated to 50–70 °C, these resins undergo considerable weight loss which is regained on exposure to air, indicating that the resins are porous and reversibly adsorb air and water vapor. The thermal characteristics of saw-dust-based composite resin reveals that the substitution by saw-dust has hardly any influence on the thermal stability. For the saw-dust-substituted resin, the capacity remains practically constant until the composition corresponds to 30–50% saw-dust. However, the sulfonated coal-substituted resin has a capacity of 1.3 meq/g, which is comparatively less than that of the unmodified resin (capacity 1.8 meq/g).

The presence of insoluble materials in the polymerization mixture may have some control on the structure of the polymeric skeleton. Seidl et al. [203] reported that the microstructure of skeletons of ion-exchange resins, based on the copolymers of styrene and DVB, can be controlled by carrying out the polymerization in presence of an inert material and by adjusting the reaction conditions and concentration of DVB. The microstructure depends on the parameter of interaction and on the molar volume of the inert material. In the case of copolymers modified by an inert component with high molar volume and interaction parameters, microstructures with small measureable surfaces and pores with relatively large radii are obtained.

The incorporation of certain materials in the polymeric matrices of ion-exchange resins can modify the ion-exchange properties, sorption characteristics, etc. of the resins thereby making them more selective in the ion-exchange process. For example, γ-manganese oxide-impregnated cellulose ion-exchange resins [204] are useful for the removal of dissolved high-molecular-weight substances, ammonium and phosphate ions from water.

An ionizable resin which is normally soluble in water can be trapped in the body of a solid insoluble crosslinked resin to give a composite resin [205]. For instance, poly(maleic anhydride) has been trapped in chloromethylated styrene-ethylvinylbenzene-DVB resin by carrying out the polymerization of maleic anhydride in the resin matrix. The resin beads on amination, followed by hydrolysis, give an ion-exchange resin having both anion-exchange and cation-exchange capacity. These ion-exchange beads are used for the separation of sodium chloride and sucrose from a solution. Magnesium chloride and zinc chloride have also been separated using these beads.

A mixture of SDVB copolymer, thermoplastic polypropylene, and dibutyl phthalate, when extruded at 170–190 °C, gives a composite material which on sulfonation yields an ion-exchange resin suitable for use as a high-temperature ion-exchange catalysts possessing high capacity [206].

Composite magnetic ion-exchange resins have been prepared in a variety of physical forms by incorporating very fine particles of magnetic materials in the crosslinked polymeric matrix of the resins [207,208]. There are three general types: (i) homogeneous resins comprising magnetic material uniformly distributed within the crosslinked ion-exchange resin, (ii) homogeneous resins comprised of magnetic material and micro ion-exchange resins within an inert crosslinked polymer, and (iii) heterogeneous resins of shell or whisker type consisting of active polymeric chains grafted onto a core of magnetic polymer formed by embedding magnetic particles within an inert crosslinked polymer.

Magnetic resins of various types can be made by bulk polymerization procedures [209]. Since grinding of the resins prepared by bulk polymerization releases or exposes the magnetic particles, magnetic resins in bead form are preferred. Granular magnetic polymers can be prepared by suspension polymerization of an appropriate monomer in which finely divided magnetic material is dispersed [210]. However, the studies on the synthesis of magnetic SDVB resins [209] adopting the above procedure reveal that complete encapsulation of the magnetic material (finely ground magnetite) is not achieved. In the most successful preparations, polystyrene is precipitated onto the surface of the magnetite. A subsequent chloromethylation reaction shows that the surface is not completely covered with polystyrene. A yield of only 16% resulted with the best product, since much of the magnetite dissolved under acidic conditions. Treatment of the surface of the magnetite with vinyl silane and other silanes is ineffective in causing complete encapsulation under suspension polymerization conditions. It has been concluded that, since a contact angle of 180° cannot be produced, some wetting of the surface by the aqueous medium is inevitable, and this is probably the reason why complete encapsulation by the polymer matrix is not achieved.

To obtain contact angles of zero, it is necessary to use a hydrocarbon suspending medium with polar reactants and the mineral suspended in an immiscible solvent with a high dielectric constant. Surfactants are added to stabilize the system, and the mineral surface is treated to ensure complete wetting by the reactants in the presence of hydrocarbon. By this method, resins of polyamine-epichlorohydrin type are made [209]. An aqueous solution of polyethylenimine is reacted partially with epichlorohydrin and the mixture is then dispersed in paraffin oil containing 1% (wt/vol) of a surfactant by stirring at $\sim$1700 r.p.m. with a high-speed stirrer. Heating at 90 °C for 16 h completes the crosslinking reaction. When very finely divided magnetic particles such as barium ferrite (2–3 μm) are added to the aqueous phase resin, beads are formed with the magnetic particles on the surface. Complete encapsulation occurs if the ferrite surface is pretreated with γ-glycidoxypropyltrimethoxysilane. Other magnetic particles could be satisfactorily encapsulated, e.g., γ-iron oxide and soft ferrites. Tetraethylenepentamine or triethylenetetramine have also been used in lieu of polyethylenimine.

Many monomers can be grafted by redox polymerization to the magnetic core [magnetic material embedded in poly(vinyl alcohol) matrix nominally 100% crosslinked with glutaraldehyde] to get whisker-type resins [211]. Acrylic acid and acrylamide give the most successful products although certain basic monomers, both strong and weak electrolyte types, result in practical grafting levels. The acrylic acid-grafted resin can be esterified with isothionic acid to introduce sulfonic acid groups or reacted with diethylamine and epichlorohydrin to form a weak-base resin. The acrylamide-grafted resin can be treated with hypochlorite to convert the amide groups to primary amino groups [212].

Synthesis of magnetic resins such as SDVB sulfonic acid resin incorporating flue dust [213] and aminated SDVB resin incorporating Fe—Si or Fe—Cr particles [214] have also been reported. Dowex 50W-X8 (100–200 mesh) in H^+ form when soaked in an aqueous solution of Fe (II) salt and then in alkali and oxidized by heating above room temperature with oxygen injection gives a resin having a ferrite coating [215]. In a similar way, KU-2F resin with magnetic properties have been synthesized [216].

Ferromagnetic phenol-formaldehyde sulfonic acid cation-exchange resin has been synthesized [217] by heating a mixture of phenolsulfonic acid and formalin containing magnesium ferrite at 65 °C for 3 h.

The properties of magnetic resins [209] depend on whether the magnetic material incorporated in the resin matrix is hard or soft. Magnetized hard ferrite (barium ferrite) resins behave as permanent magnets; they flocculate strongly once magnetized but are difficult to demagnetize. Soft ferrite (Ferrox cube) resins flocculate in the presence but not in the absence of an applied magnetic field and retain little or no magnetism when the magnetic field is withdrawn. Resins which incorporate γ-iron oxide possibly have unique intermediate properties. After magnetization they retain substantial magnetism and so flocculate outside the initial applied field. Although they can be partly demagnetized, complete demagnetization is difficult.

The flocculating properties [209] of resins from barium ferrite and γ-iron oxide are similar. Even though the magnetized resin is flocculated, the flocs break up readily when stirred so that the particles react more rapidly [218] than standard-sized resin beads which are 300–1200 μm in diameter (Fig. 2). The shell configuration of whisker-type resins also enhances reaction rates. Because of the magnetic flocculation phenomenon the magnetized beads separate out much more rapidly than the unmagnetized ones (Fig. 3).

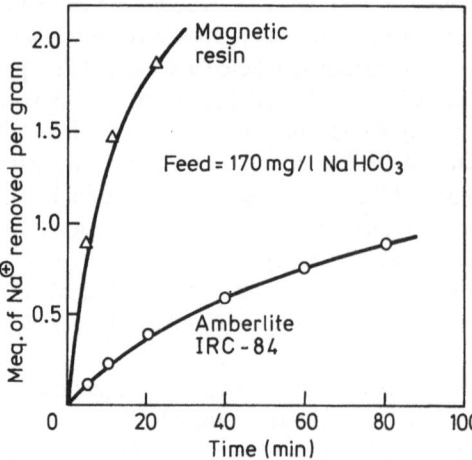

Fig. 2. Reaction rates of magnetic and commercial carboxylic acid resins with sodium bicarbonate solution [218]

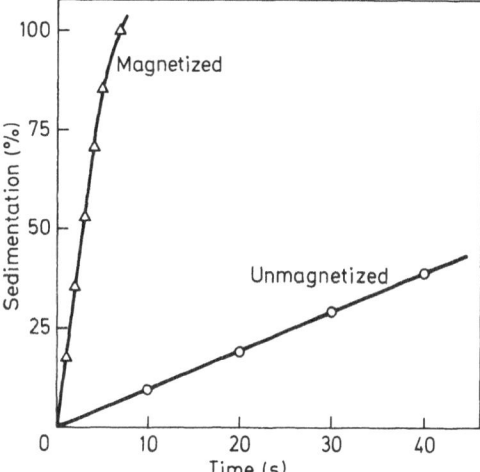

Fig. 3. Sedimentation rates of magnetized and unmagnetized forms of magnetic carboxylic acid resin [218]

The electron micrographs of unmagnetized and magnetized forms of crosslinked polyethylenimine resin beads are shown in Figs. 4a and b, respectively [209,218]. When magnetized they become individual magnets and flocculate strongly to form rings and networks in which the magnetic forces are neutralized. The resulting magnetically flocculated bed has a very high void volume and is characterized by an extensive system of channels (Fig. 4b). The high void volume of the magnetically flocculated material is responsible for another very important property: since the flocs collapse under shear, the resin can be pumped directly without attrition even from a settled bed. This can be achieved with an equipment of peristaltic type and it is a distinct practical advantage over the conventional resins which cannot be handled in such a manner. The unique features of magnetic resins make possible their use in a continuously operated mode in which the resin bed is fluidized by a rapid upflow of water.

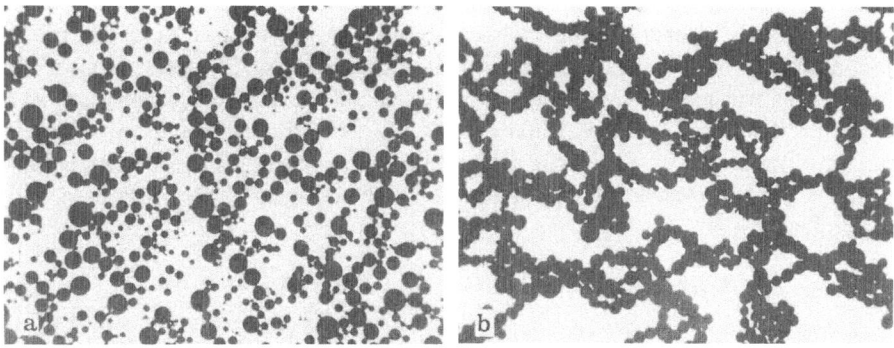

Fig. 4a. Electron micrograph of magnetic ion-exchange resin bead, containing γ-Fe_2O_3 in the unmagnetized form [209,218]; b. Electron micrograph of magnetic ion-exchange resin bead, containing γ-Fe_2O_3 in the magnetized form [209,218]

Table 7. Comparison of typical dealkalization results obtained with the fluidized bed of whisker-type magnetic resin and a conventional contactor of Asahi type [218]

	Fluidized bed of whisker-type magnetic resin	Asahi-type resin
Product: flow (l/min)	120.0	116.0
Ca^{2+} (mg/l)	65.0	46.0
Na$^+$ (mg/l)	160.0	170.0
HCO$_3^-$ (mg/l)	10.0	10.0
Effluent: flow (l/min)	5.6	8.6
Ca^{2+} (mg/l)	2,700.0	1,300.0
Na$^+$ (mg/l)	310.0	240.0

The whisker-type carboxylic acid resin obtained by grafting poly(acrylic acid) onto an inert magnetic core containing a high portion of γ-Fe$_2$O$_3$ finds application in the standard ion-exchange procedure known as dealkalization for treating hard alkaline water. A comparison of typical dealkalization results (Table 7) obtained with the fluidized bed of this type of magnetic resins and a conventional continuous contactor of the Asahi type [218, 219] using the same feed water, shows that calcium removal by this process is almost as efficient as conventional methods and the concentration of the effluent is roughly doubled.

Bolto and Dixon [220] advocated a system which involves a simple pipeline for the adsorption reaction, in which the magnetized resin and raw water are pumped through a pipe at such a flow rate that turbulence results and the flocs are broken apart so that the reaction is complete when the mixture leaves the pipe. The slurry flows into a settling vessel, whereupon the magnetic flocs reform and rapid sedimentation ensues.

5 Thermally Regenerable Ion-Exchange Resins

Thermally regenerable ion-exchange resins or Sirotherm[1] resins refer to a novel ion-exchange process which makes use of hot water as regenerant [221]. A comprehensive review (1977) by Bolto and Weiss [222] has discussed the developments in this field.

The thermally regenerable ion-exchange resins are polymers of weak electrolyte type, a mixture of polymers containing weakly basic and weakly acidic groups [223]. These resins will adsorb significant quantities of salt at ambient temperature and release it on washing with hot water (70–90 °C). The adsorption step involves the transfer of protons from carboxylic acid groups to amino groups to form the cation-exchange and anion-exchange sites followed by the co-ion entry into both basic and acidic sites.

$$\overline{R'CO_2H} + \overline{R''NR_2} + Na^+Cl^- \underset{hot}{\overset{cold}{\rightleftharpoons}} \overline{R'CO_2^-Na^+} + \overline{R''NR_2H^+Cl^-} \quad (1)$$

---

[1] Sirotherm is the ICI Australia Limited Trade Mark for thermally regenerable ion-exchange resins and associated plant

The equilibrium is temperature sensitive with both types of groups showing weaker electrolytic behaviour on heating. The driving force behind this equilibrium is the large increase, about 30 fold in the ionization of water on heating from 25 °C to 70 to 90 °C. The additional protons and hydroxyl ions supress the ionization of weak electrolyte resins.

The research effort in perfecting a resin system capable of efficient performance in a thermally regenerable ion-exchange process can be divided into two distinct phases [224–229]: (i) selection of mixtures of weakly basic and weakly acidic resins as suitable systems and extensive study of the relationship between the chemical structure, basicity, acidity, and thermal stability of ion-exchange resins of the weak electrolyte type, and (ii) efforts put forth to increase the rate of salt uptake and capacity of thermally regenerable systems.

To aid in the selection of an optimum pair of resins, measurements of equilibrium properties of the individual resins have been made in the form of titration curves, which give the loading of ions adsorbed by the resins as a function of pH. When plotted on the same graph, the intersection of two curves gives the resin loading under the conditions of titration. By determining the curves at the temperatures used for adsorption and regeneration, the fraction of the resins' total sites available for thermal regeneration can be forecast [230]. For a system containing crosslinked poly(methacrylic acid) resin (A) and crosslinked poly(vinylbenzylamine) derivative (B) containing a mixture of primary, secondary, and tertiary amino groups, only 8 % (Fig. 5; portion CD) of the resins capacity is utilized at full equilibrium. Certain weakly basic resins titrate to give a remarkable plateau, the outcome of which is that the proportion of exchange sites available for thermal regeneration is greatly magnified. This is shown in Fig. 6, which depicts the curves for a crosslinked poly(acrylic acid) resin (A') and a crosslinked poly(vinylbenzyldiethylamine) resin (B'). This arises because of the more advantageous geometry of the overlapping curves and also due to the greater shift of the curves with temperature. To obtain this effect, the basic resin must be essentially homofunctional. It is also necessary for the basic resin to contain secondary or tertiary amino groups and for the resin backbone and the substituent groups on the nitrogen to be nonpolar in character. By optimizing the adsorption pH

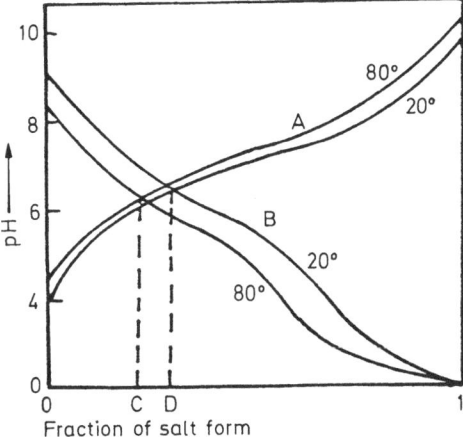

Fig. 5. Titration curves of a poly(methacrylic acid) resin (A) and a substituded polyvinylbenzylamine resin (B) in 1760 mg/l NaCl [230]

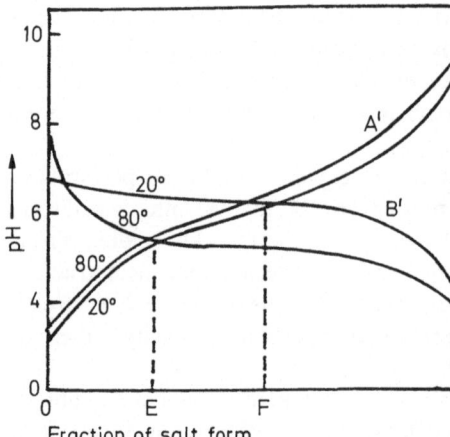

Fig. 6. Titration curves of a poly(acrylic acid) resin (A') and polyvinylbenzyldiethylamine resin (B') in 1760 mg/l NaCl [230]

and the ratio of the acidic to the basic sites, the basic resin's capacity available for thermal regeneration can be improved.

It is, however, much more difficult to obtain plateau behaviour in carboxylic acid resins, since they contain a significantly higher concentration of exchange sites and are inherently more polar [231]. A weak plateau is produced by incorporating many nonpolar residues in the polymer, but the lowering of the concentration of active sites makes such an approach impractical. The advantage of a plateau-type acidic resin can be understood from Fig. 6. Curves of lower gradient at the appropriate acidity level would mean a more effective utilization of the resins in a thermally regenerable system.

Since the salt uptake process takes place at near neutral pH levels where the proton concentration is very low, the rate of salt adsorption by the resin system is extremely slow [232]. The rate-limiting step in thermally regenerable ion-exchange resins is the transfer of protons from the acidic to the basic sites. Rapid reaction rates in these resins have been achieved through the use of different methods.

5.1 Normal-Sized Beads Incorporating Short Diffusion Paths

Bolto et al. [233] have constructed normal-sized beads by including both the required types of groups within one particle so that the diffusion paths for proton transfer are much shorter. Amphoteric resins and snake-cage resins can be considered for use as thermally regenerable resins for the reason stated above, however, their practical disadvantage is their lower performance due to the interaction between carboxyl and amino groups which are in very close proximity to one another.

The rates of salt uptake by mixed-bed normal-bead resins, mixed-bed microresins, and snake-cage resins are compared in Fig. 7. The rate for the original mixed bed of normal-sized resins is the slowest. A slight but inadequate improvement occurs if a macroporous amine resin is used instead of a gel-type resin. Rapid rates have been achieved for the mixed-bed microresins of size 10–20 μm and the snake-cage resin of normal bead size comprising a linear poly(acrylic acid) within a macroporous amine

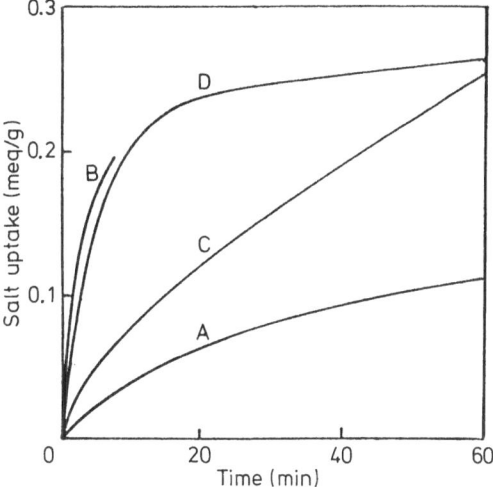

Fig. 7. Rates of salt uptake by weak electrolyte resin systems. (A) mixed bed of De-Acidite G and Zeo-Karb 226 (300–1200 μm); (B) mixed bed of De-Acidite G and Zeo-Karb 226 (10–20 μm); (C) mixed bed of Amberlite IRA-93 and Zeo-Karb 226 (300–1200 μm); (D) snake-cage resin from Amberlite IRA-93 and poly(acrylic acid) (300–1200 μm) [233]

resin. The mixed bed of the microresins is, however, very difficult to handle and requires high pressure for adequate flow through packed beads, and it is difficult to selectively backwash to remove accumulated colloids. In the case of snake-cage resins the increase in the rate of salt uptake is at the expense of seriously lowered capacities for salt uptake arising from interaction between sites of opposite character.

5.2 Plum Pudding Resins

One solution to the problem of making thermally regenerable resins rapidly reacting without loss of capacity is to group together [232] the basic and acidic sites into large domains containing one type of site which would minimize internal salt formation between the acidic and basic groups. If the two types of domains can then be held together with an inert binder, composite beads of normal size can be produced. This principle is utilized in the preparation of 'plum pudding resins' which are normal-sized beads containing micro particles of basic and acidic ion-exchange resins embedded in a matrix permeable to water and salt.

Plum pudding resins are prepared [232] from micro-bead particles of commercial resins, De-Acidite G[2] (1–10 μm) and Zeo-Karb 226[2] (5–10 μm) and a variety of polymers such as cellulosics, ionically crosslinked polysalts, or poly(vinyl alcohol) cross-

---

[2] Products of Permutit Company, London. De-Acidite G: weak-base resin in free form; Zeo-Karb 226: weak-acid resin in free form

linked with aldehydes as the matrix material. The resin is prepared by dispersing in an oil a slurry of the micro ion-exchange resins in acetone solution of the cellulosic and stripping of acetone to obtain the composite beads. Similar composite beads can be made from the polysalts obtained by mixing aqueous solutions of cationic and anionic polyelectrolytes. The polyelectrolytes employed are polyvinylbenzyltrimethyl-ammonium chloride, polyethyleniminium chloride, and sodium polystyrene sulfonate each of varying molecular weight and degree of substitution which dissolve in con-centrated aqueous solutions of inorganic salts containing small amounts of acetone. The solution of polysalts is used in bead preparation, in a procedure similar to that with cellulosics with the matrix being deposited on the removal of acetone. A water-soluble polymer such as poly(vinyl alcohol) can also be employed as the matrix, provided that covalently bonded crosslinks are introduced with the help of dialdehydes through an acid-catalyzed reaction to insolubilize the polymer in the bead-forming step.

Figure 8 suggests that the rates of salt uptake by plum pudding resins containing matrices of different structures are indeed much faster than the same for conventional weak electrolyte resins [232]. The rate is quite dependent on the nature of the matrix material, being slower when the matrix is ethylcellulose or a polysalt made from polyvinylbenzyltrimethylammonium chloride and sodium polystyrene sulfonate. The best result is obtained with a crosslinked poly(vinyl alcohol) matrix, the rate being much more superior to the mixed-bed system and of similar magnitude to that displayed by the strong electrolyte resins in common usage in water-treatment in-dustry. The adsorption rates are also influenced by the amount of matrix present in the composite beads containing polysalt and crosslinked poly(vinyl alcohol) matrices

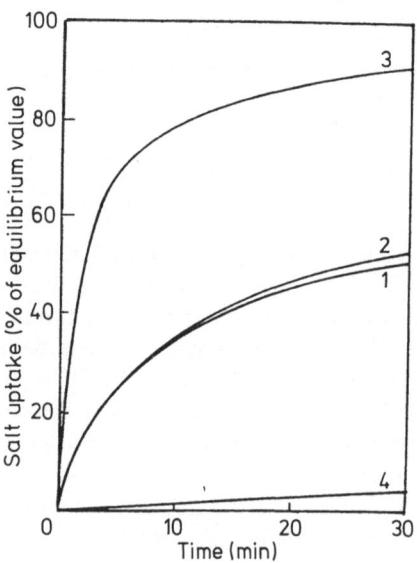

Fig. 8. Rate of salt uptake from 0.02 N salt solution by thermally regenerable plum pudding resins containing the same De-Acidite G and Zeo-Karb 226 plums, and 30 % matrix by weight, as a function of the nature of matrix. Particle size 14–52 mesh, BSS, base resin to acid resin ratio 2:5. (1) ethyl cellulose; (2) polysalt; (3) crosslinked poly(vinyl alcohol) (4) normal De-Acidite G and Zeo-Karb 226 mixed bed [232]

(Table 8 and 9). In both instances the rates slacken as the matrix content is increased suggesting that the diffusion of ions through the matrix is the rate-determining step. For a fixed level and type of matrix, the adsorption rate increases with the decrease in the size of the composite bead (Table 10). The faster reaction of the smaller bead is the confirmation that the matrix is providing the main diffusion barrier.

Table 8. The influence of the amount and type of polysalt[a] matrix on the rate of salt uptake by thermally regenerable plum pudding resins [232]

Cationic polyelectrolyte		Anionic polyelectrolyte		Amount of matrix weight %	Adsorption half-time, minutes
Molecular weight	Degree of substitution	Molecular weight	Degree of substitution		
500,000	0.57	300,000	0.61	16	5.0
20,000	0.74	3,000,000	1.00	19	6.3
500,000	0.57	300,000	0.61	28	29.0
500,000	0.74	300,000	0.77	28	31.0
500,000	0.74	300,000	0.77	28	28.0
500,000	0.74	300,000	0.77	32	32.0
500,000	0.57	300,000	0.61	43	35.0

[a] Polysalts polyvinylbenzyltrimethylammonium chloride and sodium polystyrene sulfonate were used. The ion-exchange plums used were De-Acidite G (1–10 µm) and Zeo-Karb 226 (5–10 µm)

Table 9. The influence of the amount and type of crosslinked poly(vinyl alcohol) matrix [232] on the rate of salt uptake by thermally regenerable plum pudding resins[a]

Crosslinking agent	Amount of matrix, weight %	Adsorption half-time, minutes
Glutaraldehyde	30	2.3
Glutaraldehyde	40	3.6
Glutaraldehyde	67	20.0
Terephthaldehyde	60	26.0

[a] Sufficient crosslinking agent to react with 20 % of the PVA was used. The ion-exchange plums used were De-Acidite G (1–10 µm) and Zeo-Karb 227 (5–10 µm). Plum pudding bead size 14–52 mesh, BSS

Table 10. The influence of the overall plum pudding bead size [232] on adsorption rates of thermally regenerable resins[a]

Bead size mesh, BSS	Microns	Adsorption half-time, minutes
14–52	300–1,200	2.0
22–36	420–700	1.0
52–80	180–300	0.4

[a] The ion-exchange plums used were Amberlite IRA 93 (1–3 µm) and Zeo-Karb 226 (5–10 µm). The plum pudding beads contained 40 % matrix by weight, prepared from crosslinked PVA

Even though the rate of adsorption of plum pudding resins are higher than those of commercially available resins the mechanical strength of the former is inferior to that of the latter. However, the mechanical strength of plum pudding beads containing poly(vinyl alcohol) as matrix can be improved by further treatment with aldehydes such as formaldehyde, glyoxal, glutaraldehyde, or tetraphthaldehyde. Nevertheless, their performance is still considerably inferior to that of commercial resins with the exception of formaldehyde-treated product even though there is an enhancement in the physical strength of the treated plum pudding resins [232].

5.3 No-Matrix Resins

No-matrix resins, i.e., resins devoid of inert matrix as described by Bolto et al. [234] constitute a novel way of enhancing the capacity of thermally regenerable resins. In the idealized structure of a no-matrix resin, active groups of one type are grouped into domains of size 0.01–5 µm, adjacent to domains of opposite character in porous particles of the conventional size 300–1200 µm. Thermally regenerable capacities up to 2.1 meq/g should be possible in the absence of any diluent.

In no-matrix resins, triallylamine is used as the self-crosslinking monomer with methacrylic or acrylic acid as the acidic monomer. The crosslinked resin obtained by polymerizing a mixture of triallylamine and an acidic monomer would be expected to have the sites arranged so that maximum interaction occurs between amine and carboxylic acid groups as shown in Fig. 9 [234]. Indeed, such particles have no thermally regenerable capacity and electron micrographs of these products show a uniform gel phase with no segragation of sites into domains. Obviously, measures are necessary to prevent such interactions. Several approaches have been made to minimize the internal salt formation in no-matrix resins; these are discussed below.

(a) *Counter-ion Route* [235]. Internal salt formation can be blocked to a degree by adding counter-ions to the polymerization mixture. These are preferably multivalent or large organic species which associate with each ionized monomer, thus minimizing the interaction between the monomers themselves. The counter-ions, if multicharged, should also encourage the sites to congregate to zones having a preponderance of the exchange site (Fig. 10). The best effect is obtained with certain

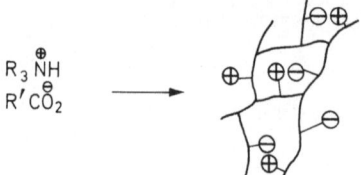

$R_3 \overset{\oplus}{N}H$
$R' \overset{\ominus}{C}O_2$

Fig. 9. Pictorial representation of the internal salt structures formed when two oppositely charged monomers are copolymerized [234]

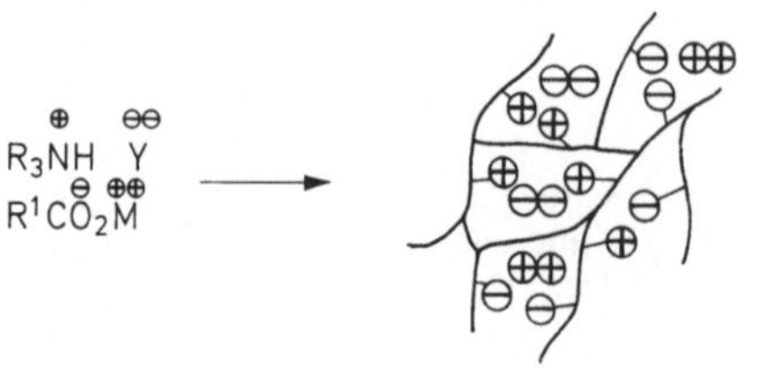

$R_3NH \quad Y$
R^1CO_2M

Fig. 10. Diagrammatic representation of counter-ion approach [235]

Table 11. The influence of counter-ions [235] on the thermally regenerable capacity of resins[a]

Added cation	Added anion	Thermally regenerable capacity, 20–80 °C (meq/g)
Nil	Nil	0.0
Na^+	Cl^-	0.1
Mg^{2+}	Cl^-	0.2
Ca^{2+}	Cl^-	0.5
Zn^{2+}	Cl^-	0.6
Mn^{2+}	Cl^-	0.4
NMe_4^+	Cl^-	0.1
$C_6H_5CH_2NMe_3^+$	Cl^-	0.5
Na^+	SO_4^{2-}	0.3
Na^+	HCO_2^-	0.0
Na^+	$CH_3CO_2^-$	0.2
Na^+	$CH_3CH_2CO_2^-$	0.2
Na^+	$CH_3(CH_2)_3CO_2^-$	0.4

[a] Resins were made from 3:2 mixtures of triallylamine and methacrylic acid at pH 4–5 in the presence of added ions

divalent cations or with bulky organic anions or cations (Table 11). 1:1 divalent salts, such as magnesium sulfate, show no particular advantage. An electron micrograph of a resin made from a mixture of hydrochloric acid and sulfuric acid salts of triallylamine and sodium acrylate reveals the acidic domains as the darkened regions with the basic domains remaining untouched. A distinct separation of the two types of sites is apparent from the electron micrograph with active zones being about 0.5 μm in size. The resin has a thermally regenerable capacity of 0.6 meq/g and is 2.5 times faster than a plum pudding resin.

The best products obtained by the counter-ion route have only about one third of their active sites available for thermal regeneration showing that substantial interaction occurs between the acidic and basic sites. The improvement by having an excess of the counter-ions is not readily achievable because of salting-out effects on the monomers and also due to the difficulty in getting a suitable solvent. In addition to these difficulties, the products obtained by the counter-ion route using γ-irradiation or chemical initiation are mechanically weak.

(b) *Precipitation Approach* [235]. In the precipitation approach a homopolymer macroradical is precipitated, onto which the second monomer is grafted. The solvent used for the precipitation of homopolymer macroradical would play a significant role. Table 12 shows the properties of the resins obtained by using different solvents.

If the precipitation takes place, an opaque product is obtained, whereas a transparent appearance denotes a material having a very high interaction of the sites of the opposite character. Many solvents cause opacity, but the resin produced may still completely lack the desired ion-exchange properties. Most of solvents have a very weak influence, but an exception arises in the case of aliphatic acids. Possibly, the solvent competes with the acidic monomer for the amine hydrochloride, thus minimizing the association of the acidic and basic monomers.

Table 12. Thermally regenerable capacity of resins[a] prepared by the precipitation approach [235)]

Solvent	Appearance of product	Thermally regenerable capacity, 20–80 °C (meq/g)
Water	Transparent	0.0
n-Butanol	Hazy	0.0
i-Hexanol	Opaque	0.1
Diethyl ether	Opaque	0.1
Dimethyl digol	Opaque	0.2
Acetone	Opaque	0.1
Methyl isobutyl ketone	Opaque	0.1
Formic acid	Opaque	0.6
Acetic acid	Transparent	0.0
Propionic acid	Opaque	0.0
n-Butyric acid	Opaque	0.2
n-Pentanoic acid	Opaque	0.3
n-Hexanoic acid	Opaque	0.3

[a] Resins were made from 2:1 mixtures of triallylamine and methacrylic acid in hydrochloric acid at pH 1

(c) *Precursor Method* [234)]. This constitutes a method of avoiding strong interaction between oppositely charged monomers in which either or both monomers are employed as electrically neutral precursors. Precursors for the acid group include esters, amides, acid chlorides, and acid anhydrides, while for primary or secondary amines they include acetyl derivatives. After polymerization the groups may be converted to their usable form by hydrolysis. This approach is shown diagrammatically in Fig. 11.

A typical no-matrix resin has been prepared by polymerizing a mixture of methyl acrylate (MA) and triallylamine (TAA) in solution using γ-irradiation. The resins have been hydrolyzed in 5N potassium hydroxide solution for 48 h at 90 °C under nitrogen atmosphere. A resin prepared in this way has a thermally regenerable capacity of 0.46 meq/g. The addition of ethylene glycol dimethacrylate (EGDMA) as a third monomer (10% by weight based on MA) brings 50% improvement of ion-exchange properties of the resin as compared to that prepared without EGDMA.

No matrix resins have also been prepared by the polymerization of methacrylamide (MAAm) or acrylamide (AAm) and TAA using γ-irradiation. The solvent used for polymerization has a profound influence on the capacity, but a smaller effect on the swelling behaviour. Ethanol, methanol, and DMF are to be preferred to just water, and ethanol is found to be the best solvent for this system giving a resin having a capacity of 1.07 meq/g when the ratio of MAAm to TAA in the polymerization mixture is 1.7.

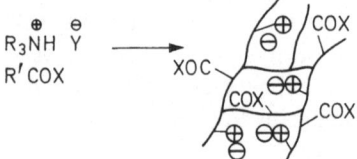

Fig. 11. Pictorial representation of no-matrix resin prepared by using electrically neutral acid precursor [234)]

With the AAm-TAA system, the best thermally regenerable capacity realized is only 0.2 meq/g as against the capacity of 1.1 meq/g for the MAAm-TAA system due to the different behaviour of acrylic and methacrylic species under irradiation conditions. The crosslinking occurs via α-hydrogen in the acrylic units, whereas the main side-reaction with the methacrylic type is the chain scission, which would greatly favour the formation of blocks of polyMAAm onto which the amine moieties could be grafted. The effect of γ-irradiation on polyacrylamide and polymethacrylamide is shown as under:

Polyacrylamide crosslinking

$$-AAm-AAm-AAm- \xrightarrow{\gamma} -AAm-AAm-AAm \qquad (2)$$
$$AAm-AAm-AAm-$$

Polymethacrylamide chain scission

$$-MAAm-MAAm-MAAm- \xrightarrow{\gamma} -MAAm-MAAm \cdot \quad \cdot MAAm \qquad (3)$$

$$-MAAm-MAAm \xrightarrow{B} -MAAm-MAAm-BBBB- \qquad (4)$$
$$B$$
$$B$$
$$-BBBMAAmMAAm-$$

Since the initiation by γ-irradiation is unsatisfactory for AAm-TAA system, the suspension polymerization procedure is adopted using isopropanol as solvent and AIBN as initiator to obtain a resin having a capacity of 0.63 meq/g. When vinylacetate (26% by weight based on AAm) is used as a third monomer, the capacity is improved from 0.63 to 0.9 meq/g. For the MAAm-TAA system the chemical initiation is not preferred since the yields of the resins obtained by this method are lower than those obtained by γ-irradiation.

The electron micrographs of no-matrix resins prepared by the precursor method show that there are discrete acidic and basic domains in which the acidic groups are in effect, embedded in a matrix of the polyamine. A possible mode of polymerization is one in which block copolymers are formed as a result of the formation, initially, of a macroradical. The marked dependence of the thermally regenerable capacity of a resin on the solvent suggests that a macroradical may be involved, since their formation and stability are also greatly dependent on the solvent. However, a study of some linear analogues suggests that this is an oversimplified picture.

(d) *Photografting* [236]. No-matrix resins in which grafting between the acidic and basic domains is controlled by the method of polymerization have been prepared by Jackson and Sasse [236] by using a prepolymer containing grafting sites which could be activated by light to initiate grafting of monomer B:

$$-BBBB-$$
$$B$$
$$AAA-X-AAAA-X-AA \xrightarrow{B} AAA-X-AAAA-X-AA \qquad (5)$$
$$B$$
$$-BBBB-$$

Prepolymer

In general, group X in the prepolymer could be a number of possible photolabile groups. Benzoin alkylethers are usually superior to other photosensitizers studied for the photopolymerization of allylamines [237]. Therefore, prepolymers are prepared in which group X is derived from an unsaturated benzoin derivative of which several suitable structures are shown:

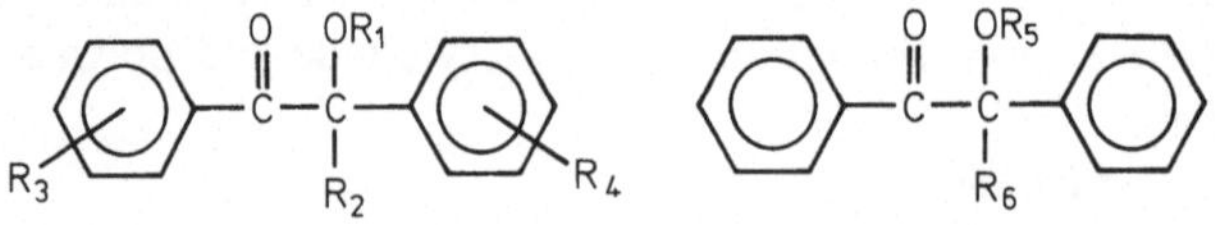

R_3 or R_4 = vinyl
R_1 and R_2 = H or alkyl

R_5 = H or alkyl
R_6 = vinyl or allyl

Allyl benzoin methyl ether (ABME) can be used as the benzoin derivative. Copolymers of methyl acrylate and ABME are prepared and irradiated with ultraviolet radiation. The homolytic cleavage of the benzoin group occurs to form two radicals, one of which is a macroradical. When the copolymer is irradiated in presence of TAA, polymerization of TAA occurs to form a graft of polyTAA onto the poly(MA-co-ABME) backbone, according to the following scheme:

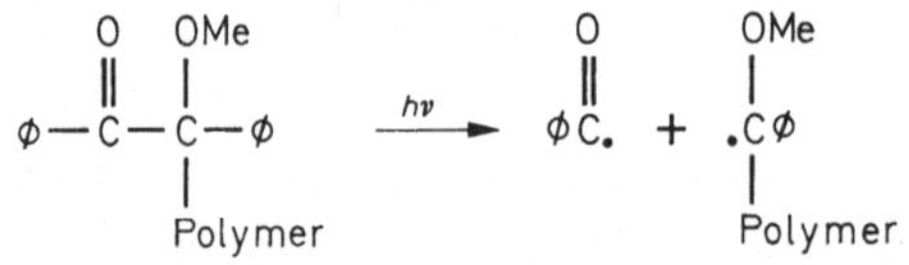

(6)

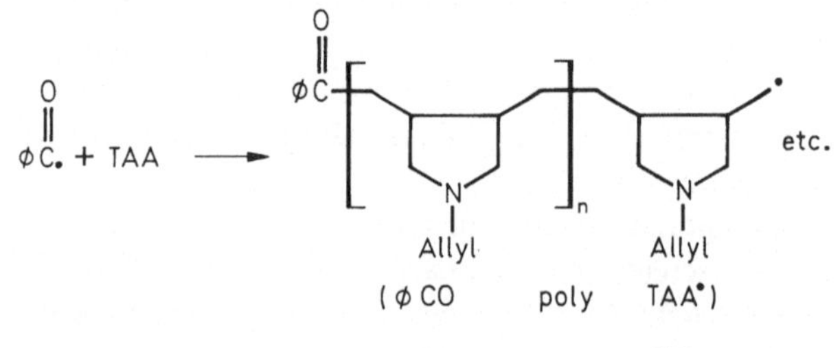

(7)

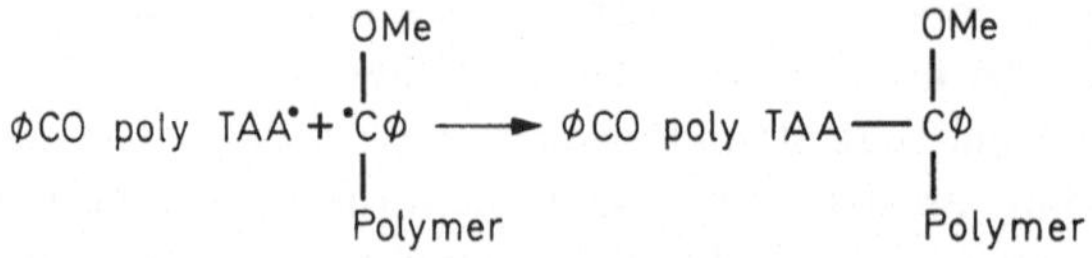

(8)

The grafting of TAA onto poly(MA-co-ABME) is carried out either by solution polymerization using dimethyl formamide as the solvent or by suspension polymerization (better yield) using dimethylformamide as the solvent and paraffin oil as the suspending medium [236].

The yield of TAA-grafted MA-ABME copolymer depends on (a) total concentration of TAA · HCl and MA-ABME copolymer, (b) suspending medium (paraffin oil better than hexane), (c) solvent (dimethylformamide superior to tetrahydrofuran or acetone), and (d) ABME amount in copolymer [236].

The best resins prepared by this method using MA-ABME' copolymer and TAA have a thermally regenerable capacity of about 0.7 meq/g and an acid-to-base ratio of 1.4.

A disadvantage of the photochemical approach is that soluble polymers are needed in order to obtain sufficient light penetration. A consequence of this could be that the final resin is composed of flexible chains, which may then result in unsatisfactorily high degrees of internal neutralization.

(e) *Use of Neutral Precursors in a Heterogeneous Emulsion* [238]. The best thermally regenerable capacity so far achieved for a resin prepared by polymerizing a homogeneous mixture of monomers by γ-irradiation is 1.1 meq/g. Nevertheless, this value is still considerably less than the theoretical value of about 2.1 meq/g, suggesting that a considerable amount of internal salt formation is still occurring. Jackson [238] has attempted the polymerization of monomers in heterogeneous emulsions rather than in homogeneous solutions. In this approach, 1–5 µm droplets of one monomer are dispersed within 0.3–1.0 mm droplets of the other monomer and the two-phase emulsion is polymerized.

In emulsions, amine hydrochloride constitutes the aqueous phase and acrylic ester the organic phase. Cetyltrimethylammonium bromide (CTAB) or span/twin (S/T)-type surfactants are used for emulsion polymerization. Solid dispersants such as talc and colloidal silica are often used to stabilize emulsions which are difficult to stabilize with usual surfactants. Hydrophilic colloidal silica (Aerosil 200) drastically increases the stability of some emulsions provided high amounts (up to 10%) of Aerosil are used. Random copolymers containing 10% hydroxyl groups can be used as polymeric dispersants for preparing w/o emulsions.

Additional crosslinkers, such as DVB in organic phase and 1,6-bis(N,N-diallyl-amino)hexanedihydrochloride (HEXA) in aqueos phase, are used in order to obtain resins having reasonable crosslinking. It is necessary to use one initiator in each phase, AIBN in the organic phase and 2,2′-azobisisobutyramidinium dichloride (amido) in the aqueous phase [238].

The thermally regenerable capacity greatly depends on the nature of the acrylic ester and on the nature of the allylamine (Table 13) [239]. The order of increasing capacity is MA/TAA, MA/DAA, MA/MDAA < EA/TAA, EA/DAA < EA/MDAA, BA/MDAA. The best thermally regenerable capacity of 1.5 meq/g is obtained with a resin prepared from EA, MDAA, and a small amount of MMA. The pronounced dependence of thermally regenerable capacities on the nature of the ester and the nature of the amine can be explained by the relative solubilities of the amine in the ester or vice versa. The greater the solubility of the one in the other, the less distinct will be the separation of the acidic and basic regions in the resin, and, hence, the thermally regenerable capacity will be lower.

Polymerization in third phase: Commercially, the preparation of beads by polymerizing a suspension of a 2-phase emulsion in a third phase appears to be more viable [238]. The third phase should ideally be one in which both acrylic esters and allylamine hydrochlorides are insoluble. However, because of the opposite solubility properties of these two monomers, one of them is invariably soluble in a given third phase. It is believed that if one phase is dispersed in the continuous phase, then that should shield the first phase from the third. However, when the two-phase system is added to a third phase, the two-phase emulsion immediately breaks up. In most cases, the two-phase emulsions also disintegrate on heating and so adding the two-phase emulsion to a heated third phase usually proves disastrous.

Jackson [238] attempted to stabilize the emulsion by carrying out the prepolymeriza-

Table 13. Thermally regenerable capacity of resins made from heterogeneous emulsion of mono-mers [239]

Acid precursor phase[a]	Amine phase[b]	Thermally regenerable capacity, 20–80 °C (meq/g)
Methyl acrylate	TAA	0.19
Methyl acrylate	MDAA	0.03
Methyl acrylate	DAA	0.17
Ethyl acrylate	TAA	0.76
Ethyl acrylate	DAA	0.78
Ethyl acrylate	MDAA	0.94
Butyl acrylate	MDAA	1.20
Ethyl acrylate (plus methyl methacrylate)	MDAA	1.50

[a] With DVB as crosslinker;
[b] As hydrochlorides. TAA = triallylamine, MDAA = methyldiallylamine, DAA = diallylamine; HEXA [1,6-(N,N-diallylamino)hexane] was also used with diallylamines

tion of the two-phase emulsion: in effect, a polymeric dispersant is formed. In addition, prepolymerization increases the viscosity of the emulsion and also reduces the solubility of the ester and amine phases in the third phase. Addition of solid dispersants such as colloidal silica and talc are helpful for bringing out effective stabilization of the partly prepolymerized mixture in the third phase.

In a modification of this approach, monomer droplets (1–5 μm) are replaced by solid particles of monomer, which are then dispersed in the other monomer [239]. For example, N-phenylmaleamic acid is used as the solid monomer and triallylamine hydrochloride is the medium in which the solid particles are dispersed. Chlorobenzene is used as the third phase. However, the thermally regenerable capacity of this resin is not satisfactory.

(f) *Encapsulation of Microparticles of one Polymer by a Matrix of a Polymer of the Opposite Type* [239]. A further reduction in the formation of internal salt structures should be possible by encapsulating microparticles of a polyacid or polyamine with a matrix of the opposite type.

Table 14. Thermally regenerable capacity of resins prepared by encapsulating microparticles of one polymer by a matrix of a polymer of the opposite type [239]

Source of basic sites	Source of acidic sites	Thermally, regenerable capacity, 20–80 °C (meq/g)
Poly(propyldiallylamine)	Ethyl acrylate plus crosslinker	0.3
Poly(diallylamine hydrochloride)	Ethyl acrylate plus crosslinker	1.2
Diallylamine hydrochloride plus crosslinker	Crosslinked poly(acrylic acid)	1.2
Diallylamine hydrochloride plus crosslinker	Crosslinked poly(ethyl acrylate)	2.1

The thermally regenerable resins prepared by polymerizing monomer of one type with a crosslinker around the microparticles of the opposite type are compared in Table 14 [239]. The thermally regenerable capacity of resins prepared from poly(propyldiallylamine) depends on whether the microparticles are present as the free base or hydrochloride form during encapsulation. The thermally regenerable capacity of the resins prepared from the free base form of the polyamine resin is much less (0.3 meq/g) than that of resins prepared from the hydrochloride (1.2 meq/g). This indicates the presence of fewer internal salt structures in the latter, endorsing that the free base form is much more swollen by ethyl acrylate than is the hydrochloride form. Microparticles of crosslinked poly(ethyl acrylate) afford a resin with a thermally regenerable capacity of 2.1 meq/g, which indicates that all internal salt structures have been eliminated.

5.4 Comparison of Thermally Regenerable Resins Prepared by Various Routes

It is evident from Table 15 that the thermally regenerable capacity of 2.1 meq/g, which is theoretically feasible in the absence of any internal neutralization, is achieved only when crosslinked poly(ethyl acrylate) microparticles are encapsulated in a crosslinked poly(diallylamine hydrochloride) matrix). The resins synthesized by other routes have capacities less than 2.1 meq/g, indicating that internal neutralization could not be avoided completely.

Table 15. Comparison of thermally regenerable capacity of resins prepared by various routes

Source of basic sites	Source of acidic sites	Preparative route	Thermally regenerable capacity, 20–80 °C (meq/g)	Ref.
Poly(triallylamine)	Crosslinked poly(acrylic acid)	Plum pudding format	1.40	[231]
Triallylamine hydrochloride	Zinc methacrylate	Counter-ion route	0.60	[235]
Triallylamine hydrochloride	Methacrylic acid in formic acid	Precipitation approach	0.60	[235]
Triallylamine hydrochloride	Methacrylamide plus crosslinker	Precursor method	1.10	[234]
Triallylamine hydrochloride	Methacrylate — allyl benzoin methyl ether copolymer	Photografting method	0.74	[236]
Methyldiallylamine hydrochloride plus crosslinker	Ethylacrylate and methyl acrylate plus crosslinker	Heterogeneous emulsion method	1.50	[239]
Diallylamine hydrochloride plus crosslinker	Crosslinked poly(ethyl acrylate)	Encapsulation method	2.10	[239]

5.5 Operational Conditions and Applications

The simple concept of thermal regeneration when applied in practice is found to involve resin equilibria which are complex. Some of the variables which influence the properties are the detailed polymeric structure of the resins, the acidity and the basicity of the functional groups, the ratio of acidic to basic groups, the resin affinities for counter-ions, the pH, the ionic strength, and temperature [223].

As the thermally regenerable resin capacity is strongly dependent on solution pH and, thus, on any water, the pH of the raw feed water must be closely controlled at a particular level, near neutrality. This is particularly important when the raw water has a high concentration of carbonate or bicarbonate alkalinity which act as a competing buffer with the dual weak-acid/weak-base resin. It can therefore be advantageous to remove the alkalinity prior to contacting the resin [223].

Sirotherm resins tend to be slowly oxidized by dissolved oxygen in raw water at the elevated temperature of regeneration [223,240]; this may be controlled by vacuum deaeration or by other techniques, such as natural gas stripping or combined coagulation and deaeration by ferrous ion.

As sirotherm resins are selective for higher valency cations (e.g., calcium and magnesium), it is difficult to remove both the cations which cause hardness in the same column [223,240]. To use the resin capacity more economically, it is desirable to operate two columns in series, the first removing calcium and magnesium salts and the second removing potassium and ammonium salts.

While desalting domestic sewage effluents for reuse in industry, certain organic anions present in the effluent to the extent of only 0.2 mg/l adsorb very strongly onto the basic component of the resin and slowly foul it. To avoid a deterioration in the performance of the system, it is necessary to pretreat the effluent by passing it through a strongly basic ion-exchange resin which acts as a trap for the removal of the offending organic compounds [231].

Typical performance data [223] achieved in a laboratory scale with a single fixed-bed column are shown in Table 16. The performance data indicate that the sirotherm resin can produce water of salinity as low as 50 to 100 mg/l dissolved salts. The operating capacity of the resins restricts the economic upper range of salinities to be treated between 2000 and 3000 mg/l. The resins can be used to treat hard water, although to remove salts of both divalent and monovalent cations the process should

Table 16. Typical 'Sirotherm' performance data [223]

Feed-water concentration TDS mg/l	Product-water concentration TDS mg/l	Throughput $m^3/day/m^3$ resin	% Recovery (yield)
540	50	408	81
1,000	500	724	81
1,000	500	420	86
1,040	350	265	73
1,750	730	330	77
2,040	1,100	330	63
3,000	1,000	160	66

be operated in two stages. The resins are suitable for deminerilization of mildly brackish water and underground water for industrial and municipal use [222,241], for the treatment of industrial and municipal effluents and cooling tower blowdown for reuse, and as a roughing stage in the production of high-quality boiled feed water.

6 Concluding Remarks

The review above highlights the use of various monomeric moieties and modification procedures involved in the synthesis and tailoring of the properties of ion-exchange resins.

The development in the field of cation-exchange and anion-exchange resins have been discussed mainly on the basis of the monomers used for the synthesis of polymer matrices. Many of the chemical systems studied do not appear to be of commercial significance at first sight. Nevertheless, enough scope exists in designing novel synthesis/modification based on such systems so as to render them more useful and economically viable.

In the literature, thermally regenerable resins have also been referred to as composite resins. However, the thermally regenerable resins are dealt with in a separate section as the developments in this new field are quite significant. Mainly, the systematic approach on the manipulation of polymerization procedures to obtain resins having the maximum thermally regenerable capacity is discussed. The Section on composite resins deals with resin systems which incorporate inert materials and magnetic particles.

While the thermally regenerable plum pudding resins can be classified as composite resins without ambiguity, the authors are of the opinion that it may not be appropriate to use such a classification for thermally regenerable no-matrix resins. Unlike the plum pudding resin, which is obtained by blending two independent resins in an inert matrix, the no-matrix resin is devoid of any inert matrix and refers to a polymer matrix in which the weakly acidic and weakly basic domains are segregated.

The developments on chelating, amphoteric and redox ion-exchange resins, and applications of ion-exchange resins which are not covered in this review will be discussed in a later publication.

Acknowledgements. The authors express their thanks to Prof. A. Ledwith for his encouragements. Thanks are also due to the authorities of the Indian Institute of Technology, Kharagpur for facilities, and the Tokyo Institute of Technology, Tokyo for providing the UNESCO fellowship to S.P.

7 References

1. Millar, J. R. et al.: J. Chem. Soc. 218 (1963)
2. Millar, J. R. et al.: J. Chem. Soc. 2779 (1963)
3. Millar, J. R., Smith, D. G., Kressman, T. R. E.: J. Chem. Soc. 304 (1965)
4. Millar, J. R. et al.: J. Chem. Soc. 2740 1964)
5. Sederal, W. L., De Jong, G. J.: J. Appl. Polym. Sci. *17*, 2835 (1973)
6. Kun, K. A., Kunin, R.: J. Polym. Sci. *C16*, 1457 (1967)
7. Abrams, I. M.: U.S. Pat. 3122514 (1964)

8. Abrams, I. M.: Ind. Eng. Chem. *48*, 1469 (1953)
9. Resindion, S. p. A.: Ital. Pat. 653389 (1963)
10. Resindion, S. p. A.: Brit. Pat. 1082635 (1967)
11. Roubinek, L., Wilson, A. G.: S. African Pat. 6604866 (1978)
12. McMaster, L. P., Gilliand, E. R.: Ind. Eng. Chem., Prod. Res. Develop. *11*, 97 (1972)
13. Hilgen, H., De Jong, G. J., Sederel, W. L.: J. Appl. Polym. Sci. *19*, 2647 (1975)
14. Millar, J. R.: J. Chem. Soc. 1311 (1960)
15. Wojaczynska, M.: Chem. Abstr. *81*, 38185 (1974)
16. Haeupke, K. et al.: Ger. (East) Pat. 71620 (1967)
17. Barrett, J. H., Clemens, D. H.: Fr. Pat. 2092354 (1971)
18. Kolarz, B. N.: Chem. Abstr. *81*, 50369 (1974)
19. Kolarz, B. N.: J. Polym. Sci., Polym. Symp. *47*, 197 (1974)
20. Haeupke, K. et al.: Brit. Pat. 1116800 (1968)
21. Haeupke, K. et al.: Ger. (East) Pat. 60888 (1968)
22. Haeupke, K. et al.: Brit. Pat. 1151480 (1968)
23. Davankov, V. A., Rogozhin, S. V., Tsyurupa, M. P.: J. Polym. Sci., Polym. Symp. *47*, 95 (1974)
24. Tsyurupa, M. P., Davankov, V. A., Rogozhin, S. V.: J. Polym. Sci., Polym. Symp. *47*, 189 (1974)
25. Davankov, V. A., Tsyurupa, M. P., Rogozhin, S. V.: Angew. Markomol. Chem. *32*, 145 (1973)
26. VEB Farbenfabrik Wolfen: Brit. Pat. 1095746 (1967)
27. Wolf, F., Frederich, K.: Ger. (East) Pat. 57703 (1967)
28. VEB Farbenfabrik Wolfen: Brit. Pat. 1521362 (1968)
29. Tsyurupa, M. P., Davankov, V. A.: J. Polym. Sci., Polym. Chem. Ed. *18*, 1399 (1980)
30. Haas, H. C., Livingston, D. L., Saunders, M.: J. Polym. Sci. *15*, 503 (1955)
31. Davankov, V. A., Tsyurupa, M. P., Rogozhin, S. V.: Angew. Makromol. Chem. *53*, 19 (1976)
32. Trushin, B. N., Tyurikov, V. K.: Vysokomol. Soedin. *B16*, 823 (1974)
33. Grassie, N., Meldrum, I. G.: Eur. Polym. J. *7*, 629 (1971)
34. Grassie, N., Gilks, J.: J. Polym. Sci., Polym. Chem. Ed. *11*, 1531 (1973)
35. Grassie, N., Gilks, J.: J. Polym. Sci., Polym. Chem. Ed. *11*, 1985 (1973)
36. Grassie, N., Flood, J., Cunningham, J. G.: Eur. Polym. J. *12*, 641 (1976)
37. Grassie, N., Flood, J., Cunningham, J. G.: Eur. Polym. J. *12*, 647 (1976)
38. Peppas, N. A., Valkanas, G. N.: Angew. Makromol. Chem. *62*, 163 (1977)
39. Peppas, N. A., Bussing, W. N., Slight, K. A.: Polym. Bull. *4*, 193 (1981)
40. Peppas, N. A., Barar, D. G.: Polym. News 7, 32 (1980)
41. Peppas, N. A., Barar, D. G.: Org. Coat. Appl. Polym. Sci. Proc. *46*, 502 (1982)
42. Davankov, V. A., Tsyurupa, M. P., Rogozhin, S. V.: Vysokomol. Soedin. *B15*, 463 (1973)
43. Negre, M., Bartholin, M., Guyot, A.: Angew. Makromol. Chem. *80*, 19 (1979)
44. Peppas, N. A., Valkanas, G. N.: J. Polym. Sci., Polym. Chem. Ed. *12*, 2567 (1974)
45. Davankov, V. A., Tsyurupa, M. P.: Angew. Makromol. Chem. *91*, 127 (1980)
46. Biswas, M., Chatterjee, S.: J. Appl. Polym. Sci. *27*, 3851 (1982)
47. Biswas, M., Chatterjee, S.: J. Appl. Polym. Sci. *27*, 4645 (1982)
48. Biswas, M., Chatterjee, S.: Angew. Makromol. Chem. *113*, 11 (1983)
49. Biswas, M., Chatterjee, S.: Eur. Polym. J. *19*, 317 (1983)
50. Biswas, M., Chatterjee, S.: J. Appl. Polym. Sci. (in press)
51. Pielichowski, J., Morawiec, E.: J. Appl. Polym. Sci. *20*, 1803 (1976)
52. Biswas, M., John, K. J.: Angew. Makromol. Chem. *72*, 55 (1978)
53. Biswas, M., John, K. J.: J. Appl. Polym. Sci. *23*, 2327 (1979)
54. Biswas, M., Packirisamy, S.: J. Appl. Polym. Sci. *27*, 149 (1982)
55. Biswas, M., Packirisamy, S.: J. Appl. Polym. Sci. *27*, 161 (1982)
56. Biswas, M., Packirisamy, S.: J. Appl. Polym. Sci. *25*, 511 (1980)
57. Biswas, M., Bagchi, S.: J. Appl. Polym. Sci. (in press)
58. Biswas, M., Mishra, G. C.: Makromol. Chem., Rapid Commun. *182*, 261 (1980)
59. Biswas, M., Mishra, G. C.: J. Appl. Polym. Sci. *26*, 1719 (1981)
60. Packirisamy, S.: Ph. D. Thesis, Indian Institute of Technology Kharagpur, India 1982
61. Biswas, M., Das, S. K.: J. Macromol. Sci., − Chem. *A16*, 745 (1981)
62. Biswas, M., Packirisamy, S.: J. Appl. Polym. Sci. *27*, 1823 (1982)
63. Armitage, G. M., Lyle, S. J.: Talanta *20*, 315 (1973)
64. Dima, M. et al.: Chem. Abstr. *61*, 16251g (1964)

65. Poinescu, Ig., Scondac, I., Dima, M.: Chem. Abstr. *61*, 16251 h (1964)
66. Scondac, I., Poinescu, Ig., Dima, M.: Chem. Abstr. *61*, 16252 a (1964)
67. Carpov, A., Cotrut, G. V., Dima, M.: Chem. Abstr. *63*, 15048 c (1965)
68. Dima, M. et al.: Chem. Abstr. *63*, 15048 h (1965)
69. Ikariya, M., Takeshita, S.: Chem. Abstr. *84*, 90908 (1976)
70. Uno, T. et al.: Chem. Abstr. *89*, 111620 (1978)
71. Poinescu, Ig.: Chem. Abstr. *71*, 39713 (1969)
72. Kuriyama, S., Yamashita, C.: Jap. Pat. 5383 (1952)
73. Ashida, K.: Chem. Abstr. *48*, 14042 (1954)
74. Daul, G. C., Reid, J. D., Reinhardt, R. M.: Ind. Eng. Chem. *46*, 1042 (1954)
75. Mirkamilov, T. M., Turaev, E.: Chem. Abstr. *83*, 164912 (1975)
76. Ferrel, R. E., Olcott, H. S., Fraenkel-Conrat, H.: J. Am. Chem. Soc. *70*, 2101 (1948)
77. Arai, K., Ogiwara, Y.: J. Appl. Polym. Sci. *20*, 1989 (1976)
78. D'Alelio, G. F.: U.S. Pat. 2340111 (1944)
79. Farbenfabriken Bayer: Brit. Pat. 719330 (1954)
80. Howe, P. G., Kitchener, J. A.: J. Chem. Soc. 2143 (1955)
81. Bodamer, G. W.: U.S. Pat. 2597437 (1952)
82. Deuel, H., Hutschneker, K., Solms, J.: Z. Electrochem. *57*, 172 (1953)
83. Galina, H., Kolarz, B.: Chem. Abstr. *82*, 86921 (1975)
84. Kolarz, B., Kuczynska, N.: Chem. Abstr. *83*, 28912 (1975)
85. Stelzner, K., Reuter, H., Rosel, E.: Ger. (East) Pat. 67583 (1969)
86. Mathieson, A. R., Shet, R. T.: J. Polym. Sci. Part A-1, *4*, 2945 (1966)
87. Telvina, A. S., Alferova, S. V., Korshak, V. V.: Chem. Abstr. *89*, 110926 (1978)
88. Kuznetsova, N. N. et al.: Ger Offen Pat. 2719723 (1978)
89. Kuznetsova, N. N. et al.: U.S.S.R. Pat. 562093 (1979)
90. Diharburg, V. A. et al.: Chem. Abstr. *69*, 19915 (1968)
91. Kuznetsova, N. N. et al.: U.S.S.R. Pat. 322332 (1971)
92. Wolf, F., Raabe, U., Richter, L.: Ger. (East) Pat. 84369 (1971)
93. Volkov, B. V., Asmakova, A. S.: Chem. Abstr. *62*, 696 g (1965)
94. Vasile'v, A. A., Gerasimyuk, T. V.: Chem. Abstr. *62*, 9310 b (1965)
95. Wolf, F., Haeupke, K.: Ger. (East) Pat. 35351 (1965)
96. Bogdanov, V. P. et al.: U.S.S.R. Pat. 194300 (1967)
97. Wolf, F., Schallert, U.: Chem. Abstr. *78*, 30611 (1973)
98. Papukova, K. P., Kuznetsova, N. N., Libel, A. N.: Chem. Abstr. *78*, 98325 (1973)
99. Kuznetsova, N. N. et al.: Chem. Abstr. *67*, 100636p (1967)
100. Petrova, O. T., Zherebtsov, I. P., Lopatinskii, V. P.: Chem. Abstr. *81*, 14054 (1974)
101. Rizaev, N. U., Sultanov, A. S., Muslimov, Kh. I.: Chem. Abstr. *73*, 46133 (1970)
102. Tursunova, D. R. et al.: U.S.S.R. Pat. 423794 (1974)
103. Askarov, M. A. et al.: U.S.S.R. Pat. 431190 (1974)
104. Nazirova, R. A., Mukhamedova, M. A.: Chem. Abstr. *85*, 161043 (1976)
105. Rizaev, N. U., Sultanov, A. S., Muslimov, Kh. I.: Chem. Abstr. *73*, 56680 (1970)
106. Biswas, M., Roy, A., Packirisamy, S.: Indian J. Tech. *18*, 259 (1980)
107. Askarov, M. A. et al.: U.S.S.R. Pat. 318596 (1971)
108. Askarov, M. A. et al.: Chem. Abstr. *76*, 141581 (1976)
109. Tsveshko, G. S., Nazirova, R. A., Dzhalilov, A. T.: Chem. Abstr. *78*, 148503 (1973)
110. Kapadia, R. N., Dalal, A. K.: Indian J. Tech. *19*, 127 (1981)
111. Askarov, M. A. et al.: Chem. Abstr. *82*, 4865 (1975)
112. Tsveshko, G. S., Nazirova, R. A., Dzhalilov, A. T.: Chem. Abstr. *83*, 132442 (1975)
113. Tadzhiev, A. T., Yunusova, D. R.: Chem. Abstr. *70*, 12347 (1969)
114. Kuzin, I. A. et al.: U.S.S.R. Pat. 467916 (1975)
115. Milusheva, A. Sh., Mirkamilova, M. S.: Chem. Abstr. *77*, 49246 (1972)
116. Askarov, M. A., Milusheva, A. Sh., Mirkamilova, M. S.: Chem. Abstr. *83*, 18158 (1976)
117. Prokopenko, V. S., Vasilenko, V. S., Milusheva, A. Sh.: Chem. Abstr. *90*, 122352 (1979)
118. Biswas, M., Kabir, G. M. A.: Angew. Makromol. Chem. *73*, 53 (1978)
119. Dasare, B. D., Krishnaswamy, N.: J. Appl. Polym. Sci. *9*, 2655 (1965)
120. Gujar, K. B., Krishnaswamy, N.: Indian J. Tech. *4*, 208 (1966)
121. Ramaswamy, R., Krishnaswamy, N.: Indian J. Tech. *10*, 185 (1972)

122. Ramaswamy, R., Krishnaswamy, N.: Indian J. Tech. *10*, 189 (1972)
123. Metha, B. J., Krishnaswamy, N.: J. Appl. Polym. Sci. *18*, 1585 (1974)
124. Metha, B. J., Krishnaswamy, N.: J. Appl. Polym. Sci. *20*, 2229 (1976)
125. Metha, B. J., Krishnaswamy, N.: J. Appl. Polym. Sci. *20*, 2239 (1976)
126. Nazirova, R. A., Pulatova, Sh. A., Askarov, M. A.: Chem. Abstr. *83*, 80060 (1975)
127. Pulatova, Sh. A., Nazirova, R. A.: Chem. Abstr. *84*, 31803 (1976)
128. Askarov, M. A. et al.: U.S.S.R. Pat. 444785 (1974)
129. Khundkar, M. H., Malek, A.: Pak. J. Sci. Ind. Res. *10*, 77 (1967)
130. Khundkar, M. H., Mahmood, A. J.: Pak. J. Sci. Ind. Res. *5*, 147 (1962)
131. Mahmood, A. J., Khundkar, M. H.: Pak. J. Sci. Ind. Res. *11*, 231 (1968)
132. Biswas, M., Packirisamy, S.: Indian J. Tech. *17*, 485 (1979)
133. Anderson, R. E.: Ind. Eng. Chem. *3*, 85 (1964).
134. Permutit, A. G.: Ger. Pat. 1234385 (1967)
135. VEB Farbenfabrik Wolfen: Ger. Pat. 1248300 (1967)
136. VEB Farbenfabrik Wolfen: Brit. Pat. 1050207 (1966)
137. VEB Farbenfabrik Wolfen: Ger. Pat. 1184959 (1965)
138. Kovaleva, M. P., Petrova, G. K.: Chem. Abstr. *71*, 102779 (1969)
139. Sabrowski, E., Hauptmann, R.: Ger. (East) Pat. 76474 (1970)
140. Egawa, H., Sugahara, K.: Chem. Abstr. *75*, 64691 (1971)
141. Egawa, H., Saeki, H.: Chem. Abstr. *75*, 64935 (1971)
142. Davankov, A. V., Trushin, B. N.: Chem. Abstr. *75*, 130374 (1971)
143. Kovaleva, M. P. et al.: U.S.S.R. Pat. 309019 (1971)
144. Trushin, B. N., Naumov, S. N., Mityukova, A. G.: Chem. Abstr. *78*, 59119 (1978)
145. Petrariu, I. et al.: Chem. Abstr. *81*, 14364 (1974)
146. Laskorin, B. N. et al.: U.S.S.R. Pat. 394391 (1973)
147. Akiyama, H. et al.: Jap. Kokai Pat. 7890179 (1978)
148. Galazzi, H.: Ger. Offen Pat. 2455946 (1975)
149. Hauptmann, R., Schwachula, G.: Z. Chem. *6*, 227 (1968)
150. Hauptmann, R., Schwachula, G., Reuter, H.: Ger. (East) Pat. 8996 (1972)
151. Rusting, Ir. N., Frielink, J. G.: Neth. Pat. Appl. 6414948 (1965)
152. Schwachula, G., Hauptmann, R., Kain, J.: J. Polym. Sci., Polym. Symp. *47*, 103 (1974)
153. Romania, Ministry of the Chemical Industry: Rom. Pat. 48095 (1967)
154. De Iong, Greet, J.: Ger. Offen Pat. 1953421 (1970)
155. Dubsky, F., Sykova, V.: Czech. Pat. 132869 (1969)
156. Romania, Ministry of the Education: Rom. Pat. 47928 (1967)
157. Warshawsky, A., Kalir, R.: J. Appl. Polym. Sci. *24*, 1125 (1979)
158. Belfer, S., Warshawsky, A.: Proc. IUPAC *26*, 718 (1979)
159. Belfer, S., Glozman, R.: J. Appl. Polym. Sci. *24*, 2147 (1979)
160. Belfer, S. et al.: J. Appl. Polym. Sci. *25*, 2241 (1980)
161. Belfer, S.: Proc. IUPAC *27*, 230 (1980)
162. Belfer, S., Warshawsky, A.: Ind. Eng. Chem., Proc. Res. Develop. *20*, 350 (1981)
163. Galitskaya, N. B., Pashkov, A. B., Lyustgarten, E. I.: Chem. Abstr. *68*, 60158 (1968)
164. VEB Farbenfabrik Wolfen: Ger. (East) Pat. 32571 (1965)
165. Kagan, E. Sh., Panyushkin, V. T.: Chem. Abstr. *71*, 62043
166. Askarov, M. A. et al.: Chem. Abstr. *83*, 11242 (1975)
167. Nabiev, M.: Chem. Abstr. *85*, 124732 (1976)
168. Prodius, L. N., Ergozhin, E. E., Zavsegolova, T. E.: Chem. Abstr. *85*, 136385 (1976)
169. Zubakova, L. B. et al.: Chem. Abstr. *75*, 141433 (1971)
170. Storey, B. T., Miller, L. M., Fries, W.: U.S. Pat. 3311572 (1967)
171. Dave, P. J. et al.: Proc. Ion-Exchange Symp. CSMCRI, India 33 (1978)
172. Strop, P.: Ger. Offen Pat. 2505350 (1975)
173. Kolomeitsev. O. P., Kuznetsova, N. N.: Chem. Abstr. *78*, 85123 (1973)
174. Miyachi, T. et al.: Jap. Pat. 7338791 (1973)
175. Kovaleva, M. P. et al.: U.S.S.R. Pat. 528310 (1976)
176. Kovaleva, M. P. et al.: U.S.S.R. Pat. 523112 (1976)
177. Dzhalilov, A. T., Yariev, O. M.: Chem. Abstr. *85*, 160925 (1976)
178. Wolf, F., Renger, P.: Chem. Abstr. *77*, 89207 (1972)

179. Luca, C. et al.: Rom. Pat. 65764 (1978)
180. Grachev, L. L., Samborskii, I. V., Chetverikov, A. F.: Fr. Pat. 2029926 (1970)
181. Chetverikov, A. F., Polikarenko, V. P., Samborskii, I. V.: U.S.S.R. Pat. 430658 (1980)
182. Grachev, L. L., Samborskii, I. V., Chetverikov, A. F.: Ger. Offen Pat. 1800308 (1970)
183. Bachmann, R., Friedrich, W.: Ger. (East) Pat. 33164 (1964)
184. Ergozhin, E. E., Menligaziev, E. Zh., Tastanov, K. Kh.: Chem. Abstr. 87, 85577 (1977)
185. Matrenkin, V. F.: U.S.S.R. Pat. 328141 (1972)
186. Samborskii, I. V. et al.: U.S.S.R. Pat. 398570 (1974)
187. Askarov, M. A. et al.: U.S.S.R. Pat. 436839 (1974)
188. Abdilakhadov, V., Nazirova, R. A., Gabrielyan, N. A.: Chem. Abstr. 84, 18148 (1976)
189. Askarov, M. A., Nazirova, R. A., Dzhalilov, A. T.: Chem. Abstr. 78, 160604 (1973)
190. Abdilakhadov, V.: Chem. Abstr. 86, 17418 (1977)
191. Fatkhullaev, E., Sukhinina, L. A., Dzhalilov, A. T.: Chem. Abstr. 93, 8839 (1980)
192. Askarov, M. A. et al.: U.S.S.R. Pat. 435257 (1974)
193. Askarov, M. A., Ibragimova, G. T., Mirkamilova, M.: Chem. Abstr. 79, 79568 (1973)
194. Ibragimova, G. T., Mirkamilova, M. S., Askarov, M. A.: Chem. Abstr. 83, 60004 (1975)
195. Ibragimova, G. T., Mirkamilova, M. S., Askarov, M. A.: Chem. Abstr. 83, 6005 (1975)
196. Askarov, M. A., Mirkamilova, M. S., Ibragimova, G. T.: Chem. Abstr. 83, 132378 (1975)
197. Askarov, M. A., Ibragimova, G. T., Mirkamilova, M. S.: Chem. Abstr. 86, 107102 (1977)
198. Askarov, M. A., Ibragimova, G. T., Mirkamilova, M. S.: Chem. Abstr. 86, 107286 (1977)
199. Renganathan, S., Krishnaswamy, N. : J. Appl. Polym. Sci. 19, 2331 (1975)
200. Vasudevan, P. et al.: J. Polym. Sci., Polym. Chem. Ed. 16, 2545 (1978)
201. Vasudevan, P., Sharma, N. L. N.: J. Appl. Polym. Sci. 23, 1443 (1979)
202. Sharma, N. L. N., Vasudevan, P.: Indian J. Tech. 17, 450 (1979)
203. Seidl, J., Malinsky, J., Krejcar, E.: Chem. Abstr. 63, 15047 (1965)
204. Joergensen, S. E.: Ger. Offen Pat. 2409951 (1974)
205. Hatch, M. J.: U.S. Pat. 3332890 (1967)
206. Chaplits, D. N. et al.: Brit. Pat. 1453336 (1967)
207. Bolto, B. A.: Ion-Exchange Process in Pollution Control (ed.) Calmon, C., Gold, H., Vol. 2, p. 213, Boca Raton, CRC Press 1979
208. Bolto, B. A.: J. Macromol. Sci., − Chem. A 14, 107 (1980)
209. Blesing, N. V. et al.: Ion-Exchange in the Process Industries, Soc. Chem. Ind., London. 371 (1970)
210. Bolto, B. A. et al.: Filtration and Separation. 11, 461 (1974)
211. Bolto, B. A., Dixon, D. R., Eldridge, R. J.: J. Appl. Polym. Sci. 22, 1977 (1978)
212. Eldridge, R. J.: Paper presented to Division of Organic Coatings and Plastics Chemistry, Symposium on Modification of Polymers, American Chem. Soc. Meeting, Hawaii, 1979
213. Acharya, H. K., Krishnaswamy, N.: J. Appl. Polym. Sci. 26, 2939 (1979)
214. Imangazieva, G. K. et al.: Chem. Abstr. 77, 75871 (1972)
215. Takeuchi, K., Fujita, K., Yamashita, H.: Japan Kokai Tokkyo Koho Pat. 78146986 (1978)
216. Svyadoshch, Yu. N. et al.: Chem. Abstr. 77, 75874 (1972)
217. Okamoto, S., Okamoto, S.: Japan Kokai Pat. 7768247 (1977)
218. Bolto, B. A. et al.: J. Polym. Sci. Polym. Symp. 49, 211 (1975)
219. Melbourne, J. D., Blesing, N. V.: Mech. Chem. Eng. Trans. MC4, 155 (1968)
220. Bolto, B. A. et al.: Progr. Water Tech. 9, 833 (1977)
221. Battaerd, H. A. et al.: Effluent Water Treatment J. 14, 245 (1974)
222. Bolto, B. A., Weiss, D. E.: Ion-Exchange and Solvent Extraction (ed.) Marinsky, J. A., Marcus, Y., Vol. 7, p. 221, New York, Marcel Dekker 1977
223. Stephens, G. K., Bolto, B. A.: Effluent Water Treatment J. 17, 116 (1977)
224. Weiss, D. E. et al.: Aust. J. Chem. 19, 561 (1966)
225. Weiss, D. E. et al.: Aust. J. Chem. 19, 589 (1966)
226. Weiss, D. E. et al.: Aust. J. Chem. 19, 765 (1966)
227. Weiss, D. E. et al.: Aust. J. Chem. 19, 791 (1966)
228. Bolto, B. A. et al.: Aust. J. Chem. 21, 2703 (1968)
229. Warner, R. E., Kennedy, A. M., Bolto, B. A.: J. Macromol. Sci., − Chem. 44, 1125 (1970)
230. Bolto, B. A.: Research Review (1974-75), p. 13, Division of Chem. Tech., CSIRO, Australia

231. Bolto, B. A.: IUPAC, Polymeric Amines and Ammonium Salts (ed.) Goethals, E. J., p. 365, New York, Pergamon Press 1980
232. Bolto, B. A. et al.: Desalination, *13*, 269 (1973)
233. Bolto, B. A. et al.: J. Macromol. Sci., — Chem. *A4*, 1039 (1970)
234. Bolto, B. A. et al.: J. Polym. Sci., Polym. Symp. *55*, 95 (1976)
235. Bolto, B. A., Siudak, R. V.: J. Polym. Sci., Polym. Symp. *55*, 87 (1976)
236. Jackson, M. B., Sasse, W. H. F.: J. Macromol. Sci., — Chem. *A11*, 1137 (1977)
237. Schildknecht, C. E.: Allyl Compounds and Their Polymers, p. 531, New York, Wiley — Interscience 1973
238. Jackson, M. B.: J. Macromol. Sci., — Chem. *A12*, 853 (1978)
239. Bolto, B. A. et al.: Desalination *34*, 171 (1980)
240. Cable, P. J., Murtagh, R. W., Pilkington, N. H.: The Chem. Engineer Issue No. 324 (1977)
241. Bolto, B. A. et al.: Desalination *25*, 45 (1978)

A. Ledwith (Editor)
Received November 16, 1984

The Miscibility of High Polymers:
The Role of Specific Interactions

D. J. Walsh and S. Rostami
Department of Chemical Engineering and Chemical Technology,
Imperial College of Science and Technology,
London SW7, UK

Polymers which are miscible with each other are so because they are of low molecular weight, because they are similar chemically and physically, or because of specific interactions between the polymers. Not surprisingly a large percentage of those mixtures which are of practical use belong to the last group. This review attempts to present a critical assessment of methods of studying homogeneous blends and some recent work on polymer mixtures with a particular emphasis on those systems which are influenced by specific interactions.

The review contains an introduction to the theory of polymer mixtures, the ways in which they can be made, how they can be studied, and how one can obtain thermodynamic data relating to them. It discusses how miscible systems having specific interactions differ from those without, evidence for the specific interactions, and how the interactions may affect the properties of blends. Finally it discusses to what extent the most widely used theories of polymer miscibility are able to deal with systems which show specific interactions.

It is shown that a lack of knowledge of the interactions in polymer blends is probably the most serious limiting factor in our understanding of polymer mixtures.

1 Introduction

At one time very few polymer pairs were believed to be miscible but over the last twenty years a host of new miscible pairs have been identified [1]. As a relatively new field, the phenomenon of miscibility and the new materials derived from it will certainly have many as yet undiscovered applications.

Many of the practical examples of miscible blends involve poly(vinylchloride) including those with butadiene-acrylonitrile copolymers [2], possibly the first put into use, and various polyacrylates and vinyl acetate copolymers [3, 4] which are extensively used in PVC formulations at present. Others involve high performance engineering plastics such as blends of polystyrene with poly(2,6-dimethyl-1,4-phenylene oxide) (Noryl®) [5]. In some cases a useful compromise or averaging of properties can be obtained whereas in others a useful combination of different desirable properties can be achieved.

Polymers were thought to be usually immiscible due to their low combinatorial entropy of mixing. Any small unfavourable heat of mixing, positive ΔH, would thus preclude miscibility. However, many pairs of polymers are now known to show specific interactions such as hydrogen bonds which result in a favourable heat of mixing. It is part of the intentions of this review to stress the importance of these specific interactions and to show that a consideration of these interactions is essential to an understanding of the phenomena and theory of polymer miscibility.

We will exclude from this review much discussion of the miscibility of low molecular weight polymers, where the miscibility is due primarily to a non-negligible entropy of mixing, except in cases where it helps to develop the ideas or relates to phenomena in high polymers. We will also exclude discussion of polymers which show strong complex formation such as in mixtures of anionic and cationic polyelectrolytes.

2 Theoretical

Central to the phenoma of miscibility is the free energy of mixing (ΔG_M) which must be negative for mixing to occur. This is a necessary but not sufficient criterion for homogeneity as the free energy of mixing can be negative across the entire composition range and phase separation still occur. With this in mind it does however serve as a useful starting point for discussion. It can be expanded as

$$\Delta G_M = \Delta H_M - T \Delta S_M \tag{1}$$

where ΔH_M and ΔG_M are the enthalpy and entropy changes on mixing and T the absolute temperature.

The combinatorial entropy of mixing is usually taken in the form of the classical Flory-Huggins theory as

$$\Delta S_M = -KT(N_1 \ln \Phi_1 + N_2 \ln \Phi_2) \tag{2}$$

where N_i and Φ_i are the molecular number and volume fraction of component i. Thus as the molecular weights of the two components tend to infinity N_1 and N_2 tend to zero and the entropy of mixing tends to zero.

The heat of mixing can be thought of as an exchange energy [2] of breaking 1/1 and 2/2 contacts and forming 1/2 contacts as

$$\Delta H_M = VZW \, \Phi_1\Phi_2/V_s \tag{3}$$

Where W is the energy change for a 1,2 contact compared to the means of 1,1 and 2,2 contacts, Z is a coordination number, V the total volume and V_s a segment volume, or in terms of the Flory-Huggins interaction parameter, χ_{12}

$$\Delta H_M = RTn_1\Phi_2\chi_{12} \tag{4}$$

where n_1 is the number of moles of component 1. Or in terms of the solubility parameters of the components (δ_1 and δ_2)

$$\Delta H_M = V(\delta_1 - \delta_2)^2 \, \Phi_1\Phi_2 \tag{5}$$

This result suggests that ΔH_M must always be positive (unfavourable) and on this basis no high polymers should be miscible (except when $\delta_1 \simeq \delta_2$).

In more advanced theories such as the equation-of-state theory developed by Flory and his co-workers [6, 7], extra contribution to the free energy of mixing are considered which take account of the possible volume changes accompanying mixing which were assumed to be zero in the simpler theories. As formulated these contributions are however also unfavourable for mixing.

Mixing can occur in three possible circumstances.
1. If the polymers are not of very high molecular weight and the combinatorial entropy of mixing is not negligible.
2. If the polymers have a very small unfavourable heat of mixing arising from a very small exchange energy. This could arise for example in the mixing of two copolymers which vary very little in composition.
3. If the polymers have a favourable heat of mixing arising from a specific interaction between them.

Most practical and interesting examples come into the last category, which shows the dominance of the heat of mixing in determining the miscibility of polymers. In one experiment low molecular weight analogues of a wide range of polymers were fairly crudely mixed in order to find favourable heats of mixing. Seven previously unstudied such pairs were found and of these six of the actual polymer pairs were found to be miscible [8].

Many polymers show a variation in miscibility with temperature. Low molecular weight polymers, having positive heats of mixing, are typically more miscible at higher temperatures and may phase separate on cooling showing upper critical solution temperature behaviour (UCST). High molecular weight polymers forming homogeneous blends are typically less miscible at higher temperatures, and may phase separate on heating showing lower critical solution temperature behaviour (LCST).

In the case of low molecular weight polymers ΔS_M is favourable and the term $T\Delta S_M$ in Eq. (1) becomes more favourable at higher temperatures. If Eqs. (2) and (3) are substituted into Eq. (1) at a series of temperatures then the plots of ΔG_M against composition (Φ_2) may have the form shown in Fig. 1. This calculation is based on the simple Flory-Huggins lattice theory and given

$$\Delta G_M = RT(n_1 \ln \Phi_1 + n_2 \ln \Phi_2 + n_1 \Phi_2 \chi_{12}) \tag{6}$$

At T_3 the polymer pair are miscible in all proportions and $\delta^2(\Delta G_M)/\delta\Phi_2^2$ is positive over all compositions. At T_1 compositions between Φ_B and Φ'_B can phase separate to reduce the overall energy to give two phases at composition Φ_B and Φ'_B where the values

$$\frac{\delta}{\delta\Phi_2}(\Delta G_M)_B = \frac{\delta}{\delta\Phi_2}(\Delta G_M)_{B'} \tag{7}$$

at T_2 the two points meet at the critical point and

$$\frac{\delta^2}{\delta\Phi_2^2}(\Delta G_M) = \frac{\delta^3}{\delta\Phi_2^3}(\Delta G_M) = 0 \tag{8}$$

Between Φ_B and Φ_S and between Φ'_B and Φ'_S a region of metastability occurs. This arises because any small fluctuation which occurs in composition produces an increase in energy which acts as a barrier to phase separation. In the inset of Fig. 1 if composi-

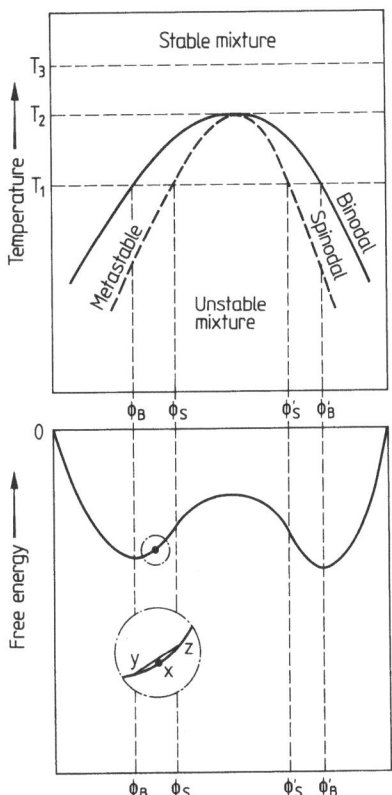

Fig. 1. The phase diagram for a system showing upper critical behaviour. At temperature T_3 mixtures are stable at all compositions whereas at T_1 phase separation can occur. The origins of this phase separation are shown in the lower diagram which shows the plot of free energy against composition at T_1. A metastable region exists between the binodal and the spinodal points (Φ_B, Φ'_B and Φ_S, Φ'_S respectively). The inset in the lower diagram shows that if composition X partially phase separates into Y and Z the average free energy is higher even though complete phase separation to Φ_B and Φ'_B produces a lower energy

tion X phase separates to a mixture of Y and Z the average energy is higher. Phase separation in this area can only occur by nucleation and growth. Between Φ_S and Φ_S' concentration fluctuations are stable and phase separation can take place spontaneously by spinodal decomposition. The mechanism of spinodal decomposition has been described by several authors [118, 119] and involves continuous changes in the composition of the phases with time while the spacing remains constant. The resulting spinodal structure is uniform and interconnected. The spinodal curve is defined by

$$\frac{\delta^2}{\delta\Phi_2^2} (\Delta G_M) = 0 \tag{9}$$

The resultant phase diagram is shown in Fig. 1. The line connecting points at various temperatures at composition Φ_B satisfying Eq. (7) is the binodal. The line connecting points at composition Φ_S satisfying Eq. (9) is the spinodal.

In practice and as described in other theories more complex phase diagrams showing evidence of bimodality can be found [9].

In the case of high molecular weight polymers the same sort of phase diagram is found but upside down, the phase separation occurs on heating and the critical point is called the lower critical solution temperature. The origins of this phase separation are more complex but can be attributed to three possible causes all three of which may in fact be operative.

1. The Equation-of-state terms as calculated by Flory and co-workers [6, 7], which take account of volume changes on mixing become more unfavourable at higher temperatures. The effect of the polymer properties and the resultant phase diagrams using this theory have been explored by McMaster [10].

2. Other unfavourable entropic contributions to the free energy exist which make the entropy of mixing negative. The $T\Delta S_M$ term in Eq. (1) is thus more unfavourable at higher temperatures. Such terms could possibly arise out of the specific interactions which themselves infer an ordering of the system. Such terms involving an empirical parameter Q_{12} have been included in modified versions of the Equation-of-state theory [11].

3. The heat of mixing could be temperature dependent. Though favourable at low temperatures it may become less so at higher temperatures. This could arise from a specific interaction which tends to dissociate at higher temperatures. There is experimental evidence that this is an important factor but as yet no theory describes it. Certain theories of non-random mixing do however have some of the features of such an effect. [12]

All of these three effects will be discussed further in later parts of this review. At this stage we will say more concerning advanced theories of polymer miscibility starting with the Equation-of-state theory of Flory and his co-workers.

2.1 Equation of State Theories

Due to the shortcomings of the classical Flory-Huggins lattice model, Flory and co-workers [6, 7] abandoned the whole concept of a lattice, and characterized each pure component by three equation of state parameters, V*, T* and P* which may be evaluated from the pure component data, density, thermal expansion coefficient and thermal pressure coefficient. In addition an interaction term, X_{12}, associated with a difference in chemical nature between the components was introduced in order to calculate the properties of the mixture. The residual quantities arising from the volume effect on mixing are derived in this theory while the combinatorial parts are borrowed from the lattice model. This theory, like other new theories of Patterson and Delmas [13], and Sanchez and Lacombe [14] benefits from the essential assumption of Prigogine [15] who divided a chain into "r-segments" and specified the number of external degrees of freedom.

The theory of Flory and his co-workers is no longer related to a fixed volume but does use a random mixing assumption. Introducing the latter concept neglects any local densification which might result from strong specific interactions. Later work [16] has shown the latter assumption to be invalid, which required an amendment to the theory by introducing a new entropy correction factor, Q_{12}. Flory's theory prescribes the following state Equation for the pure components and their mixtures.

$$\frac{\tilde{P}\tilde{v}}{\tilde{T}} = \frac{\tilde{v}^{1/3}}{\tilde{v}^{1/3} - 1} - \frac{1}{\tilde{T}\tilde{v}} \tag{10}$$

If subscript one is used to denote parameters of component one, subscript two of component two, and the parameters without or with 12 subscripts relate to the blends, then the reduced quantities are defined with i = 1, 2 or 12.

$$\tilde{v}_i = v_{i, sp}/v^* \tag{11}$$

$$\tilde{T}_i = T/T_i^* \tag{12}$$

$$\tilde{P}_i = P/P_i^* \tag{13}$$

From these equations it is obvious that $v_1^* = v_2^* = v^*$. $\tilde{v}_i$ is normally obtained from the thermal expansion coefficient of component i [6, 7].

$$\tilde{v}_i = \left[\frac{3 + 4\alpha_i T}{3 + 3\alpha_i T} \right]^3 \tag{14}$$

$\tilde{T}_i$ is obtained from $\tilde{v}_i$.

$$\tilde{T}_i = \frac{\tilde{v}_i^{1/3} - 1}{\tilde{v}_i^{4/3}} \tag{15}$$

P_i^* (of pure components only) is obtained from the thermal pressure coefficient, γ_i:

$$P_i^* = \gamma_i T \tilde{v}_i^2 \tag{16}$$

and P* of the blend is given by:

$$P^* = \Phi_1 P_1^* + \Phi_2 P_2^* - \Phi_1 \theta_2 X_{12} \tag{17}$$

where

$$\Phi_2 = \frac{m_2 v_{2sp}^*}{m_1 v_{1sp}^* + m_2 v_{2sp}^*} , \Phi_1 = 1 - \Phi_2 \tag{18}$$

$$\theta_2 = \frac{(s_2/s_1)\, \Phi_2}{(s_2/s_1)\, \Phi_2 + \Phi_1} , \theta_1 = 1 - \theta_2 \tag{19}$$

and Φ_i, θ_i and m_i are volume, segmental and weight fractions of component i.
T^* of the mixture is related to P^* of the mixture by:

$$T^* = \frac{P^*}{(\Phi_1 P_1^*/T_1^* + \Phi_2 P_2^*/T_2^*)} \tag{20}$$

On this basis the enthalpy change on mixing is given by:

$$\Delta H = v_{sp}^*[\Phi_1 P_1^*/\tilde{v}_1 + \Phi_2 P_2^*/\tilde{v}_2 - P^*/\tilde{v}] \tag{21}$$

with

$$v_{sp}^* = m_1 v_{1,\,sp}^* + m_2 v_{2,\,sp}^* \tag{22}$$

The residual entropy of mixing is given by:

$$TS^R = -3Tv_{sp}^*\left[\frac{\Phi_1 P_1^*}{T_1^*}\, \ln \frac{\tilde{v}_1^{1/3} - 1}{\tilde{v}^{1/3} - 1} + \frac{\Phi_2 P_2^*}{T_2^*}\, \ln \frac{\tilde{v}_2^{1/3} - 1}{\tilde{v}^{1/3} - 1} - \frac{1}{3}\, \Phi_1\theta_2 Q_{12}\right] \tag{23}$$

and the residual Gibbs free energy changes on mixing is:

$$G^R = \Delta H - TS^R \tag{24}$$

Differentiating the enthalpy equation with respect of N_1 will give the partial molar heat of mixing of component i, which in turn is related to the enthalpy interaction parameter, χ_H:

$$\Delta \bar{H}_1 = \bar{H}_1 - H_1^0 = (\partial \Delta H_M/\partial N_1)_{\tilde{T},\, \tilde{P},\, N_2}$$

$$= (\partial \Delta H_M/\partial N_1)_{N_2,\, \tilde{T},\, \tilde{v}} + (\partial \Delta H_M/\partial \tilde{v})_{N_2,\, \tilde{T},\, N_1} \cdot (\partial \tilde{v}/\partial N_1)_{N_2,\, T,\, v}$$

$$= P_1^* V_1^*\left[(\tilde{v}_1^{-1} - \tilde{v}^{-1}) + \frac{\chi T}{\tilde{v}}\cdot\frac{\tilde{T}_1 - \tilde{T}}{\tilde{T}}\right] + \frac{V_1^* X_{12}}{\tilde{v}}\cdot\theta_2^2(1 + \alpha T) \equiv RT\chi_H\Phi_2^2 \tag{25}$$

Applying similar procedure for the entropy Equation gives $T\bar{S}_1^R$ which is similarly related to the entropy interaction parameter, χ_s:

$$T\bar{S}_1^R = -P_1^* V_1^*\left[3\tilde{T}_1\, \ln \frac{\tilde{v}_1^{1/3} - 1}{\tilde{v}^{1/3} - 1} - \frac{\alpha T}{\tilde{v}}\cdot\frac{\tilde{T}_1 - \tilde{T}}{\tilde{T}}\right] + \frac{V_1^*\theta_2^2}{\tilde{v}}(\alpha T X_{12} + T\tilde{v}Q_{12})$$

$$\equiv -RT\chi_s\Phi_2^2 \tag{26}$$

The partial molar residual chemical potential of component one is therefore:

$$\Delta\mu_1^R = P_1^* V_1^* \left[3\tilde{T}_1 \ln \frac{\tilde{v}_1^{1/3} - 1}{\tilde{v}^{1/3} - 1} + \tilde{v}_1^{-1} - \tilde{v}^{-1} \right] + \frac{V_1^* \theta_2^2}{\tilde{v}} (X_{12} - T\tilde{v}Q_{12})$$

$$\equiv RT\chi_t \varphi_2^2 \tag{27}$$

where χ_t is the total interaction parameter, i.e. $\chi_t = \chi_H + \chi_s$.

The last three Equations serve as definitions of χ_H, χ_s and χ_t respectively. The values of X_{12} and Q_{12} are supposed to be composition independent, whereas the composition dependence of χ_t, χ_H and χ_s are given as:

$$\chi_t = \chi_{t,1} + \chi_{t,2}\Phi_2 + \chi_{t,3}\Phi_2^2 + \ldots \tag{28}$$

$$\chi_H = \chi_{H,1} + \chi_{H,2}\Phi_2 + \chi_{H,3}\Phi_2^2 + \ldots \tag{29}$$

$$\chi_s = \chi_{s,1} + \chi_{s,2}\Phi_2 + \chi_{s,3}\Phi_2^2 + \ldots \tag{30}$$

The volume change of mixing is approximated on the basis of the assumption that it is only enthalpy dependent and the entropy contribution to the volume change of mixing is ignored.

$$\frac{\Delta V^M}{V^0} = \frac{\tilde{v} - \tilde{v}^0}{\tilde{v}^0} = \frac{\tilde{v}}{\tilde{v}^0} - 1 \tag{31}$$

where

$$\tilde{v}^0 = \Phi_1 \tilde{v}_1 + \Phi_2 \tilde{v}_2$$

The spinodal, binodal and critical point Equations derived on the basis of this theory will be discussed later. When the theory has been tested it has been found to describe the properties of polymer blends much better than the classical lattice theories [17, 18]. It is more successful in interpreting the excess properties of mixtures with dispersion or weak attraction forces. In the case of mixtures with a strong specific interaction it suffers from the results of the random mixing assumption. The excess volumes observed by Shih and Flory [19] for C_6H_6-PDMS mixtures are considerably different from those predicted by the theory and this cannot be resolved by reasonable alterations of any adjustable parameter. Hamada et al. [20], however, have shown that the theory of Flory and his co-workers can be largely improved by using the number of external degrees of freedom for the mixture as:

$$C = C_1 \Phi_1 + C_2 \Phi_2 - C_{12} \Phi_1 \theta_2 \tag{32}$$

where C_{12} is the parameter characterizing the deviation from additivity. And also

$$v^* = v_1^* \Phi_1 + v_2^* \Phi_2 + 2 v_{12}^* \Phi_1 \Phi_2 \tag{33}$$

where

$$v^* = \{(v_1^{*1/3} + v_2^{*1/3})/2\}^3 \tag{34}$$

The predicted values of ΔH_M and $\Delta V_M/V^0$ for the mixtures studied showed reasonable agreement with experimental results.

Introduction of the entropy correction term containing the Q_{12} parameter as an empirical parameter limits the usefulness of the theory of Flory and his co-workers. A consistent correction for the entropy and volume dependancies of this theory should start from a completely new form of the partition function.

There has been an attempt by Renuncio and Prausnitz [12] to introduce a partition function by modifying the energy of interaction to take into account the non-random mixing of two components. They introduced two site fractions between segment i and j as θ_{ij} and θ_{ji} where

$$\theta_{ii} + \theta_{ji} = 1. \qquad \theta_{ij} + \theta_{jj} = 1, \qquad \theta_{ij} \neq \theta_{ji} \tag{35}$$

On this basis for $\theta_{ji} > \theta_{jj}$, i–J contact is favourable, whereas for $\theta_{ji} < \theta_{jj}$ it is unfavourable for mixing. Canovas et al. [21] have derived the following Equation of state by using the non-randomness concept of mixing.

$$\frac{\hat{P}\tilde{v}}{\hat{T}} = \frac{\tilde{v}^{1/3}}{\tilde{v}^{1/3} - 1} - \frac{1}{\hat{T}\tilde{v}} + \frac{A}{\hat{T}^2\tilde{v}^2} + \frac{C}{\hat{T}\tilde{v}} \tag{36}$$

where A and C are composition dependent [21].

This Equation will reduce to the Equation of Flory and his co-workers for $A = C = 0$. The third term on the right hand side of this Equation arises due to the volume dependence of the local composition. The fourth term is a consequence of the volume dependence of the combinatorial contribution. Theories of non-random mixing, although very attractive in that they correct the Equation of state for combinatorial and non-combinatorial factors, is unlikely to be a great improvement over Flory's theory. By using a new partition function it has considered the contribution of the entropy in the interaction term, and hence removed the Q_{12} factor. It, however, assumes $v_1^* = v_2^* = v^*$ and neglects the effect of internal degrees of freedom. An early test of this model [22] has shown a good agreement between predicted and experimental χ values, but gave a positive volume change of mixing for mixture of PDMS with some solvents, while their experimental values have been shown to be negative. This theory is more complicated than Flory's theory and more application of it needs to be documented before any further comments are possible. Early ideas of non-random mixing involving molecules with orientation dependant forces have been discussed by Tompa [124] and Baker [125].

Another useful and simpler theory is the Lattice-Fluid (LF) Theory developed by Sanchez and Lacombe [14, 23, 24]. This theory has much in common with the Flory-Huggins theory but differs in one important respect in that it allows the lattice to have some vacant sites and to be compressible. Thus the compressible lattice theory is capable of describing volume changes on mixing as well as LCST and UCST behaviours. As with the theory of Flory and his co-workers, X_{12} (which is proportional to the change in energy that accompanies the formation of a 1–2 contact from a 1–1 and a 2–2 contact) is obtainable from experimental values of heats of mixing.

The excess properties and interaction parameters are defined in a similar way as in the theory of Flory and co-workers, whereas this theory shows that miscibility for high molecular weight polymers can only be predicted when the heat of mixing of the two components is negative. Some comparisons of these theories are given elsewhere [25].

A disadvantage of the (LF) theory is the prediction of ΔV^M from the close-packed densities. Most of the hard core densities, ϱ^*, predicted by the (LF) theory are about 10 % smaller than their known crystalline densities, which is most probably due to the packing factor of the lattice. There have been few applications of this theory to a real mixture, but from the work done by Sanchez [24] it seems that the introduction of an entropy correction factor into the model is inevitable if it is going to be applied to a system with specific interactions.

Many variations on these and similar theories have been developed and their relative merits have been discussed [1,9,25]. We believe that most theories suffer because they do not address themselves to the important problem of the specific interaction directly. In our own work we have used the Equation-of-state theory of Flory and co-workers. Although it cannot fully describe systems with specific interactions it does have the merits of being moderately easy to use, of allowing volume changes on mixing, and of using parameters which are mostly obtainable either by experimental measurement or calculation.

3 Ways of Making Miscible Blends

Three main ways exist for making homogeneous blends of polymers: mechanical mixing, mixing in a common solvent, and *in situ* polymerisation. Mixing in a common solvent has been the commonest method in academic studies though this method, including the recovery of the blend by solvent evaporation or precipitation of the polymers in a non-solvent, would have very limited industrial application.

In this section we will describe the uses and limitations of the three methods. We also find that the kinetic and thermodynamic properties of the polymers and blend, including the specific interactions, control the blend forming process.

3.1 Mechanical Mixing

Mechanical mixing of two polymers to form a blend is probably the method of greatest practical importance. Various factors however tend to lead to inhomogeneity even in miscible systems. The low diffusion rates of high polymers means that very long mixing periods would be required to produce complete homogeneity. The low thermal stability of many polymers means that such prolonged mixing is not possible. Secondly, many polymer mixtures exhibit phase separation on heating. It may be that attempts at melt mixing take place above the phase separation temperature of a blend. In practical terms, however, complete homogeneity may not be necessary or even desirable and many commercial blends are prepared in this way.

Many polymeric plasticisers, impact modifies, and processing aids for PVC are incorporated into the PVC by mechanical mixing [4]. Many of these, including butadiene-acrylonitrile copolymers [13], ethylene-vinyl acetate copolymers [26, 27], chlorinated polyethylenes [28] and various polyacrylates and copolymers [3] are known to be miscible or partially miscible with PVC. In the case of polymeric plasticisers homogeneity may be desirable but optimum impact modification is often achieved by a heterogeneous structure. Processing aids may in some cases rely on phase separation at the temperatures of processing for their efficacy.

Blends of polystyrene and poly(2,6-dimethyl-1,4-phenylene oxide) (PPO) can be mixed in the melt as both polymers have reasonable thermal stability. There has however been much discussion as to whether the blends are truly one phase. Some techniques suggest homogeneity while others suggest a heterogeneous structure. On balance it appears that the two polymers are in fact thermodynamically miscible in all proportions but completely efficient mixing is difficult to achieve [29].

It seems possible that complete homogeneity may be close to impossible to achieve by mechanical mixing of two high molecular weight polymers. It will however remain to be of the greatest practical importance in the preparation of commercial blends.

3.2 Mixing in a Common Solvent

As stated earlier this method is the commonest method of preparing homogeneous blends in academic studies. It is however not without its pitfalls. Two phase blends can be formed by the evaporation of solvent from solutions of polymers which are themselves thermodynamically miscible.

In the case of polystyrene blends with poly(vinyl methyl ether) two phase behaviour was found for blends from various chlorinated solvents whereas single phase behaviour was found for blends from toluene [30, 31]. The phase separation of mixtures of these polymers in various solvents has been studied and the interaction parameters of the two polymers with the solvents measured by inverse gas chromatography [32]. It was found that those solvents which induced phase separation were those for which a large difference existed between the two separate polymer-solvent interaction parameters. This has been called the $\Delta\chi$ effect [33] (where $\Delta\chi = \chi_{12} - \chi_{13}$). A two phase region exists within the polymer/polymer/solvent three component phase diagram as shown in Fig. 2. When a dilute solution at composition A is evaporated, phase separation takes place at B and when the system leaves the two phase region, at overall

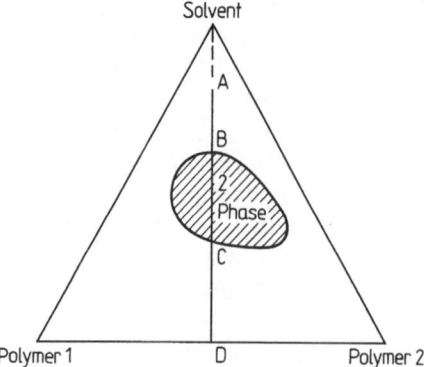

Fig. 2. A hypothetical phase diagram for two polymers with a common solvent showing a two phase region. When solvent is evaporated from a mixed solution at composition A it enters the two phase region at B. When it leaves at C the phase separated regions have grown to such a size that remixing does not easily occur and the resultant blend at D is inhomogeneous

composition C, the phases have grown too large and the rate of diffusion in the mixture is too low for the polymers to remix and the resultant polymer blend, of overall composition D, is two phase.

Computer simulations of the binodal in the phase diagrams of polymer/polymer solvent mixtures have been carried out and show the effect of the various interaction parameters on the shape of the phase diagram [33].

Other examples of solvent effects in casting blends include epoxy resin/copolyester/tetrachloroethane [34], polyethersulphone/poly(ethylene oxide)/cyclohexanone [35], and mixtures of PVC with various polyacrylates in solvents such as THF [3]. One particular pair of polymers PVC/poly(ethyl acrylate) appear to be miscible but no suitable solvent has been found as yet [36]. Homogeneous blends can only be prepared by *in situ* polymerisation though it is possible that miscibility is enhanced by small amounts of graft copolymer which is inevitably formed by this technique.

Coprecipitation of polymers from a common solvent into a non-solvent is an alternative way of making blends. This method is useful when the common solvent has a very high boiling point. It however produces finely divided or coarse particular samples which must be pressed before testing by many techniques. This hot pressing could result in phase separation.

3.3 In situ Polymerisation

In situ polymerisation is the polymerisation of one monomer in the presence of another polymer. It has been used extensively in the preparation of two phase blends, for example high impact polystyrene, but has only recently been applied to the preparation of homogeneous blends. It does have many advantages for the preparation of blends where the base polymers are not very thermally stable and have a high T_g, or show phase separation on heating which precludes the use of mechanical mixing. As described earlier the method of mixing in a common solvent does not guarantee homogeneity even for a miscible blend and would not be industrially viable except in special circumstances.

In situ polymerisation does not however guarantee homogeneous blends as two phase regions can exist within the polymer/polymer/monomer three component phase diagram. In the case of vinyl chloride polymerisation with solution chlorinated polyethylene, the vinyl chloride has limited solubility in both poly(vinyl chloride) and chlorinated polyethylene. The phase diagram has the form shown in Fig. 3 [28]. The limit of swelling of vinyl chloride in the chlorinated polyethylene is A and the highest concentration of PVC prepared by a 'one-shot' polymerisation is B.

In the case of vinyl chloride polymerisation in poly(butyl acrylate) these materials are completely miscible but a two phase region exists within the phase diagram as shown in Fig. 4 [37]. Polymerisation from A to B produces a homogeneous blend whereas from E to F produces a two phase structure. Composition B can be reswollen to C with vinyl chloride which can then be polymerised to D to produce a homogeneous blend. This route avoids the two phase region in the phase diagram and in principle all compositions of polymer blend can be prepared in a series of steps.

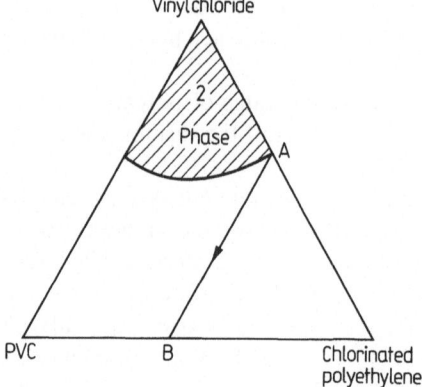

Fig. 3. The three component phase diagram for vinyl chloride, PVC and a chlorinated polyethylene is A and the highest concentration of PVC in a homogeneous blend prepared by a one-shot polymerisation is B

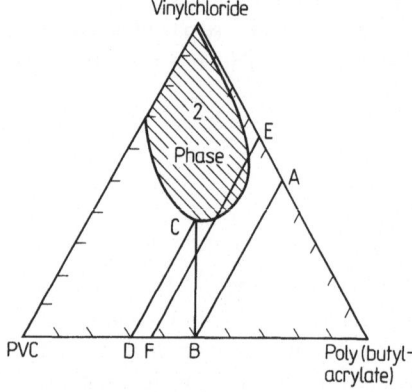

Fig. 4. The three component phase diagram for mixtures of vinyl chloride, PVC and poly(butyl acrylate). A polymerisation from E to F passes through the two phase region and an inhomogeneous blend results. A polymerisation from A to B, followed by reswelling with vinyl chloride to C and repolymerisation to D, avoids the two phase region and produces a homogeneous blend

An interesting observation is that PVC blends prepared without passing through the two phase region of the phase diagram show no evidence of PVC crystallinity [38].

Homogeneous blends can also be prepared by *in situ* polymerisation of butyl acrylate in PVC but a much larger two phase region exists and many steps are necessary for some compositions [39].

The phase diagram has been simulated for the PVC/poly(butyl acrylate)/vinyl chloride system [36]. Binodals are usually difficult to simulate but in this case the critical point is close to the apex of the triangle and the tie lines radiate from the apex. This means that the binodal approximates to a line joining compositions where the chemical potential is equal to that of the pure monomer and is easily calculated. It is a paradox that a large negative (favourable) interaction parameter between the two polymers would result in a phase diagram with a larger two phase region getting closer to the polymer/polymer axis. As the interaction between the two polymers becomes more favourable they have a greater tendency to exclude the monomer.

In situ polymerisation may have wide applicability in preparing blends of polymers such as engineering plastics which have few solvents. Because the solvents interact strongly with the polymers they are more likely to induce phase separation. This has been shown to be true for PVC which also has few solvents.

4 Evidence for Miscibility and Phase Diagrams

The optical, mechanical, electrical, morphological and thermodynamic properties of various polymer mixtures are often used as evidence for establishing miscibility. The methods have been extensively reviewed by MacKnight et al. [40] and Olabisi [1]. In this section we will attempt only to discuss the applicability of some of the methods to various types of blends.

4.1 Optical Clarity

Optical clarity is usually the first indication that two polymers are miscible, though it is not itself a sufficient proof of homogeneity. It can be deceptive if the refractive indices of the blend components happen to be close to each other, if the blend has phase separated into two separate layers, or if a two phase structure exists with domains much smaller than the wave length of light.

A clear homogeneous blend of two high molecular weight polymers may phase separate and become cloudy on heating. It therefore transmits light and its scattering intensity increases. This can be used as a method of detecting the cloud point of a blend. The usefulness of this technique can be seen from experiments on a homogeneous blend of ethylene-vinyl acetate copolymer (45 wt.-% vinyl acetate) and chlorinated polyethylene (52 wt.-% chlorine) at 50/50 w/w composition. Transmitted light was detected by a photodiode, set in line with the incident light beam, whilst the sample was heated at a constant rate of 0.175 °C/min. The intensity of transmitted

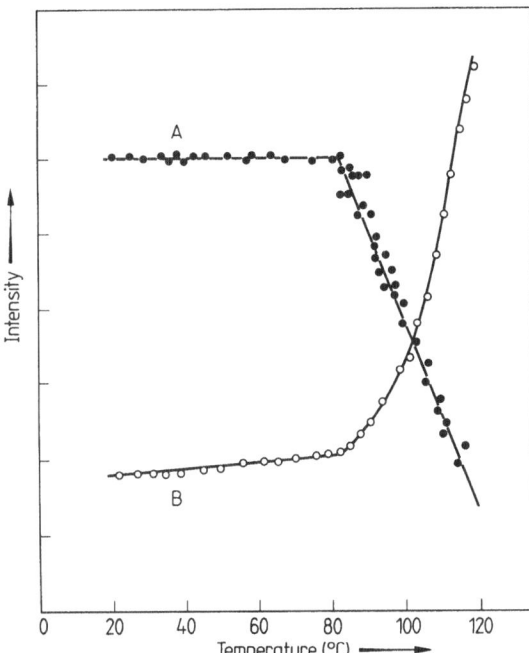

Fig. 5. Plots of A, transmitted light intensity and B scattered light intensity (arbitrary units) against temperature for an initially homogeneous blend of an ethylene vinyl acetate copolymer with a chlorinated polyethylene. At the phase separation temperature there is a drop in transmitted intensity and a rise in scattered intensity

light dropped rapidly at about 83 °C (as shown in Fig. 5) where the blend phase separated. Similar results obtained by detecting the forward scattering intensity at an angle of 45° showed a sharp increase in the scattered intensity as the blend started to phase separate. The heating rate in this case was 0.2 °C/min. This is also shown in Fig. 5 for comparison. The first deflection in a plot of scattering intensity against temperature is taken as the cloud point of the blend. The full cloud point curves, measured in this way, for the above blend and for blends containing another ethylene-vinyl acetate copolymer having 40 wt.-% vinyl acetate are shown in Fig. 6.

This technique only works well when studying blends of two rubbery materials or occasionally for blends of rubbery and glassy materials. If the mobility of the two polymers is not sufficiently high then phase separation may not proceed to give phases large enough to scatter appreciable light until a sample has been heated for a considerable time at temperatures above the phase separation temperature. Scattering may, however, still be observed with X-rays or neutron beams which have shorter wavelengths. Light scattering also has been used successfully in determining cloud points for poly(butyl acrylate)-chlorinated polyethylene blends [41], for poly(ethylene oxide)-poly(ether sulphone) blends [35], and for polystyrene-poly(vinyl methyl ether) blends [42].

This technique has also been used for solutions of two polymers in a common solvent. If these measurements are extrapolated to zero solvent concentration, then results are obtained which are compatible with those obtained for solvent free blends [43]. This could be a useful technique for blends of polymers which have low mobility.

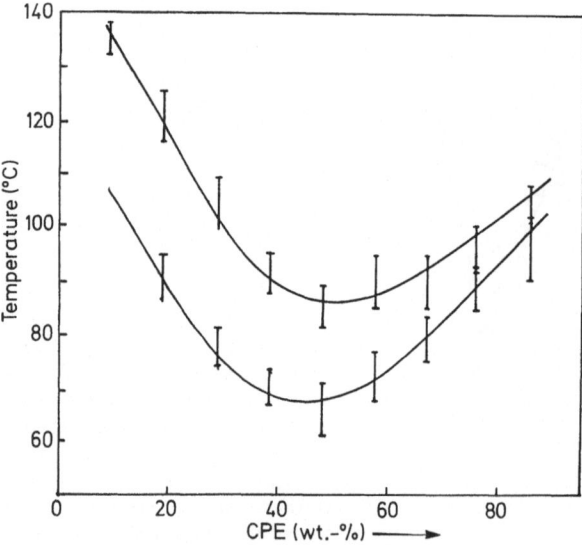

Fig. 6. The cloud point curves obtained using turbidimetry for a chlorinated polyethylene (52% chlorine) with ethylene-vinyl acetate copolymers having A, 40% vinyl acetate and B, 45% vinyl acetate. A higher vinyl acetate content is found to give a larger temperature range of miscibility

4.2 Optical and Electron Microscopy

Both optical and electron microscopy are widely used in studies of polymer blends. Phase contrast microscopy is preferably used in polymer blend studies due to its high resolving power for materials with similar refractive indices. In this type of microscopy the beam for the interference diffraction maxima of the light passing through a specimen is split into two parts by a beam splitting prism. Each part contains the full object information. These beams recombine in an interferometer and by shearing one beam vertically against the other the two waves hit each other sheared, and interference will occur depending on the phase difference of the components. This method produces coloured pictures of the specimen, with a homogeneous blend having one colour over the observed field, and a two phase blend showing two different colours. One advantage of this technique is that it avoids the possibility of artifacts due to staining which is a problem in electron microscopy. However, it has a limited resolution depending on the polymer mixture and the microscope used. We have used this method successfully to study miscibility and phase transition of several blends. An example of the results obtained for a blend of ethylene vinyl acetate co-polymer (40 wt.-% VA) with chlorinated polyethylene (43 wt.-% Cl) before and after phase separation is given in Fig. 7.

The resolving power of a light microscope is limited by the wave nature of the light. The minimum spacing which can be resolved by a good microscope is of the order of the wavelength of the radiation source. For better resolution a larger angle of acceptance of the lens and a shorter wavelength radiation are required. Transmission

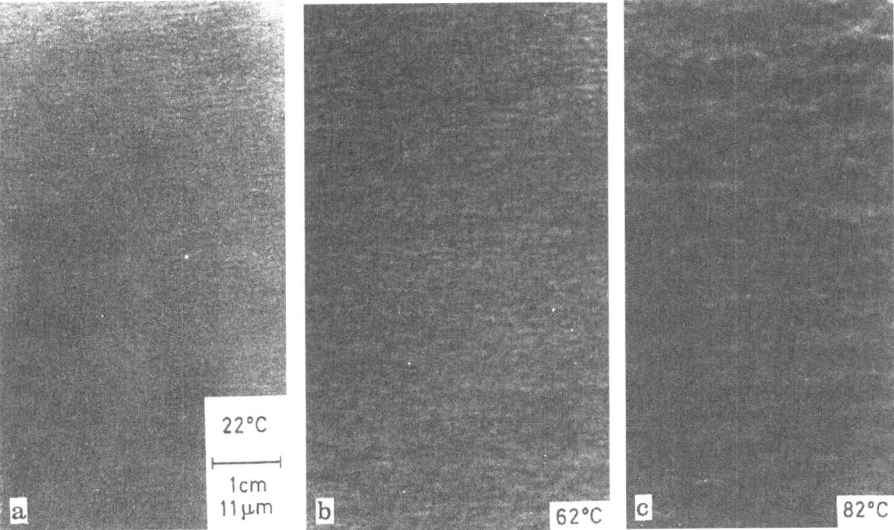

Fig. 7 a–c. Phase contrast microscope pictures of a blend of ethylene-vinyl acetate copolymer (40 % vinyl acetate) with chlorinated polyethylene (43 % chlorine) before and after phase separation. Since both polymers are elastomers the mobility is quite high. The original pictures are coloured red and green. These black and white pictures have enhanced contrast to make the phase separation clear

Electron Microscopy (TEM) with a resolution of 10–20 Å has proved a powerful tool for studies of polymer-polymer miscibility and phase separation.

For electron microscopy the problem always exists that the image may not be a perfect reconstruction of the object. The reasons for this arise from preparation of the sample, replication methods and damage due to the electron beam. The quality of the image produced in microscopy is judget by the reasolution and contrast. Staining techniques using osmium tetraoxide create a high contrast in systems containing double bonds. A reasonable contrast is produced with polymers containing heavier atoms, such as chlorine, without any staining. Electron micrographs are shown in Fig. 8 for poly(ether sulphone)-poly(ethylene oxide), 80/20 blends which have previously been heated at various temperatures and quenched, they show the development of the microstructure of the blends on phase separation [35]. This technique has also been used for blends of poly(vinyl chloride) with poly(butyl acrylate) [37] and for ethylene-vinyl acetate copolymers with chlorinated polyethelene [45].

4.3 Glass Transition Temperature Measurements

A miscible blend which behaves as one homogeneous phase shows a single glass transition temperature, T_g, which is generally between the T_g's of the individual polymers. Plots of T_g against composition for blends with a strong specific interaction show marked deviation from linearity. The extent of the departure depends on the

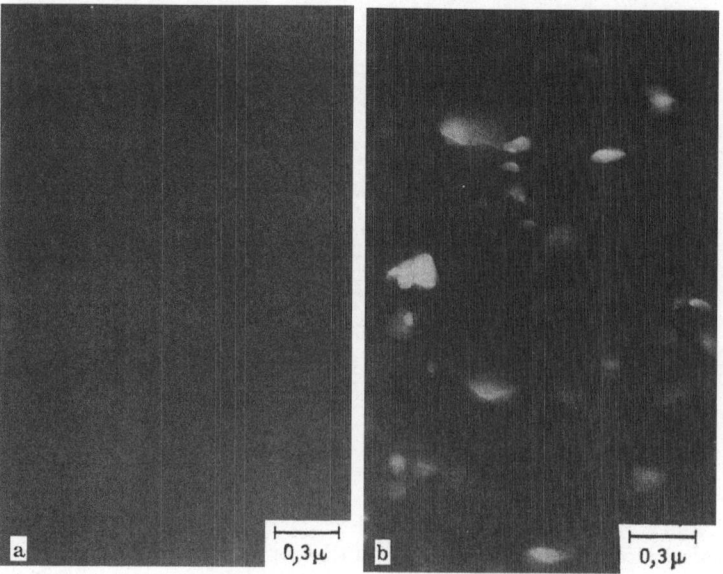

Fig. 8a and b.

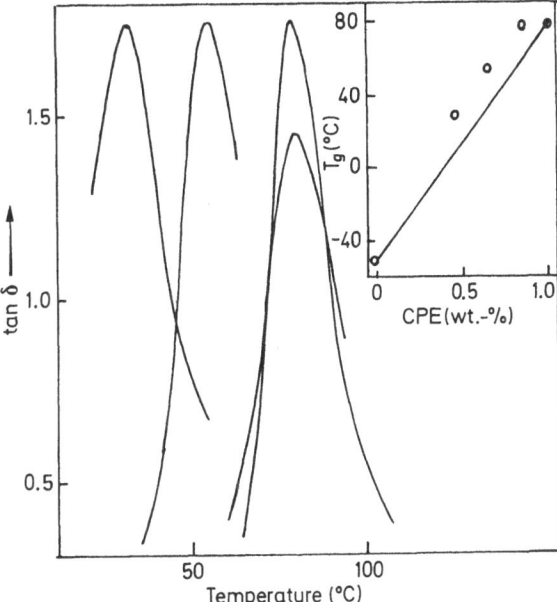

Fig. 9. Dynamic mechanical analysis: plots of tan δ against temperature for chlorinated polyethylene (52 % Cl) (4) and blends with poly(butyl acrylate) containing (3) 84.7 % PBA, (2) 64.1 % PBA, and (1) 46.1 % PBA. The inset shows a plot of T_g against weight percent chlorinated polyethylene where there is a marked deviation from linearity indicative of a specific interaction

strength of the specific interaction. An example of this is shown in Fig. 9 for blends of poly(butyl acrylate) with chlorinated polyethylene. In this case the blend requires a higher activation energy than its additivity value in the form of heat to allow chain movements. A review of this subject and of the relations between T_g and chemical structure of blends has been given by Cowie [46].

For miscible blends many attempts have been made to correlate the T_g with the blend composition as is frequently done with random copolymers. Several miscible blends studied by Hammer [26] and Hichman and Ikeda [47] exhibit a composition dependence of T_g which can be described by the simple Fox relationship.

$$\frac{1}{T_{g_b}} = \frac{m_1}{T_{g_1}} + \frac{m_2}{T_{g_2}} \tag{37}$$

where m_i is the weight fraction of polymer i with T_{g_i}.

A logarithmic form of this equation is given by Pochan et al. [48]. Other expressions include the Wood equation [49], the Kelley-Bueche expression [50], the Gordon-Taylor equation [51], and the DiMarzio-Gibbs equation [52]. None of these Equations directly take into account the specific interactions within a blend.

There are various techniques available for the determination of glass transition temperatures as are outlined below.

Dynamic mechanical analysis involves the determination of the dynamic properties of polymers and their mixtures, usually by applying a mechanical sinusoidal stress [53]. For linear viscoelastic behaviour the strain will alternate sinusoidally but will be out of phase with the stress. The phase lag results from the time necessary for molecular rearrangements and this is associated with the relaxation phenomena. The energy loss per cycle, or damping in the system, can be measured from the loss tangent defined as:

$$\tan \delta = \frac{E''}{E'}$$

where E' is the storage modulus and E'' is the loss modulus.

A peak in the value of $\tan \delta$ occurs around the glass transition temperature.

The usual instruments used for these measurements have a limited number of frequencies available. It is therefore usual to keep the frequency constant while lowering or raising the temperature to match the molecular motions of the blend to the set frequency.

A plot of $\tan \delta$ versus temperature gives a strong indication of blend homogeneity when a single T_g is detected or inhomogeneity when two T_g's are detected. An example of this method in studying blends of ethylene-vinyl acetate copolymers (45 wt.-% Ac) with chlorinated polyethelene (52 wt.-% Cl) at a constant frequency of 11 Hz is shown in Fig. 10. Similar results obtained for blends of chlorinated polyethylene (44 wt.-% Cl) with chlorinated polyethylene (62 wt.-% Cl) are shown in Fig. 11.

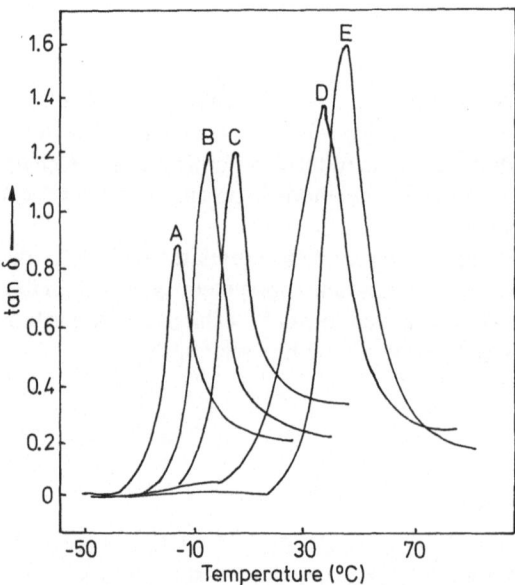

Fig. 10. Plots of $\tan \delta$ against temperature for blends of ethylene-vinyl acetate copolymer (45 % vinyl acetate) with chlorinated polyethylene (52 % Cl) showing a single composition dependent glass transition temperature: A EVA45 = 100; B EVA45:CPE = 20:80; C EVA45:CPE = 40:60; D EVA45: CPE = 80:20; E CPE = 100

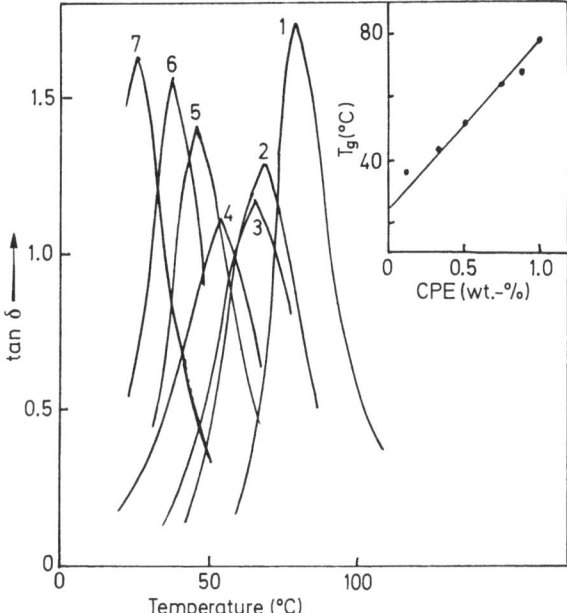

Fig. 11. Plots of tan δ against temperature for blends of chlorinated polyethylene (44 % Cl) with chlori-nated polyethylene (66 % Cl): 1 pure CPE (66 % Cl); 2 87.0 %; 3 73.5 %; 4 50.0 %; 5 25.0 % and 6 11.0 % of CPE (66 % Cl) in CPE (44 % Cl); 7 pure CPE (44 % Cl). The inset shows a plot of T_g against weight percent of the chlorinated polyethylene with the higher chlorine content, there is no marked deviation from linearity which is the expected result for systems with no specific interaction

Dynamic mechanical analysis can also be used to estimate the position of the phase diagram. This is possible by heating the blend (for a LCST behaviour) to a certain temperature and freezing in its structure by quenching it below the T_g's of both com-ponents. The normal test on the quenched blend reveals its homogeneity or heter-ogeneity at the temperature of heat treatment. By repeating this heating and quenching procedure for several temperatures an approximate phase separation temperature can be obtained. This method is particularly useful for blends with small refractive index differences between the two components, or blends which undergo degradation on prolonged heating which may be necessary when using another technique. An example of this method used for a 40 wt.-% poly(butyl acrylate) in poly(vinyl chloride) is shown in Fig. 12. The existence of two T_g's at 135 °C indicates that the blend undergoes phase separation at a temperature between 125–135 °C. Similar studies on blends of ethylene vinyl acetate copolymers with chlorinated polyethylenes are discussed elsewhere [54].

The dynamic mechanical method, however, suffers from two important limitations. Firstly, it is not sensitive for blends having less than 10 wt.-% of one component, as the response is dependent on the fraction of the stress born by each phase. Secondly, instruments have poor resolution for blends of materials with close T_g's. Different machines obviously have different sensitivities in both these respects.

Dielectric relaxation measurement in similar to dynamic mechanical measurements, except that it exploits the dipole electrical properties of the blend. It is, therefore,

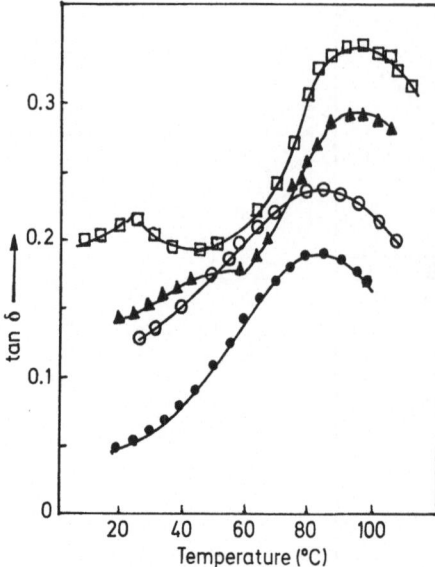

Fig. 12. Plots of tan δ against temperature for a 40/60 blend of PBA/PVC heated to various temperatures; ● room temperature, ○ 125 °C (tan δ + 0.05), ▲ 135 °C (tan δ + 0.1), □ 140 °C (tan δ + 0.15). This serves to identify the phase separation temperature of the blend as between 125 °C and 135 °C

not applicable to a blend of polymers without dipole moments. Inclusions can cause low frequency conductivity in the blend while measuring the glass transition temperature. This is more pronounced when the heating scan programme is used. This effect was first discovered by Maxwell-Wigner and Sillars and is fully described elsewhere [55]. The advantage of dielectric over dynamic relaxation measurements is the range of frequencies available. The use of dielectric relaxation in studying blends has been described by Wetton et al. [56].

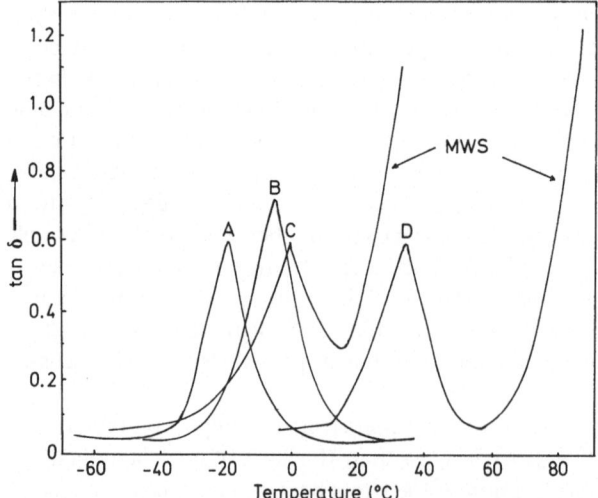

Fig. 13. Plots of tan δ against temperature dielectric relaxation: for EVA45, CPE3 and their mixtures at 37 Hz. A single composition dependent glass transition temperature is found, indicative of miscibility. The MWS effect in pure CPE3 and 30% EVA45 in CPE3 is also shown. A 100% EVA45; B 70% EVA45; C 30% EVA45; D 100% CPE3

Application of this method for miscibility studies of blends of ethylene-vinyl acetate copolymer (45 wt.-% Ac) with chlorinated polyethylene (52 wt.-% Cl) at a constant frequency of 37 Hz is demonstrated in Fig. 13. The method has shown single T_g's for the blends studied. The Maxwell-Winger-Sillars conductivity effect which appears after the glass transition temperatures is also shown in this Figure.

An alternative form of spectra can be obtained by recording the short-circuit current during warming up after the blend has been polarized at a constant d.c. field above the glass transition temperature. This method is referred to as the thermo-stimulated current depolarization or dielectric depolarization current method. The scheme of this method is shown in Fig. 14. The blend is heated up to a polarization temperature, T_p, where it is polarized for a known length of time, stage B. The current is switched off and the elctrically orientated blend structure is frozen in (stages C and D). Using a temperature program unit the polarized blend is heated gradually to a point when the dipoles discharge a new disorientation current (E). The rate of disorientation is a maximum near the glass transition temperature of the blend.

The depolarization method is less popular in T_g measurements, firstly due to the difficulty of sample preparation, secondly to the fact that the Maxwell-Wigner-Sillars effect interferes with the depolarization of the dipoles, and thirdly the detrapping of the charges above the transition temperature hinders the detection of the actual depolarization.

Differential scanning calorimetry, DSC, and differential thermal analysis, DTA show similar traces for T_g measurements although the property being measured is different. DSC measures the amount of heat required to increase the sample temperature over that required to heat up a reference material, normally an empty pan, to the same temperature. The variation in power necessary to maintain this level during a transition is monitored. DTA measures the difference in temperature between the sample and the reference material when both are heated at the same temperature rate. These techniques require a small amount of specimen, about 15 mg, and have

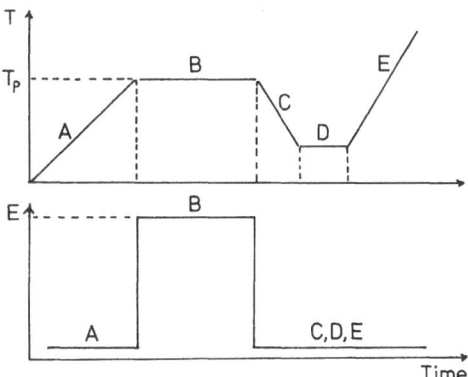

Fig. 14. The thermostimulated current depolarisation method; a schematic diagram showing temperature and applied field as a function of time. The sample is heated (A) to the polarisation temperature T_p where it is polarised for a time (B). The current is switched off and the oriented structure frozen in (stages C and D). The polarised blend is then heated gradually to the point where the dipoles discharge a disorientation E. The rate of disorientation is a maximum near T_g.

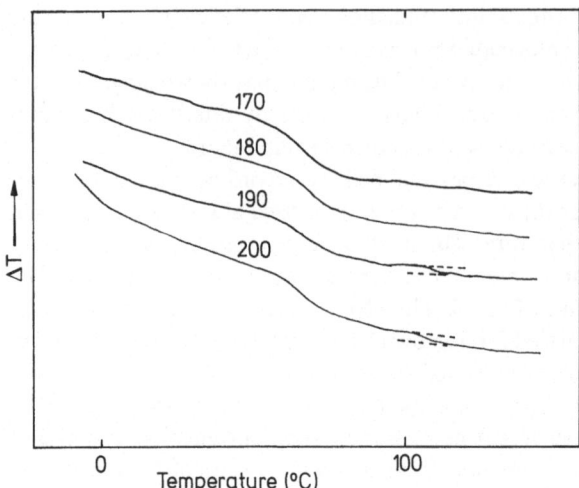

Fig. 15. Differential thermal analysis traces of a blend of chlorinated polyethylene with poly(methyl methacrylate) obtained by heating at a rate of 10 °C/min. The blend was kept at the quoted temperatures for 10 min. and was quenched prior to scanning. The appearance of two T_g's after treatment at 190 °C is indicative of phase separation at this temperature

sophisticated controlled rates of heating and cooling. More detailed descriptions of these methods can be found elsewhere [40, 57]. The relation between DSC results and polymer thermodynamics has been described by Flynn [58].

The utility of the DSC for studying polymer-polymer miscibility has been demonstrated for poly(vinyl chloride)/nitrile rubber [59], poly(vinyl methyl ether)/polystyrene [31] and poly(2,6-dimethyl 1,4-diphenylene oxide)/poly(styrene-co-chlorostyrene) [49]. It has also been particularly useful for measuring the melting point depressions of crystalline polymers in blends [60, 61] in order to calculate the interaction parameter as will be discussed later.

DSC and DTA can in principle also be used to study the phase diagram of a mixture by the same heating and quenching procedure as described earlier for dynamic mechanical measurements. The phase diagram of poly(methyl methacrylate)-chlorinated polyethylene (52 wt.-% Cl) has been obtained in this way by using DTA. The traces obtained for several blends of this system at various temperatures are shown in Fig. 15, where a single T_g defines a homogeneous mixture and two T_g's give an indication of phase separation in the blend at the temperature of heat treatment.

4.4 Other Methods

The optical clarity, morphological structure and glass transition properties of blends have been used to obtain information about the homogeneity and phase diagram of mixtures. Other properties such as the radius of gyration as measured by scattering methods [62] and the frequency shift of interacting functional groups as detected by Fourier transform infrared spectroscopy [63, 64] have also been used to give evidence of miscibility and phase separation of polymer-polymer mixtures. In the former case it has been shown [62] for a mixture deuterated polystyrene in poly(vinyl methyl ether) that the polystyrene coils are expanded, with respect to the unperturbed state, at room temperature. The coil expansion reduces as the temperature increases indicative of a reduction in the solvent power of the poly(vinyl methyl ether) for polystyrene. This trend was similarly found for mixtures of deuterated polystyrene in poly(2,6-dimethyl-1,4-diphenylene oxide), deuterated poly(methyl methacrylate) in polystyrene acrylonitrile and deuterated poly(methyl methacrylate) in poly(vinyl chloride). In each case the difference in chain dimensions from that found in the pure polymer gives of miscibility. These methods are, however, used primarily to study the interactions in polymer mixtures and will be discussed in this context later in this review.

Other methods, such as small angle X-ray scattering (SAXS), small angle neutron scattering (SANS) and pulse-induced critical scattering (PICS), are less attractive for blends of high molecular weight polymers, mainly due to their instrumental limitations. SAXS is sensitive to differences in electron density and hence usually requires a heavy atom in the blend to give a larger scattering cross section for X-rays after phase separation. Measurements using this method have been described by Stein et al. [65] for blends of poly(ε-caprolactone) with poly(vinyl chloride). Their experimental finding indicates that the PCL chains expanded as much as 35–50 % in the PVC matrix, which is a strong evidence for miscibility of this pair. Application of SAXS to mixtures of poly(p-iodostyrene) in polystyrene indicated an immiscible system whereas poly(chlorostyrene) in polystyrene was found to be miscible [79].

SANS, on the other hand, measures the difference in the neutron scattering cross Section, for example, of a deuterated chain dispersed in a matrix of protonated blend or vice versa. This method has been used to study the miscibility of deuterated poly(2,6-dimethyl-1,4-diphenylene oxide) with polystyrene in the temperature range of 104 to 273 °C [66]. It was shown that the mixture was thermodynamically homogeneous over the range of temperatures studied. Similar studies of blend miscibility by SANS have been published elsewhere [62]. The necessity to use labelled chains makes this method less attractive as few deuterated polymers are easily obtainable. There is also always the possibility that deuteration may affect the miscibility. Application of Pulse Induced Critical Scattering (PICS) for polymer-polymer mixtures with LCST behaviour is severely limited by the slow diffusion rate of the high molecular weight polymers. It is mainly designed to study polymer-solvent mixtures and only recently has been applied to low molecular weight polymer-polymer mixtures showing a UCST [58, 67]. PICS determines the spinodal directly and if it was found to be possible to use this or a similar technique on high polymers in the same way as has already been done for oligomers showing UCST behaviour, then this would be an appreciable advance since other methods generally only give the cloud point which is rather indeterminate in position between the binodal and spinodal.

5 Measurement of the Interaction Parameter

There are several different ways of obtaining an estimate of the interaction between two polymers. These include heat of mixing measurements, inverse gas chromatography, solvent vapour absorption, various scattering techniques and viscosity measurements. In the case of crystalline materials measurement of melting point depression by a technique such as D.S.C. can also be used. In this section we shall not attempt to describe the techniques in detail. We shall instead describe the applicability and limitations of the techniques and discuss the significance which may be given to the results.

5.1 Heat of Mixing Measurements

The direct measurement of the heats of mixing of two polymers is not possible. Several authors have attempted to measure the heats of mixing in the presence of a solvent, using Hess's law to extract the heat of mixing of the polymers [68]. In order to do this one needs to measure the heats of dissolution of the base polymers and of the blend in a common solvent. Some workers have said that this technique should be confined to rubber samples since glasses are not at equilibrium and results depend on the history of the glassy samples [120,121]. In principle this is a very attractive method but in practice it has not been particularly successful, possibly due to accumulation of errors in the series of experiments involved. Other workers have preferred to use oligomers or low molecular weight analogues, normalising the results with respect to the interacting segments of the actual polymers [44,69]. This may not always give reliable results due to differences between the high polymers and the analogues used. These differences could arise from chain end effects, steric differences or density differences. Low molecular weight materials often have a lower density. The cohesive energy density is strongly dependent on the density so that in systems where the dispersive forces are important the use of low molecular weight analogues may be less reliable. When dispersive forces are small and specific interactions strong the low molecular weight materials may be more representative of the polymers themselves. Measurements of heat of mixing of analogous materials in specially built microcalorimeters have been the most successful. The design of several of these has been described in the literature [70,71].

A combination of heat of mixing data and the state parameters of the pure components can be used with Eq. (21) to obtain the interaction term X_{12}. It should be noted that X_{12} is obtained rather than $\bar{X}_{12}$ as the heat of mixing does not also depend on the entropy correction term, Q_{12}.

If the heat of mixing is determined over a range of composition, then a result may be obtained as shown in Fig. 16 for mixtures of sec-octyl acetate with a chlorinated hydrocarbon (52 wt.-% Cl). One could use the value for example for a 50/50 mixture to obtain an interaction parameter or obtain a best fit of the results to Eq. (21). In our experience, for systems which have a strong specific interaction the deviation of these experimental results from Eq. (21) is always of the same form. ΔH values are larger than expected near to 50/50 compositions and smaller than expected at extremes of composition. This gives rise to a composition dependence of X_{12}. We believe that this arises because a system with specific interactions need not follow Eq. (21). The heat of mixing instead would also depend on the concentrations of interacting groups and the equilibrium extent of formation of interactions.

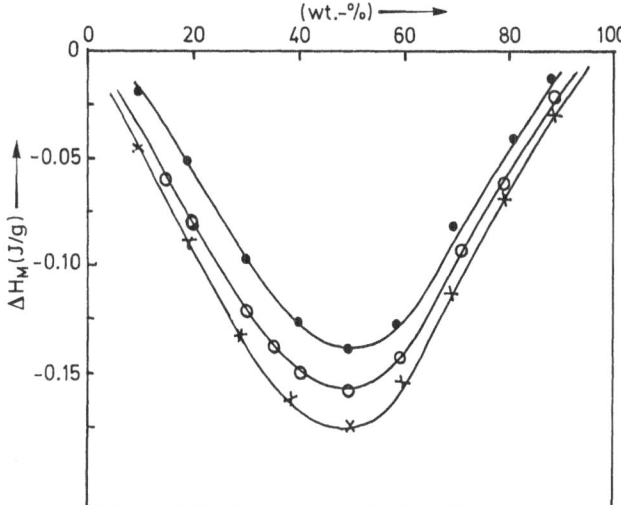

Fig. 16. Heat of mixing measurements for mixtures of secondary octyl acetate with a chlorinated paraffin (used as analogues for an ethylene-vinyl acetate copolymer and chlorinated polyethylene) at 64.5 °C ($\times$), 73.1 °C ($\bigcirc$) and 83.5 °C ($\bullet$) as a function of composition. The heat of mixing is very temperature dependent and the composition dependence is not as would be expected from the Equation-of-state theory. This will result in a temperature and composition dependence of the interactional parameter X_{12}

5.2 Inverse Gas Chromatography (I.G.C.)

Conventional gas liquid chromatography determines the retention of an unknown sample in the moving phase with a known stationary phase. The inverse method, however, determines the property of the stationary phase using a known volatile solute in the moving phase. The volatile molecules are referred to as probe molecules.

Since the initial work of Smidsrod and Guillet [72] numerous investigators have used I.G.C. to determine physicochemical parameters characterising the interaction of small amounts of volatile solutes with polymers [1]. Baranyi [73] has shown that infinite dilution weight fraction activity coefficients, interaction parameters and excess partial molar heats of mixing can be readily determined with this technique. Partial molar heats and free energies of mixing, and solubility parameters of a wide variety of hydrocarbons in polystyrene and poly(methyl methacrylete) have been determined [74]. The temperature dependence of the interaction parameter between two polymers has also been studied [45,32].

When studying polymer mixtures one needs to prepare packed columns containing an inert support coated with each of the base polymers and with the mixture, which therefore must be capable of being evaporated onto the support from a common solvent. Using a solvent probe one can then measure the specific retention volumes of the solvents when passed as vapour through the columns, i.e. $V^0_{g_2}$, $V^0_{g_3}$ and $V^0_{g_{23}}$

for each of the polymers and the blend respectively. The interaction parameter between two polymers can be obtained from the following Equation [45]

$$\frac{\chi_{23}}{V_2} = \frac{1}{V_1} \left\{ \frac{1}{\varphi_3} \ln \frac{V_{2SP}}{Vg_2^0} + \frac{1}{\varphi_2} \ln \frac{V_{3SP}}{Vg_3^0} - \frac{1}{\varphi_2\varphi_3} \ln \frac{W_2 V_{2SP} + W_3 V_{3SP}}{Vg_{23}^0} \right\} \quad (38)$$

where χ_{23} is the interaction parameter between the two polymers, and V_i, V_{sp}, w_i and φ_i are the molar volume, specific volume, weight fraction and volume fraction of component i respectively. The calculated interaction parameter when using different probes tends to vary more than can be explained by experimental error and various specific solvent effects have been invoked to explain this. The interactions of solvents with the single stationary phase, quantified by interaction parameters, activity coefficients or partial molar heats of mixing can also give indications of the sorts of groups which are capable of forming specific interactions with the polymers, and are hence of relevance to understanding interactions in the polymer blends themselves.

Although I.G.C. experiments are easy and fast one must question the quantitative reliability of the technique. There is a problem in removing the effects of both surface adsorption and diffusion limitation in the stationary phase [75,76]. It has not been definitely proven that any choice of loading weight or extrapolation procedure can remove both these effects. A further limitation of the technique is that it can only be used to give information at temperatures at least 50 °C above the T_g's of the homogeneous systems being studied in order to reduce the effects of surface adsorption and diffusion limitations.

It should be pointed out that since I.G.C. measures the total free energy of the interaction, any value of the Flory-Huggins interaction parameter which is derived will be a total value including combinatorial and residual interaction parameters as well as any residual entropy contributions. Similarly when using Equation-of-state theory one will obtain $\tilde{X}_{12}$ rather than X_{12}. The interactions are measured at high polymer concentration and are therefore of more direct relevance to interactions in the bulk state but this does not remove problems associated with the disruption of intereactions in a blend by a third component.

5.3 Solvent Vapour Sorption

Measurement of the adsorption of solvent vapours by films of polymer and of polymer mixtures can be used to obtain information about interactions within mixtures. The theoretical basis for the technique is similar to I.G.C. but, being a static rather than dynamic method it is in principle easier to reduce the effects of surface adsorption and diffusion limitation inherent in I.G.C. The technique is however not so fast and easy as I.G.C., but in principle gives information about the interaction between two polymers over the whole range of solvent composition.

The chemical potential of the solvent at pressure P is given by

$$\Delta\mu_1 = \frac{RT}{M_1} \ln P/P_s (= \Delta\mu_1^{comb} + \Delta\mu_1^{residual}) \quad (39)$$

where M_1 is the molecular weight of the solvent and P_s the saturation vapour pressure. Knowing $\Delta\mu_1$, the χ value is readily obtainable.

The method has been used to study the thermodynamics of various polymer blends [68]. The results of vapour sorption have been compared with I.G.C. for PVC, polystyrene, and poly(methyl methacrylate) with various solvents and the interaction parameters have been found to agree within experimental error [77].

There are a number of difficulties in performing the experiment due to the long period of time required to ensure the establishment of equilibrium conditions. The time taken to reach equilibrium increases as the vapour pressure increases. This long time requires an accurate temperature control which is more difficult at higher temperatures. The necessary temperature control limits the range of temperature covered by the method. The experimental errors are higher in the regions of low and high concentration of solvents due to limited accuracy of the method in measuring the solvent concentration and the pressure (which becomes closer to the saturation value).

5.4 Light, X-ray and Neutron Scattering

The theory of the scattering of radiation by large molecules has been reviewed in several recent publications [78, 83]. Thermodynamic data can be obtained using scattering techniques via the Flory-Huggins interaction parameter, χ_{12}, or the second osmotic virial coefficient, A_2. At low values of the scattering vector, Q (where $Q = \dfrac{4\pi}{\lambda} \sin \dfrac{\theta}{2}$, λ is the wavelength of the incident radiation and θ is the scattering angle) and for low concentrations of polymer 2 in component 1, the scattered intensity of the mixture minus 'solvent' (polymer 1) scattering, $I(Q)$, obeys the following Equation:

$$\frac{KC_2}{I(Q)} = \frac{1}{\bar{M}_w} \left| 1 + \frac{R_z^2 Q^2}{3} \right| + 2A_2 C_2 \tag{40}$$

where C_2, $\bar{M}_w$ and R_z are the concentration, the weight average molecular weight and the Z average radius of gyration of polymer 2.

K is a constant containing all the "optical" parameters for light, "electron density fluctuations" for X-rays and "scattering length difference" for neutrons. The exact definitions of K are given as:

$$K = \frac{2\pi^2 \left(\dfrac{\delta n}{\delta c} \right) n^2}{N_A \lambda^4} \qquad \text{for light} \tag{41}$$

$$K = \frac{P(\Delta z)^2 \, dN_A}{a^2} \cdot \left(\frac{e^2}{mc^2} \right)^2 \qquad \text{for X-rays} \tag{42}$$

$$K = \left| b_2 - \frac{V_2}{V_1} b_1 \right|^2 \frac{N_A}{m_2} \qquad \text{for neutrons} \tag{43}$$

where $(\delta n/\delta C)$ is the refractive index increment of the solute. e^2/mc^2 is the electron radius, P is the total energy per unit time irradiating the sample, d is the sample thickness, a is the distance between the sample and the plane of registration, ΔZ is the number of effective mol. electrons per gram, B_i, V_i and m_i are the scattering length, molar segmental volume and the segmental molecular weight of polymer i, and N_A is Avogadro's number.

The osmotic second virial coefficient, A_2, is obtained by plotting the right hand side of Eq. (40) against the concentration of polymer 2. The slope of the line thus obtained gives A_2. The value of A_2 is angle dependent and normally is extrapolated to zero angle. It is therefore dependent on the range over which the extrapolation has been performed and on the extrapolation procedure itself. Derivation of A_2 from neutron scattering data at several temperatures has been performed by Schmitt et al [62] and Maconnachie et al [66]. The Flory-Huggins χ_{12} parameter at low concentration of polymer 2 is then obtained by using the following relation:

$$\chi_{12} = \frac{1}{P_1} \left| \frac{1}{2} - \frac{V_1 M_2^2}{V_2^2} A_2 \right| \tag{44}$$

with P_1 being the degree of polymerisation of component one. For systems showing LCST behaviour the value of A_2 is positive, zero or negative below, at or above the θ temperature of the mixture respectively. The magnitude of A_2 is indicative of the strength of the interaction between the two polymers at a dilute concentration range. The Flory-Huggins interaction parameters, χ_{12}, using small angle neutron scattering, for several deuterated binary mixtures have been found by Schmitt [62]. A similar study of mixtures of deuterated polystyrene with poly(2,6-dimethyl-1,4-phenylene oxide) by small angle neutron scattering at temperatures in the range of 104 to 273 °C has been reported [66]. The results obtained for the latter mixture show that A_2 and χ_{12} decrease as the temperature increases. This is indicative of an LCST behaviour for the mixture investigated. The experimental results obtained so far using SANS are at very low concentrations of one component in another [62]. A new form of Eqs. (43) and (44) proposed recently [122] would allow measurements of an interaction parameter over the whole range of concentrations. This treatment in principle can be used to analyse data from SAXS and SALS to obtain information about the interactions in a mixture. No information about the expansion or contraction of the chains can be obtained by this treatment and the authors prefer to use Eq. (43) to interpret their SAXS or SALS scattering measurements due to this limitation. Several examples of this treatment have been given elsewhere [79, 83, 96].

5.5 Viscosity Measurements

Polymer-solvent interactions have been examined by viscometric studies of polymer-solvent-non-solvent mixtures in dilute solution [84–86]. The Fox-Flory model which relates the molecular parameters of the unperturbed dimension and the linear expansion coefficient to the total sorption parameter has been used. The latter can be obtained by the simultaneous solution of several Equations when the intrinsic viscosities of the mixtures are known. This method is in an early stage of development and provides a new way of measuring interaction parameters. The feasibility of this technique being used for computing binary and ternary interaction functions from intrinsic viscosity data alone has been discussed [86]. It could potentially be used to obtain information concerning polymer mixtures.

5.6 Melting Point Depression

The depression of the melting point of a crystalline polymer in blends can be used to measure the Flory-Huggins interaction parameter of the mixture. The melting points on heating, or crystallisation points on cooling, can be measured using techniques such as differential scanning calorimetry or by turbidity measurements.

For very high molecular weight polymers the interaction parameter χ'_{21} (segmental) can be obtained using a simplified version of that derived by Flory [87] as used by Nishi and Wang [61]

$$\frac{1}{T_m} - \frac{1}{T_m^0} = -\frac{R\chi'_{12}\varphi_1^2}{\Delta H_u} \tag{45}$$

where T_m^0 is the melting point of the pure crystalisable polymer and T_m its value in the blend, ΔH_u is the heat of fusion, and φ_1 the volume fraction of the amorphous component. A plot of $1/\varphi_1(1/T_m - 1/T_m^0)$ against φ_1 should give a straight line from the slope of which the interaction parameter can be found. In practice the plot does not go through the origin as expected, and this has been attributed to the effect of morphological changes in melting point depression. A generalised form of the above expression for non-infinite molecular weight polymers has been given which takes account of this factor [88]

$$\frac{\Delta H_u(T_m^0 - T_m)}{\varphi_1 R T_m^0} - \frac{T_m}{m_1} - \frac{\varphi_1 T_m}{2m_2} = \frac{C}{R} - b\varphi_1 \tag{46}$$

This assumes that the morphological contribution is solely proportional to φ_1 with a proportionality constant C. The left hand side of the above expression is plotted against φ and the slope b is related to the interaction parameter.

In practice it is very difficult to measure the true thermodynamic equilibrium melting point depression in blends due to problems of inhomogeneity and diffusion limitation in heating measurements and of supercooling during cooling measurements and results are dependent on the rate of heating or cooling. Along with the above theoretical assumptions this does bring into question the reliability of this technique. It is also limited in that it only gives one measure of the interaction parameter at the temperature of melting.

As well as the above quoted studies this method has also been used to study the interaction between poly(vinylidene fluoride) and poly(methyl methacrylate) [61], between poly(ethylene oxide) and the hydroxy ether of bisphenol A [60], and between poly(ethylene oxide) and a poly(ether sulphone) [35]. The above equations have also been reformulated in terms of the equation-of-state theory [123] to obtain the interaction energy, $\bar{X}_{12}$, which is concentration independent rather than the Flory-Huggins χ parameter which is composition dependent.

5.7 Other Methods

Interaction parameters can also be calculated from values of the expansion coefficients of polymer blends [66] using Equation-of-state theories, or from values of the isothermal compressibility of the mixture [89]. They can also be obtained from measures of volume changes on mixing. The measure of a cloud point diagram itself can in principle be used to calculate an interaction parameter though the converse is usually done in that spinodal curves are simulated using interaction parameters. This will be discussed in a later section of this review.

Methods based on radiationless transfer between chromophores in polymers can give information on local conformation and structure in blends but they are unable to give direct values of interaction parameters [90, 91].

6 Miscible Systems with no Specific Interactions

As discussed earlier two polymers may be miscible in the absence of any specific interaction if they have a heat of mixing which, though positive, is so small that it is outweighed by the small favourable entropy of mixing. The commonest example of this is the case of copolymers with differing composition with respect to the two components. Any copolymers must show some variation in composition and yet they are normally considered to be homogeneous. Many commercial copolymers are polymerised to high conversion under far from azeotropic conditions so that very large variations must exist. It would not be surprising if phase separation occurred under some circumstances.

One example in this category is the case of one polymer in two stereoregular forms [92–95]. Other examples are of two polymers which are chemically very similar such as poly(methyl acrylate) with poly(vinyl acetate) [96]. A series of systems which have been studied in some detail are various mixtures of chlorine containing polymers. Blends of chlorinated PVC with PVC have been studied [96]. It has been suggested that at 65.2 % wt.- % chlorine they are miscible and at 67.5 wt.- % they are not. Chlorinated polyethylene with 45 wt.- % chlorine has also been found to be miscible with PVC [75]. In this case it was suggested that phase separation occurs on heating.

It should be pointed out that in both these cases the degree of chlorination differs from PVC by around 10 %. By any estimate, the heat of mixing in these cases should be quite unfavourable. For example, an estimate based on solubility parameters and using group contribution gives for PVC ($\delta = 19.28$ $J^{1/2}cm^{-3/2}$) and chlorinated polyethylene (45 wt.- % Cl) ($\delta = 18.77$ $J^{1/2}cm^{-3/2}$), hence for a 50/50 mixture ΔH is $+0.065$ J per cm^3 of mixture. Together with unfavourable equation-of-state terms and a small combinatorial entropy contribution these mixtures would not be expected to be miscible.

In another study various chlorinated polyethylenes having different degrees of chlorination were examined [17]. The miscibilities are shown in Fig. 17. It can be seen that some mixtures differing in chlorination by 16 % were miscible whereas others differing by 7 % were immiscible. The phenomena could be explained to some extent

62.1	49.8	48.0	45.6	44.7	44.1	36.9	35.7	27.4	24.5	22.4	21.0	16.5	wt.-% Cl	
20	17	3	18	15	19	1	4	2	5	11	12	8	CPE	
−			−	−	−	+	+		+	+	+	+	20	62.1
	−		A	−		+	+		+				17	49.8
		−		−		+	+		+		C		3	48.0
			−	−	−	+	+			+	+	+	18	45.6
				−	−	+	+			+	+	+	15	44.7
					−	+	+	+	+	+	+	+	19	44.1
						−	−	+	+	+	+		1	36.9
							−	−	−	−	+	D	4	35.7
								−	B	−	−	+	2	27.4
									−	−	−	+	5	24.5
										−	−	+	11	22.4
											−	+	12	21.0
												−	8	16.5

Fig. 17. The miscibility of chlorinated polyethylenes with others having a different chlorine content, (−) miscible (+) immiscible. Group C includes mixtures of rubbery polymers with glassy polymers and group D is generally mixtures of rubbery polymers with the largest differences in chlorine content. These two groups give the largest (unfavourable) equation-of-state contributions to the free energy of mixing

by differences in Equation-of-state contributions, but most miscible blends had a calculated unfavourable free energy of mixing, using Flory's Equation-of-state theory when $Q_{12} = 0$.

It is possible that the explanation of this phenomena lies in the size of the unfavourable free energy of mixing. For a moderately long piece of a polymer chain, around 25,000 mol. wt., the free energy translates into a value which is of the same order as kT, the ambient energy level. We might therefore expect very large equilibrium fluctuations in composition. The small free energies would also give only a very small driving force for the kinetic process of phase separation (or mixing); in effect the polymers might tend to stay wherever they find themselves.

It is the authors' experience that, when phase separation takes place in such systems, for example with PVC/chlorinated polyethylene, it is a rather indeterminate process. It appears to take place at different temperatures using different techniques for establishing phase separation and for different heating rates. The process may more resemble a gradual increase in equilibrium phase separation extent over a wide temperature range and time.

In the case of polymers which are miscible due to specific interactions the free energies involved are an order of magnitude greater, their change with temperature is also greater, and phase separation phenomena are much more well defined processes.

7 Miscible Systems Having Specific Interactions

Most common and practically important miscible polymers owe their miscibility to a specific interaction of some sort between the two polymers. The commonest of these is the postulated weak hydrogen bond between halogen containing polymers and oxygen contaning polymers, e.g.

$$\text{>C=O} \ldots \text{H–}\overset{|}{\underset{|}{\text{C}}}\text{–Cl}$$

This is responsible for the miscibility of various polyesters [70], polyacrylates [3] and vinyl acetate copolymers with PVC [4]. Another postulated interaction which has not been studied so much is that between ether groups and aromatic rings which may be responsible for the miscibility of polystyrene and poly(methyl vinyl ether). Interactions probably also exist between other groups and aromatic moeties. However, some interactions can at present only be inferred from favourable heats of mixing found for low molecular weight analogues without much being really understood at a molecular level.

7.1 The Effect of Specific Interactions on T_g

The first indication of a strong specific interaction in a blend may be the measurement of a much higher glass transition temperature than would be expected from a weighted mean of the two polymers as has been previously discussed in Section 4. This arises because, if the chains interact together, their mobility will be decreased. A wide range of polyacrylates blended with PVC show a large positive deviation from linearity in plots of T_g against composition [1]. Another example of this effect has already been illustrated in Fig. 9.

7.2 The Effect of Specific Interactions on Cloud Point Curves

As might be expected it is often found that polymers which contain a higher concentration of such groups as are involved in the specific interaction are more miscible as evidenced by being miscible over a wider range of temperatures. Blends of chlorinated polyethylene with ethylene-vinyl acetate copolymers show phase separation on heating. If one increases the degree of chlorination of the chlorinated polyethylene, or the vinyl acetate content of the ethylene-vinyl acetate copolymer, the cloud point moves to a higher temperature. Fig. 18 shows the minimum in the cloud point curve as a function of these two variables.

The miscibility of PVC with several polyeters has also been studied by Ziska et al. [98]. Their experimental findings indicate that a negative interaction parameter exists between PVC and polyesters when the ratio of methylene groups to ester groups, $3 < (CH_2)_x/COO < 12$, and it is more negative for linear polyesters compared to branched ones at the same value of CH_2/COO.

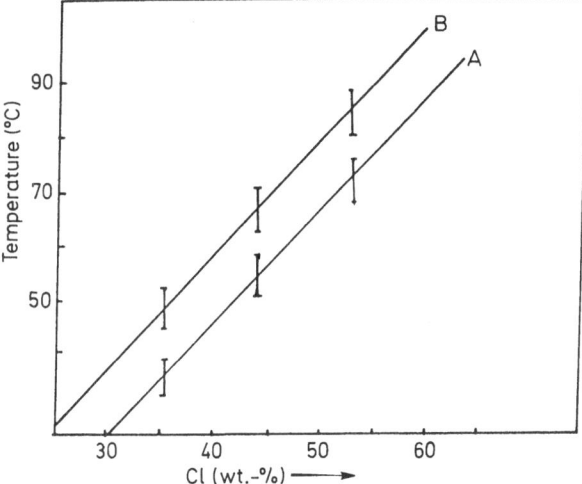

Fig. 18. A plot of the minimum of the cloud point curve against chlorine content for blends of chlorinated polyethylene with ethylene-vinyl acetate copolymers having A (40% vinyl acetate), B (45% vinyl acetate). It is seen that the higher concentrations of interacting groups, CHCl or C=O, give a larger temperature range of miscibility

In the case of PVC with various polyacrylates (and polymethacrylates) the choice of polyacrylate would change the concentration of interacting groups as measured by the weight fraction of carbonyl groups ($w_{c=0}$). Table 1 shows the miscibility of polymethacrylates with PVC as prepared both by *in-situ* polymerisation and by solvent casting from butan-2-one[3] (solvent casting from THF produces two phase blends). All the lower methacrylates are miscible with PVC but poly(octyl methacrylate) is found to be immiscible. Poly(hexyl methacrylate) is miscible at room temperature but phase separation on heating. Thus those polymers with a higher concentration of interacting groups appear to be more miscible.

In the case of the miscibility of polyacrylates with PVC the situation is more complicated as shown in Table 2. The higher acrylates are immiscible with PVC as expected but so is poly(methyl acrylate). Poly(ethyl acrylate) is a case where miscible blends

Table 1. The miscibility of polymethacrylates with PVC when prepared by in situ polymerisation and by solvent casting, (+) miscible, (−) immiscible

	In situ	Solvent cast
Poly(methyl methacrylate)	+	+
Poly(ethyl methacrylate)	+	+
Poly(n-propyl methacrylate)		+
Poly(n-butyl methacrylate)	+	+
Poly(n-pentyl methacrylate)		+
Poly(n-hexyl methacrylate)	+ (shows LCST)	+
Poly(iso-octyl methacrylate) (2-ethyl hexyl methacrylate)	−	−

Table 2. The miscibility of polyacrylates with PVC, when prepared by in situ poly-merisation and by solvent casting, (+) miscible, (−) immiscible

	In situ	Solvent cast
Poly(methyl acrylate)	−	−
Poly(ethyl acrylate)	+	−
Poly(n-propyl acrylate)	+ (shows LCST)	+
Poly(n-butyl acrylate)	+ (shows LCST)	+
Poly(n-pentyl acrylate)	uncertain results	
Poly(n-hexyl acrylate)	−	−
Poly(iso-octyl acrylate) (2-ethyl hexyl acrylate)	−	−

can be prepared by *in-situ* polymerisation but not by solvent casting, presumably due to interference by the solvents in the specific interaction. All the polyacrylates are less miscible than the polymethacrylates probably due to a weaker interaction arising from inductive or steric effects.

7.3 Evidence from Heat of Mixing Measurements

We have already described how the heats of mixing dominate the miscibility of poly-mers and so it was obvious to examine the heats of mixing of the above acrylates with PVC in order to understand the reasons for their miscibility or immiscibility. Fig. 19 shows the heats of mixing of various oligomeric polyacrylates and polymethacrylates with a chlorinated hydrocarbon which was used as an analogue for PVC. One can see

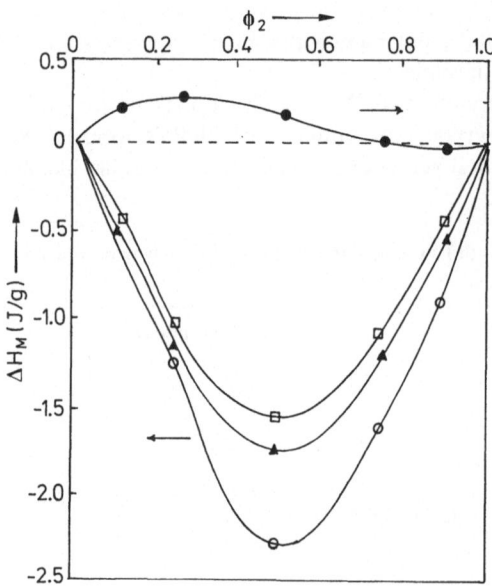

Fig. 19. The heats of mixing of a chlori-nated paraffin (as an analogue of PVC) with oligomers of (○) poly(butyl acryl-ate), (△) poly(hexyl methacrylate), (□) poly(butyl methacrylate), and (●) poly-(methyl acrylate) plotted against volume fraction chlorinated paraffin. That for poly(methyl acrylate) is positive (un-favourable) at most compositions which corresponds to the fact that PVC is immiscible with poly(methyl acrylate)

that poly(butyl acrylate), poly(hexyl methacrylate) and poly(butyl methacrylate) all show favourable (negative) heats of mixing but that poly(methyl acrylate) shows a small unfavourable heat of mixing thus accounting for the high polymer being immiscible with PVC. One should also note the concentration dependence of ΔH which shows a marked deviation from Eq. (21) as was described and discussed for other systems earlier.

The heats of mixing of a wide range of polyacrylates with chlorinated paraffin were measured at 50/50 compositions. The results are shown plotted against the weight fraction of carbonyl groups ($w_{C=O}$) in Fig. 20. They show unfavourable heats of mixing at both very high and very low values of $w_{C=O}$. Some immiscible polymers show favourable heats of mixing in the analogues but this can be explained by differences in the polymers and analogues which probably arise mostly from density differences.

One is still however left with the problem of why, since poly(methyl acrylate) contains the highest carbonyl group concentration of all polyacrylates, does it show an unfavourable heat of mixing? The answer lies in the neglection of contributions from dispersive forces. If one assumed that the dispersive and specific contributions could be separated,

$$\Delta H_{TOTAL} = \Delta H_{DISPERSIVE} + \Delta H_{SPECIFIC} \tag{47}$$

then the dispersive contributon could be written in the most simple way in terms of the cohesive energy densities or solubility parameters as in Eq. (5) as,

$$\Delta H_{DISPERSIVE} = V(\delta_1 - \delta_2)^2 \, \varphi_1 \varphi_2 \tag{48}$$

and the contribution from the specific interaction would depend in some way on the concentration of carbonyl groups in the polyacrylate, most simply as

$$\Delta H_{SPECIFIC} \alpha [C=O] \tag{49}$$

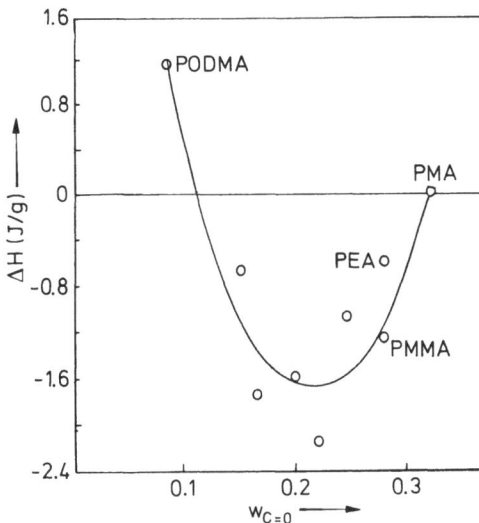

Fig. 20. The heats of mixing of 50/50 mixtures of a chlorinated paraffin with various oligomeric polyacrylates, plotted against the weight fraction of C=O groups in the polyacrylate. Those polyacrylates with very low and very high C=O group content are positive (unfavourable) which corresponds to the immiscibility of the respective polymers

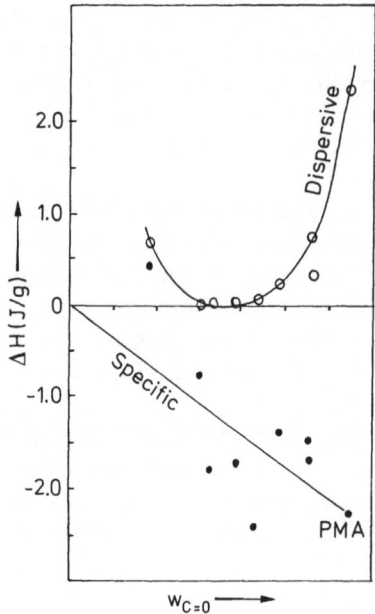

Fig. 21. The estimated contributions of the dispersive forces and the specific interaction to the heat of mixing for a series of oligomeric polyacrylates with a chlorinated paraffin (as an analogue for PVC) plotted against the weight fraction of carbonyl groups in the polyacrylate. The dispersive contribution is estimated using solubility parameter theory and the specific interaction obtained by difference from the experimental heats of mixing. The results explain why poly-(methyl acrylate) which has the highest concentration of interacting groups has an unfavourable heat of mixing and the respective polymers are immiscible

One can then calculate the dispersive contribution from tabulated solubility parameters, or from solubility parameters calculated from group contributions. When the calculated dispersive contribution is subtracted from the observed heats of mixing the contribution of the specific interaction should remain. The results are shown in Fig. 21. There is considerable scatter which is not surprising considering the assump-

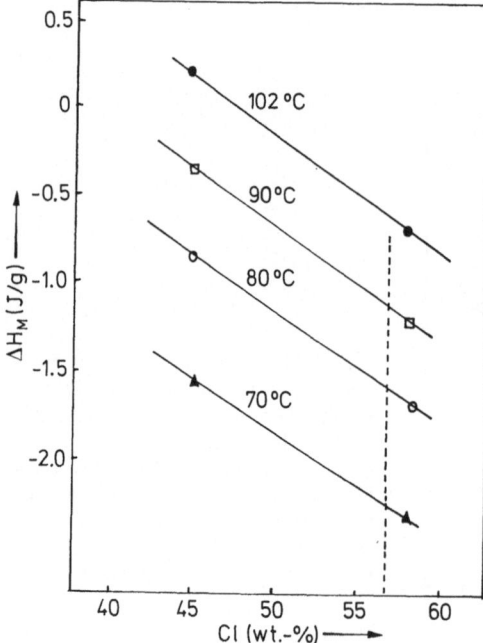

Fig. 22. The heats of mixing of chlorinated paraffins (as analogues for PVC) with poly(butyl acrylate) at a series of temperatures. The heat of mixing is a function of the chlorine content of the paraffins as expected and is also strongly dependent on temperature. A linear extrapolation to the chlorine content of PVC is also shown

tions made, but the results support the qualitative explanation above. The poly-(methyl acrylate) does in fact have a large contribution from the specific interactions but this is outweighed by a large unfavourable contribution from the dispersive forces.

One might expect that the effect of the specific interaction in blends would reduce as the temperature increases due to dissociation of the interacting pair. This dissociation of the specific interaction on heating should show up as a reduction in the heat of mixing at higher temperature. A low molecular weight poly(butyl acrylate) was mixed with two chlorinated paraffins [41], which can be used as analogues for PVC. The results are shown in Fig. 22. The chlorinated paraffin with the higher chlorine content shows a more favourable (negative) heat of mixing but both show a decrease in the favourable heats of mixing at higher temperature, becoming positive (unfavourable) in one case. This, in energy terms, is a very large effect and supports the view that the dissociation of the specific interaction can be the largest single factor resulting in phase separation on heating.

7.4 Evidence from Infra-Red Spectroscopy

The most direct way of looking at the specific interactions is by using spectroscopic measurements. Infra-red spectroscopy is the technique which has been most commonly used to study mixtures involving polymers. Studies of blends of PVC with polycaprolactone showed shifts of 4–$6\ cm^{-1}$ in the carbonyl band of polycaprolactone relative to the pure polymer [99, 100], but this Figure should be treated with caution as the peak probably consists of the sum of a shifted and an unshifted peak and it is difficult to say what the frequency of the shifted peak would be, or what fraction of the carbonyl groups are, or can be, involved in the interaction. Frequency shifts have also been shown to exist in blends of poly(methyl methacrylate) with poly(vinylidene fluoride) [63].

In the case of blends involving PVC with other polymers such as polycaprolactone there has been some controversy about the exact nature of the interaction. It had originally been assumed that the interaction involved the methine hydrogen of PVC but mixing studies of THF with various chlorinated hydrocarbons showed that the heats of mixing were dependent on the number of chlorines present independent of whether they were present as CCl_2 or CCl_3 groups etc. [101]. It was concluded that in this case the chlorine atoms were involved in the interaction. Unfortunately infra-red spectra due to C—Cl compounds are difficult to interpret as the bands are very dependent on the environment of the groups. Coleman eventually showed that the α hydrogen atoms of PVC were certainly involved in interactions with caprolactone by an experiment involving infra-red studies of blends of selectively deuterated PVC [102].

Evidence from spectral studies for interactions other than the above hydrogen bonds is not very plentiful. Polystyrene/poly(2,6 dimethyl-1,4-phenylene oxide) blends have been studied by infra-red and ultraviolet spectroscopy [103, 104]. Interactions involving the aromatic rings of the two polymers were proposed. Studies of low molecular weight ethers with aromatic compounds have shown evidence for specific interactions and this has recently been extended to blends of polystyrene with poly(methyl vinyl ether) [105].

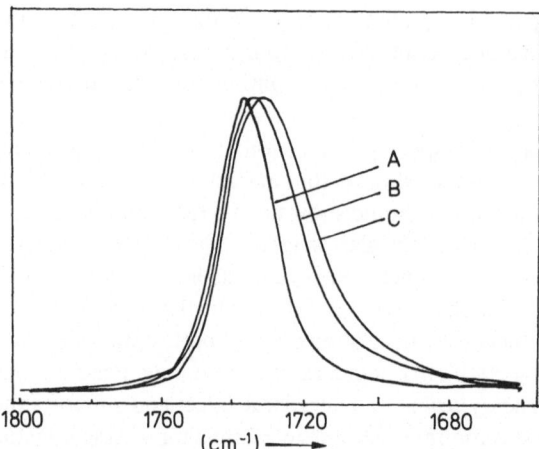

Fig. 23. A part of the infra red spectra showing the carbonyl band absorption for (A) an ethylene-vinyl acetate copolymer and its blends with (B) 40% and (C) 80% chlorinated polyethylene. The peak is shifted due to a specific interaction between the carbonyl and the methine hydrogen of chlorinated polyethylene. The shifted peaks are actually a combination of a shifted and an unshifted peak at different ratios

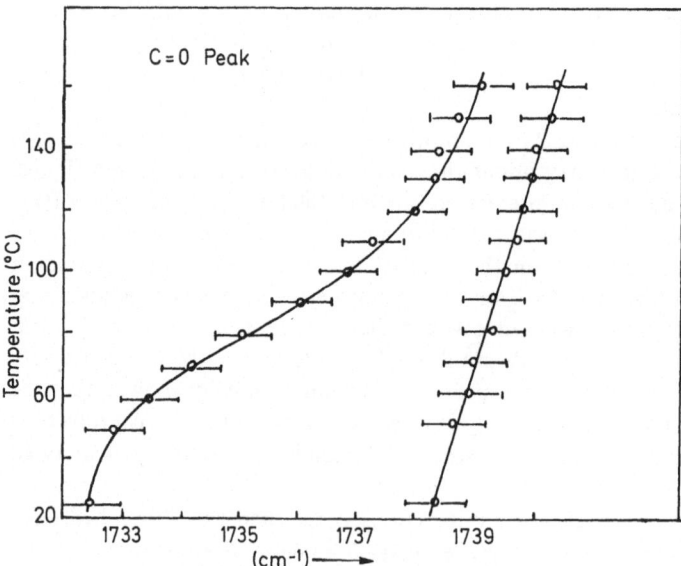

Fig. 24. A plot of the frequency of maximum infra red carbonyl absorption against temperature for an ethylene-vinyl acetate copolymer and a blend containing 80% chlorinated polyethylene. The relative shift is reduced at higher temperature which is attributed to a dissociation of the specific interaction between the polymers

Since many polymers phase separate on heating, studies of changes in infra-red spectra on heating were expected to give interesting information. Blends of chlorinated polyethylene with ethylene-vinyl acetate copolymers which phase separated at around 80 °C for 50/50 compositions were examined [106]. The infra-red carbonyl adsorption of EVA and two blends are shown in Fig. 23. The gradual shift with increasing chlorinated polyethylene content confirms that these peaks are each the sum of two peaks, one shifted and one unshifted. The blends were also examined as a function of temperature and the change in the position of the maximum in the peaks is shown in Fig. 24 with a corresponding EVA peak position for comparison. This particular 80/20 composition is not expected to phase separate until around 110 °C, but a gradual reduction of the shift is apparent over most of the temperature range studied.

This gradual reduction in the infra-red shift we attribute to a dissociation of the specific interaction on heating. In energy terms the size of this effect is very large compared to other contributions to the free energy such as the combinatorial entropy and the equation-of-state terms. In effect the main reason for the phase separation of this blend on heating is the weakening of the specific interaction to the point where other unfavourable contributions to the free energy outweigh it. This has important theoretical consequences which will be discussed later.

8 Simulation of Phase Diagrams

The thermodynamic definition of the spinodal, binodal and critical point were given earlier by Eqs. (9), (7) and (8) respectively. The variation of ΔG_M with temperature and composition and the resulting phase diagram for a UCST behaviour were illustrated in Fig. 1. It is well known that the classical Flory-Huggins theory is incapable of predicting an LCST phase boundary. It has, however, been used by several authors to deal with ternary phase diagrams [33, 36]. Other workers have extensively used a modified version of the classical model to explain binary UCST or ternary phase boundaries [106–113]. The more advanced equation-of-state theories, such as the theory of Flory and his collaborators, are capable of predicting both upper and lower critical phase diagrams. In using these theories the phase boundaries are simulated by means of a computer. Using the generalised version of the theory, McMaster [10] simulated the spinodal and binodal curves of a hypothetical polymer-polymer mixture. By applying the spinodal, binodal and critical condition, described earlier, to the chemical potential of mixtures we have calculated simpler forms of McMaster's equation which are presented below. The polydispersity of the samples were ignored in our derivation.

8.1 The Phase Diagram Equations

The spinodal Equation can be obtained by applying the spinodal condition (Eq. 9) to the chemical potential of a polymer in a mixture, (obtained from Eq. (27) by including the Flory-Huggins combinatorial chemical potential) as has been described previously [44, 69] to give:

$$\frac{\partial(\Delta\mu_1/RT)}{\partial\Phi_2} = -1/\Phi_1 + (1 - r_1/r_2) + (P_1^*V_1^*/RT_1^*)(-D/(\tilde{v} - \tilde{v}^{2/3}))$$

$$+ \overset{*}{P_1}\overset{*}{V_1}D/RT\tilde{v}^2 + PV_1^*D/RT + V_1^*X_{12}\, 2\theta_2^2\theta_1/RT\tilde{v}\Phi_1\Phi_2$$

$$- V_1^*X_{12}D\theta_2^2/RT\tilde{v}^2 - V_1^*Q_{12}\, 2\theta_2^2\theta_1/R\Phi_1\Phi_2 \tag{50}$$

where

$$D = \frac{\partial\tilde{v}}{\partial\Phi_2} \tag{51}$$

the derivation of which is given elsewhere [44, 69].

In the simulation of the spinodal curve, using the Equation above, the Q_{12} parameter can be used as an adjustable parameter to match the minimum temperature of the simulated spinodal to the minimum of the experimental cloud point temperature. Then the full spinodal curve with constant values of X_{12} and Q_{12} can be calculated. If both X_{12} and $\bar{X}_{12}$ are known from different measurements then in principle the spinodal can be calculated without the need for adjustable parameters.

The critical point Equation is derived from the spinodal Equation using the critical condition given earlier. The differential of the spinodal Equation with respect to Φ_2 is zero at the critical point. This calculation is perfectly feasible but so far no one has made any attempt to use these Equations to predict the critical point. In the past it has usually been approximated using the simple Flory-Huggins expression for the critical point.

The binodal Equation can be derived from the chemical potentials of the components using the binodal condition quoted earlier. The resulting Equations are much more complicated and more difficult to solve than those for the spinodal. The simulation of binodal and spirodal curves in a practical manner is shown elsewhere [126].

8.2 The Equation of State Parameters

In order to simulate the phase boundaries the following state parameters of the pure components are required.

a. The specific volume, $v_{sp} = 1/d$
b. The thermal expansion coefficient, $\alpha = (1/v)(\delta v/\delta T)_P$
c. The thermal pressure coefficient, $\gamma = (\delta P/\delta T)_V$

and the following binary parameters:

d. The surface per unit of core volume ratio, S_2/S_1.
e. The interaction term, X_{12}.

The entropy parameter, Q_{12}, is also needed but is generally used as an adjustable parameter.

The specific volumes of the pure components are normally obtained by equal density titration or by use of a density gradient column.

The thermal expansion coefficients are obtained either from density measurements or more accurately by dilatometry as described by Orwoll and Flory [114].

The thermal pressure coefficients of the pure polymers can be estimated from the ratio of thermal expansion coefficient, α, to isothermal compressibility, β_T, as $\gamma = \alpha/\beta_T$. In the case where γ is not available from the literature it can be calculated from solubility parameters which themselves are related to the cohesive energy density (C.E.D.) and hence to the strength of the internal pressure of the structural molecules [115]. The binary parameter, S_2/S_1 is obtainable from the method of group contribution given by Bondi [116]. It can alternatively be calculated by casting shadows of models of the molecules for various orientations, where the area for the monomer unit is estimated from the area of the projections.

The X_{12} parameter is best obtained by fitting the equation for to the experimental heats of mixing of analogous materials as reported elsewhere [44,69]. It can also be obtained from any other binary quantity such as the second virial coefficient, the thermal expansion coefficient of mixture, or the volume change on mixing. X_{12} is assumed to be independent of temperature but as we described in the previous section this may not be valid. At present there is no way of predicting the temperature variation and one can only use empirical expressions or assume a constant value most appropriate for the temperature range of interest.

The temperature dependance of v_{sp}, α and γ at atmospheric pressure can also be allowed for [44,66].

8.3 Simulation of Model Phase Diagrams

McMaster simulated binodal and spinodal curves for hypothetical polymer pairs with various values of the Equation-of-state parameters [10]. We have also simulated many hypothetical spinodal curves using the equations presented in the previous section and some of these are presented in Figs. 25 and 26. Various other workers have also calculated theoretical curves. An assessment of the effect of changes in the various properties is presented below.

a) Polymer-polymer miscibility is decreased (reduction in range of temperature) by increasing the difference between the α values of the two polymers. Curve 1 of Fig. 25 shows the result for a small difference in α values whereas no difference gives complete miscibility.

b) Changes in γ values have less effect than changes in α values. The size of the effect depends on the "α effect". It is often stated that changes in $\gamma_1 - \gamma_2$ alter the extent of miscibility but in practice it is the average absolute value of the α values which affect this most, higher γ values reducing the range of miscibility (curve 2 of Fig. 25). The difference between the γ values and its sign affect the shape of the curve. If γ_2 is higher than γ_1 this tends to move the minimum in the curve to lower volume fractions of component 2 (see curve 3 of Fig. 25).

c) Increasing molecular weights of the polymers decreases the miscibility. Changes in one molecular weight relative to the other moves the curve to one side in the way that would be expected from a simple Flory-Huggins treatment.

d) Introducing a negative or positive X_{12} makes the polymers more or less miscible respectively. The curves also become much flatter with larger X_{12} values as this composition independent factor dominates the miscibility (see curves 1 and 2 of Fig. 26).

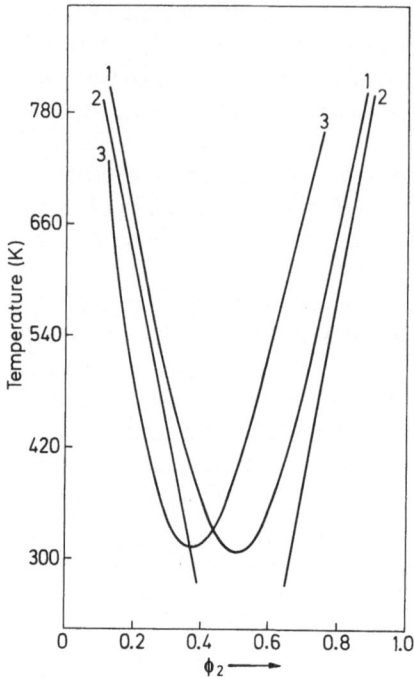

Fig. 25. Hypothetical simulated spinodal curves for the phase separation of two polymers on heating, illustrating the effect of the thermal expansion coefficient (α) and thermal pressure coefficient (γ). The curves are all simulated using values of $S_2/S_1 = r_2/r_1 = 1$; $X_{12} = Q_{12} = 0$; $V_1^* = 100,000$ cm^3 $\times$ mol^{-1}. With $\gamma_1 = \gamma_2 = 1$ (J cm^{-3}K^{-1}) and $\alpha_1 = \alpha_2$ the polymers will always be miscible but if $\alpha_1 = 5 \times 10^{-4}$ K^{-1} and $\alpha_2 = 4.7 \times 10^{-4}$ phase separation occurs with a spinodal as shown (1). Increasing both γ values to 1.1 moves the phase diagram to lower temperatures (2). Making $\gamma_1 = 0.8$ and $\gamma_2 = 1,2$ (i.e. keeping the sum the same but increasing the difference) moves the spinodal to one side in composition

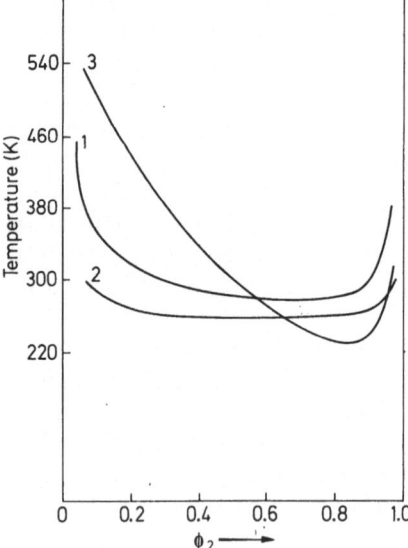

Fig. 26. Hypothetical simulated spinodal curves for the phase separation of two polymers on heating illustrating the effect of the interaction parameter (X_{12}); the non-combinatorial entropy parameter (Q_{12}) and the ratio of surface areas per unit volume S_2/S_1. The curves are all simulated using values of $\gamma_1 = \gamma_2 = 1$ (J cm^{-3}K^{-1}); $r_2/r_1 = 1$; $V_1^* = 100,000$ cm^{-3}mole^{-1} $\alpha_1 = 4 \times 10^{-4}$, $\alpha_2 = 4 \times 10^{-4}$K^{-1}. If $X_{12} = -0.6$ J cm^{-3}, $Q_{12} = 0$, and $S_2/S_1 = 1$, the curve is much flatter than those in the previous figure. If there is a larger (favourable) X_{12}, say -1.2 and this is balanced by an unfavourable $Q_{12} = -0.0023$ J cm^{-3}K^{-1}, then the curve is much flatter as these parameters swamp the effect of other terms (2). If $X_{12} = -0.6$, $Q_{12} = 0$ and S_2/S_1 is now set at 1.5 then the spinodal curve will be skewed to one side as shown (3)

e) The values of S_2/S_1 has little effect in the absence of X_{12} (or Q_{12}) terms. As can be seen from the spinodal equation it operates mostly on these terms. When X_{12} terms are present it operates to lean the curve over to one side or the other. If S_2/S_1 is above the median value of 1 then the curve leans down to high values of volume fraction component 2, often with a minimum at very high volume fractions of component 2 (see curve 3 of Fig. 26).

f) A more negative value of Q_{12} makes the polymers less miscible. In systems where a large negative X_{12} is compensated by a large negative Q_{12} the curve can become extremely flat.

g) When two polymers are polydisperse the spinodal curve remains unchanged if the weight average chain length is used [17, 18].

h) Bimodality in the spinodal curve can arise in simulations involving the variation of state parameters of pure components and binary quantities.

8.4 Simulation of Actual Phase Diagrams

Many authors have tried to simulate phase diagrams of the upper critical type for low molecular weight polymers using modified forms of classical theories [107–112]. This is a much easier task than for high molecular polymers showing lower critical temperature behaviour. The cloud points are much easier to measure due to the high mobility of low molecular weight polymers. Upper critical solution temperature phase type exist for relatively non-polar polymers where there are no specific interactions and the theories available are more applicable. Also it is possible to use pulse induced critical scattering (PICS) [109] to determine the position of the spinodal curve directly rather than a cloud point curve which can be anywhere between the binodal and spinodal. A full discussion of this work is outside the limit of this review but the reader is directed to several reviews and papers published by others [9, 107–112].

Olabisi [100] used the Equations developed by McMaster to attempt to simulate the LCST phase diagram of mixtures of PVC with poly(caprolactone). He obtained a value of X_{12} using the inverse gas chromatography technique and assumed Q_{12} to be zero. The simulated curves did not come close to reality but considering the

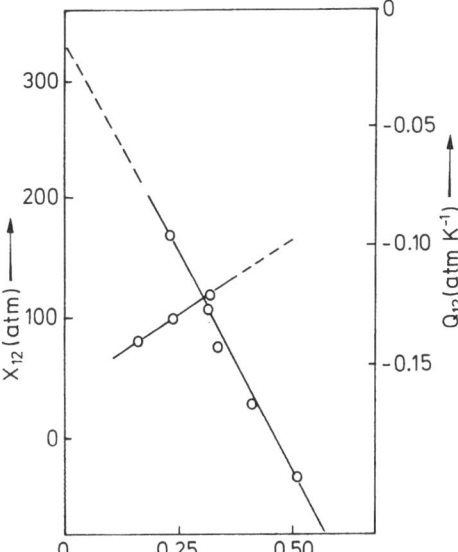

Fig. 27. A plot showing the values of X_{12} calculated from heat of mixing data for chlorinated paraffins with oligomeric poly-(methyl methacrylate). This is more favourable for higher chlorine content. Also shown are the Q_{12} values which had to be used to simulate the spinodal curves shown in the following figure

approximation made and the fact that only S_2/S_1 ratio was used as an adjustable parameter this is not too surprising.

In one study of blends, using the spinodal equation shown in this review, for chlorinated polyethylene with poly(methyl methacrylate) [44] it was possible to show LCST behaviour for high molecular weight polymers and also study the UCST behaviour for low molecular weight analogues where the low molecular weight analogue had a lower degree of chlorination. X_{12} values were found from heat of mixing studies on oligomers to be negative (favourable) and become more negative for chlorinated hydrocarbons with higher chlorine contents indicating a stronger specific interaction as shown in Fig. 27. Spinodals of the UCST curves of oligomers were simulated using Q_{12} as an adjustable parameter and the results are shown in Fig. 28. The fact that the cloud point curve may be closer to the binodal than the spinodal may explain the differences in shape. The Q_{12} values required for the simulation are also shown in Fig. 27. The LCST curves for the high molecular weight polymers were also simulated as shown in Fig. 29. The X_{12} value was -30 atm and the Q_{12} value -0.06 atm K^{-1} used was not far distant from that extrapolated from Fig. 27.

Simulation studies have also been carried out on blends of chlorinated polyethylene with poly(butyl acrylate). The results are shown in Fig. 30. It was found in this case (in a similar way to the previous example) that with the value of X_{12} obtained from heat of mixing studies at 70 °C on oligomers (-94 atm) and with the value of Q_{12} necessary to match the spinodal to the minimum of the cloud point (-0.235 atm/K) the resulting spinodal was very flat bottomed and lay outside the cloud point curve, an impossible situation. To match the spinodal to the cloud point curve a much smaller value of X_{12} (and correspondingly Q_{12}) must be chosen. This discrepancy could have resulted from differences between the low molecular weight materials used for heat

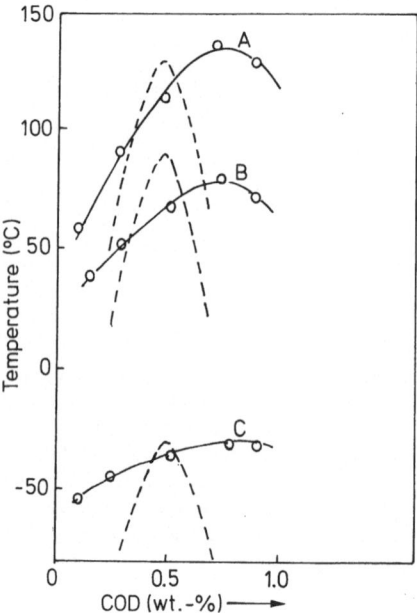

Fig. 28. The experimental cloud point curves (solid lines) and simulated spinodal curves (dotted) for an oligomeric poly(methyl methacrylate) with chlorinated paraffin (octadecane) with (A) 17.4% Cl, (B) 24.6% Cl and (C) 33.4% Cl. The cloud points are probably closer to the binodals than the spinodals which might explain the difference in shape. The chlorinated paraffin with the highest chlorine content is found to be the most miscible

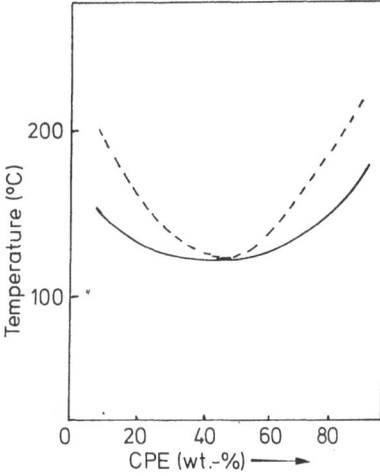

Fig. 29. The experimental cloud point curve and simulated spinodal for a mixture of poly(methyl methacrylate) and chlorinated polyethylene using a value of $X_{12} = -30$ atm and $Q_{12} = -0.06$ atm K^{-1} showing that the LCST can be simulated using data which is broadly compatible with that used in simulating a UCST in the previous figure

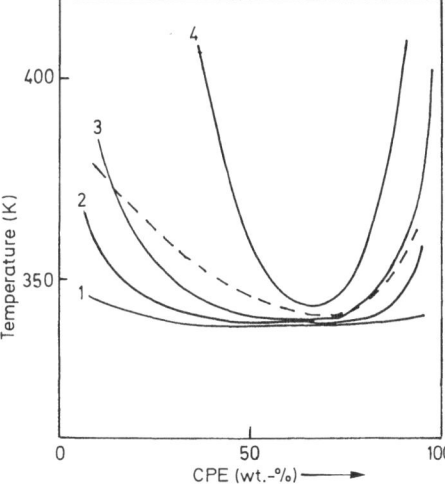

Fig. 30. The experimental cloud point curve (dotted line) and simulated spinodals for blends of poly(butyl acrylate) with a chlorinated polyethylene. The initial value of $X_{12} = -94$ atm obtained from heat of mixing data, and the adjusted $Q_{12} = -0.235$ atm K^{-1} give a curve which is too flat-bottomed (1). By adjusting X_{12} (and using an appropriate Q_{12}) a closer fit can be obtained; (2) $X_{12} = -30$ $Q_{12} = -0.076$; (3) $X_{12} = -10$ $Q_{12} = -0.026$; (4) $X_{12} = -1$ $Q_{12} = -0.0034$

of mixing determination and the polymers themselves or from the temperature dependance of X_{12} and Q_{12}.

Studies on mixtures of chlorinated polyethylenes with ethylene-vinyl acetate copolymers show similar results. Some of these simulated spinodals are shown in Fig. 31. Again the X_{12} values obtained from low molecular weight analogues are too large to give a satisfactory spinodal curve. The discrepancy may again be due to differences between low molecular weight analogues and the polymers and/or to the variation of X_{12} with temperature. In this case we know from FT infra-red studies that the specific interaction dissociates over the temperature range of interest. This would result in an X_{12} value which becomes smaller at higher temperatures and this would account very satisfactorily for the discrepancy in the simulated spinodals.

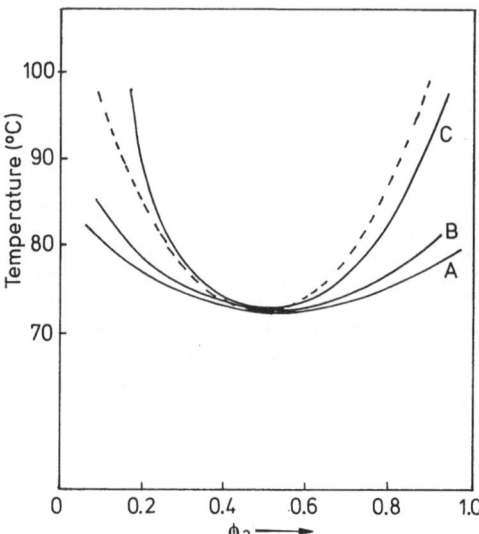

Fig. 31. Experimental cloud point curve (dotted line) and simulated spinodals, for blends of an ethylene-vinyl acetate co-polymer with chlorinated polyethylene. The initial curve (A) using an X_{12} (-4.2 J cm^{-3}) value calculated from heat of mixing data and an adjusted Q_{12} (-0.0108 J cm^{-3} K^{-1}) was too flat bottom-ed. By adjusting X_{12} (and the appropriate Q_{12}) a closer fit could be obtained; (B) $X_{12} = -2.63$, $Q_{12} = -0.00678$; (C) $X_{12} = -0.5$, $Q_{12} = -0.00138$

The importance of simulations of the phase diagrams is that they are one of the few direct tests of any theory and hence of any model used to describe the systems. The Equation-of-state theory has the advantage of being able to predict LCST be-haviour but in most real systems this is only possible by including the non-combi-natorial entropy correction Q_{12}. This may have a physical significance in the entropy change accompanying a specific interaction, but its use as an adjustable parameter is obviously unsatisfactory. Ideally one would have independent measures of both X_{12} and $\bar{X}_{12}$, preferably over a range of temperature, and hence have no need of adjustable parameters.

Finally there is the problem that the theory was not intended to fully describe systems with a specific interaction. If a specific interaction varies with temperature then X_{12} will not be constant. An ideal theory and model would predict this behaviour. The inclusion of yet another adjustable parameter to describe the temperature depend-ence of X_{12} would not be desirable. In systems where the specific interaction does not vary over the temperature of study the Equation-of-state theory may give a satis-factory description of the system. This underlines the importance of experimental evidence which gives direct information about the specific interactions.

9 Conclusion

We have concluded that in most systems of miscible pairs of high polymers specific interactions are responsible for the miscibility, and an understanding of these specific interactions is vital for an understanding of the behaviour of the blends.

In those miscible, high polymer pairs showing no specific interaction the polymers are usually very similar chemically. Other polymers which are not similar chemically are nevertheless miscible due to specific interactions between dissimilar groups in the two polymers. The specific interaction shows itself in many ways; from the favourable heats of mixing, from predictable changes when the concentrations of interacting groups are changed, from favourable interaction parameters determined by various techniques, by the T_g of the blends which are higher than expected by additivity, and from direct observation of spectroscopic changes in the mixed polymers. Some of these techniques also show that the specific interactions tend to dissociate on heating; and the heats of mixing and interaction parameters become less favourable and the spectra revert to those of the pure components. This has important implications for the theories of polymer miscibility. None of the theories available address themselves directly to the problem of the specific interaction.

It is possible to simulate the spinodal curves of the phase diagram of polymer pairs using the Equation-of-state theory developed by Flory and co-workers. It is only, however, possible to do this using the adjustable non-combinatorial entropy parameter, Q_{12}. Another problem arises in the choice of a value for the interaction parameter X_{12}. This is introduced into the theory as a temperature independent constant whereas we know that in many cases the heat of mixing, and hence X_{12} is strongly temperature dependent. The problem arises because X_{12} was never intended to describe the interaction between two polymers which are dominated by a temperature dependent specific interaction.

10 References

1. Olabisi, O., Robeson, L. M. and Shaw, M. T.: "Polymer-Polymer Miscibility", Academic Press, N.Y. (1979)
2. Emmett, R. A.: Ind. Eng. Chem., 36, 730 (1979)
3. Walsh, D. J. and McKeown, J. G.: Polymer, 21, 1330 (1980)
4. Adler, K. and Paul, K. P.: Kunststoffe, 7, 70 (1980)
5. Cizek, E. P.: US Patent 3,383,435 (to General Electric Co.), (1968)
6. Flory, P. J., Orwell, R. A. and Vrij, A.: J. Am. Chem. Soc., 86, 3507, 3515 (1964)
7. Flory, P. J.: J. Am. Chem. Soc., 87:9, 1833 (1965)
8. Newton, A. B.: Paper presented at "MACRO GROUP" meeting, Belgrave Square, London (1983)
9. Onclin, M. H., Kleintjens, L. A. and Koningsveld, R.: Makromol. Chem. Suppl., 3, 197 (1979)
10. McMaster, L. P.: Macromolecules, 6, 760 (1973)
11. Eichinger, B. E. and Flory, P. J.: Trans. Faraday Soc., 64, 2035–2065 (1968)
12. Renuncio, J. A. R. and Prausnitz, J. M.: Macromolecules, 9, 898 (1976)
13. Patterson, D. and Delmas, G.: Discussion Faraday Soc. No. 49, 98 (1970)
14. Sanchez, I. C. and LaComb, R. H.: Macromolecules, 11, 1145 (1978)
15. Prigogine, I.: "The Molecular Theory of Solutions", Holland Pub. (1975)
16. Chahal, S. R., Kao, W. P. and Patterson, D.: Faraday Trans., 1, 1834 (1973)
17. Chai, Z.: Ph. D. Thesis, Imperial College, 1982
18. Rostami, S.: Ph. D. Thesis, Imperial College, 1983
19. Shih, H. and Flory, P. J.: Macromolecules, 5, 758, 761 (1972)
20. Hamada, F., Shiomi, T., Fujiswa, K. and Nakajima, A.: Macromolecules, 13, 729 (1980)
21. Canovas, A. A., Rubio, R. G. and Renuncio, J. A. R.; J. Polym. Sci. Phys. Ed., 20, 784 (1982)
22. Canovas, A., Rubio, R. G. and Renuncio, J. A. R.: J. Polym. Sci. Phys. Ed., 21, 841 (1983)
23. Sanchez, I. C. and LaComb, R. H.: J. Phys. Chem., 80, 2352, 2568 (1976)
24. Sanchez, I. C.: J. Macromol. Sci. Phys. Ed., B17 (3), 565 (1980)
25. Sanchez, I. C.: Chapter 3 of "Polymer Blends", Paul and Newman Ed., Academic Press, 1978

26. Hammer, C. F.: Macromolecules, 4, 69 (1971)
27. Nielsen, L. E.: J. Polym. Sci., 42, 357 (1980)
28. Doubé, C. D. and Walsh, D. J.: Polymer, 20, 115 (1979)
29. Shultz, A. R. and Beach, B. M.: Macromolecules, 7, 902 (1974)
30. Davis, D. D. and Kwei, T. K.: J. Polym. Sci. Phys. Ed., 18, 2337 (1980)
31. Bank, M., Leffingwell, J. and Thies, C.: Macromolecules, 4, 43 (1971)
32. Robard, A., Patterson, D. and Delmas, G.: Macromolecules, 10, 706 (1977)
33. Hsu, C. C. and Prausnitz, J. M.: Macromolecules, 7, 320 (1974)
34. Aharoni, S. M.: Macromolecules, 11, 277 (1978)
35. Walsh, D. J. and Singh, V. B.: to be published
36. Walsh, D. J. and Cheng, G. L.: Polymer, to be published
37. Walsh, D. J. and Cheng, G. L.: Polymer, 23, 1965 (1982)
38. Walsh, D. J., Higgins, J. S., Doube, C. D. and McKeown, J. G.: Polymer, 22, 168 (1981)
39. Walsh, D. J. and Sham, C. K.: to be published
40. MacKnight, W. J., Karasz, F. E. and Fried, J. R.: Chapter 5 of "Polymer Blends", Paul and Newman Ed., Academic Press, 1978
41. Chai, Z. and Walsh, D. J.: Makromol. Chem., 184, 1549 (1983)
42. Reich, S. and Cohen, Y.: J. Polym. Sci. Phys. Ed., 19, 1255 (1981)
43. Cowie, J. M. G.: Private Communication
44. Chai, Z., Ruona, S., Walsh, D. J. and Higgins, J. S.: Polymer, 24, 263 (1983)
45. Walsh, D. J., Higgins, J. S., Rostami, S. and Weeraperuma, K.: Macromolecules, 16, 391 (1983)
46. Cowie, J. M. G.: J. of Macromole. Sci. Phys. Ed., B18, 569 (1980)
47. Hichman, J. J. and Ikeda, R. M.: J. Polym. Sci. Phys. Ed., 11, 1713 (1973)
48. Pochan, J. M., Beatly, C. L. and Pochan, D. F.: Polymer, 20, 879 (1979)
49. Fried, J. R., Karasz, F. E. and MacKnight, W. J.: Macromolecules, 11, 150 (1979)
50. Prest, W. M. and Porter, R. S.: J. Polym. Sci. Phys. Ed., 10, 1693 (1972)
51. Gordon, M. and Taylor, J. S.: J. Appl. Chem., 2, 493 (1953)
52. Dimarzio, E. A. and Gibbs, J. M.: J. Polym. Sci., XL, 121 (1959)
53. Murayama, T.: "Dynamic Mechanical Analysis of Polymeric Materials", Elsevier (1978)
54. Walsh, D. J., Higgins, J. S., Rostami, S.: Macromolecules, 16, 387 (1983)
55. Hedvig, P.: "Dielectric Spectroscopy of Polymers", Adam Higler, Bristol (1977)
56. Wetton, R. E. et al.: Macromolecules, 11, 158 (1978)
57. Rabek, J. E.: "Experimental Methods in Polymer Chemistry", Wiley (1980)
58. Flynn, J. M.: Thermochim, Acta, 8, 69 (1974)
59. Zabrzewski, G. A.: Polymer, 14, 347 (1973)
60. Kimura, M., Porter, R. S. and Salee, G.: J. Polym. Sci. Polym. Phys. Ed., 21, 367 (1983)
61. Nishi, T. and Wang, T. T.: Macromolecules, 8, 909 (1975)
62. Jeleniö, J., et al.: Makromol. Chem. 185, 129 (1984)
63. Coleman, M. M., et al.: J. Polym. Sci. Polym. Phys. Lett. Ed., 15, 745 (1977)
64. Varnell, D. F. and Coleman, M. M., Polymer, 22, 1324 (1981)
65. Khambata, F. B., et al.: J. Polym. Sci. Polym. Phys. Ed., 14, 1391 (1976)
66. Maconnachie, A., et al.: to be published
67. Koningsveld, R. and Kleintjens, L. A.: Br. Polym. J.; 9, 212 (1977).
68. Tager, A. A., Scholokhovich, T. T. and Bessanov, Yu. S.: Eur. Polym. J., 11, 321 (1975)
69. Rostami, S. and Walsh, D. J.: Macromolecules, 17, 315 (1984)
70. Cruz, C. A., Barlow, J. W. and Paul, D. R.: Macromolecules, 12, 276 (1979)
71. Chong, C. L.: Ph. D. Thesis, Imperial College, 1980
72. Smidsrod, O. and Guillet, J. E.: Macromolecules, 2, 272 (1969)
73. Baranyi, C. D.: Macromolecules, 14, 683 (1981)
74. Baranyi, C. D. and Guillet, C. D.: Macromolecules, 11, 228 (1977)
75. Doube, C. D. and Walsh, D. J.: Europ. Polym. J., 17, 63 (1980)
76. Braun, J. M., Guillet, J. E.: Macromolecules, 9, 617 (1976)
77. Tait, P. J. and Abushihada, M. A.: Polymer, 18, 810 (1977)
78. Huglin, M. B.: "Light Scattering from Polymer Solutions", Academic Press, 1972
79. Russell, T. P. and Stein, R. S.: J. Polym. Sci. Phys. Ed., 20, 1594 (1982)

80. Richards, R. G.: "An Introduction to Physical Properties of Large Molecules in Solutions", Cambridge University Press, 1980
81. Gilmer, J., Goldstein, J. S. and Stein, R. S.: J. Polym. Sci. Polym. Phys. Ed., 20, 2219 (1982)
82. Maconnachie, A. and Richards, R. W.: Polymer, 19, 739 (1978)
83. Glater, O. and Kratky, O.: "Small Angle X-Ray Scattering", Academic Press, 1982.
84. Battiglione, V., Morcellet, M. and Loucheux, C.: Macromol. Chem., 181, 485 (1980)
85. Apostolopoulos, M., Morcellet, M. and Loucheux, C.: Macromol. Chem., 183, 1293 (1982)
86. van Den Berg, J. W. A. and Altena, F. W.: Macromolecules, 15, 1447 (1982)
87. Flory, P. J.: "Principle of Polymer Chemistry", Cornell University, 1954
88. Kwei, T. K. and Frisch, H. L.: Macromolecules, 11, 1267 (1978)
89. Aıcrat, A., Tardajos, G. and Diaz Pena, M.: J. Solution Chemistry, 12, 41 (1983)
90. Frank, C. W. and Gashgari, M. A.: Macromolecules, 12, 163 (1979)
91. Amrani, F., Hung, Ju. M. and Morawetz, H.: Macromolecules, 13, 649 (1980)
92. Yeh, G. S. Y. and Lambert, S. L.: J. Polym. Sci., Part A2, 10, 1183 (1972)
93. Krause, S. and Roman, N.: J. Polym. Sci., Part A3, 1631 (1965)
94. Liu, H. Z. and Liu, K. J.: Macromolecules, 1, 157 (1968)
95. Liquori, A. M., et al.: Nature (London), 206 358 (1965)
96. Hughes, L. J. and Britt, G. E.: J. Appl. Polym. Sci., 5, 337 (1961)
97. Carmain, B., Villoutreix, G. and Berlot, R.: J. Macromol. Sci. Phys., 14 (2), 307 (1977)
98. Ziska, J. J., Barlow, J. W. and Paul, D. R.: Polymer, 22, 918 (1981)
99. Coleman, M. M., Zarian, J.: J. Polym. Sci. Phys. Ed., 17, 837 (1979)
100. Olabisi, O.: Macromolecules, 9, 316 (1975)
101. Pouchly, J. and Biros, J.: Polymer Lett. 7, 463 (1969)
102. Varnell, D. F.: Ph. D. Thesis, Pennsylvania State University (1982)
103. Wellinghoff, S. T., Keonig, J. L. and Baer, E.: J. Polym. Sci., Phys. Ed., 15, 1913 (1977)
104. Lefebvre, D., Jasse, B. and Monnerie, L.: Polymer, 22, 1616 (1981)
105. Yang, H., Hazhoannou, G. and Stein, R. S.: J. Polym. Sci., Phys. Ed., 21, 159 (1983)
106. Moskala, et al.: to appear in Polymer
107. Koningsveld, R., Kleintjens, L. A.: J. Macromol. Sci. Phys., B17, 144 (1970)
108. Koningsveld, R. and Kleintjens, L. A.: Macromolecules, 4, 637 (1971)
109. Kleintjens, L. A., Koningsveld, R. and Gordon, M.: Macromolecules, 13, 303 (1980)
110. Koningsveld, R., Kleintjens, L. A. and Schaffelers, H. M.: Pure Appl. Chem. 39, 1 (1974)
111. Koningsveld, R. and Kleintjens, L. A.: Br. Polym. J., 9, 212 (1977)
112. Koningsveld, R., Kleintjens, L. A. and Onclin, M. H.: J. Macromol. Sci. Phys. B18 (3), 363 (1980)
113. Koningsveld, R., Kleintjens, L. R.: J. Polym. Sci. Part C, 39, 281 (1972)
114. Orwoll, R. A. and Flory, P. J.: J. Am. Chem. Soc., 89, 6814 (1967)
115. van Krevelen, D. W.: "Properties of Polymers", Elsevier, 1972
116. Bondi, A.: J. Phys. Chem., 68, 411 (1964)
117. ten Brinke, G., Eshuis, A., Roerdink, E. and Challa, G.: Macromolecules, 14, 867 (1981)
118. van Aartsen, J. J.: Europ. Polym. J., 6, 919 (1970)
119. Snyder, H. L., Meakin, P. and Reich, S.: Macromolecules, 16, 757 (1983)
120. Struminskii, G. V. and Slonimskii, G. L.: Rubber Chem. Tech., 31, 250 (1958)
121. Ichihara, S., Komatsu, A. and Hata, T.: Polym. J., 2, 640 (1971)
122. Warner, M., Higgins, J. S. and Carter, A. J.: Macromolecules, 16, 1931 (1983)
123. Walsh, D. J., Rostami, S. and Singh, V. B.: Makromol. Chem. 186, 145 (1985)
124. Tompa, H.: "Polymer Solutions", Butterworths Scientific Publications, 1956
125. Baker, J. A.: J. Chem. Phys., 20, 1526 (1952)
126. Rostami, S. and Walsh, D. J.: Macromolecules, June (1985)

K. Dušek (Editor)
Received May 17, 1984

Flammability of Polymeric Materials

Roza M. Aseeva and Gennadiy E. Zaikov
Institute of Chemical Physics, USSR Academy of Sciences,
117334 Moscow, Kosyginastr. 4,
Moscow, U.S.S.R.

This is a state-of-the-art review of ignition, combustion and extinction in polymeric systems. Factors affecting the flammability of polymers are discussed. The relationship between the chemical nature of the polymer and its flammability is analyzed assuming an important role of pyrolysis and gasification reactions occurring in the complex combustion process. Thermodynamic data on the chemical reactions responsible for the combustion of high molecular weight compounds are considered. The action mechanisms of the three major flame retardants for polymer, halogen-, phosphorus- and metal-containing compounds, as well as their relative inhibiting effects are discussed.

Foreword

The burning behavior of polymeric material depends mainly on

— its chemical composition
— its mass, density, shape and specific surface area
— its interaction with other products in composites
— the type, intensity and duration of any exposure to heat and/or a source of ignition.

The burning behavior of a polymer cannot, therefore, be regarded as an intrinsic property of the material. Rather do the major variables in the case of a fire — flammability, propagation of the fire, production of heat, smoke and toxic or corrosive decomposition products — depend on all the factors mentioned above.

Nevertheless, it can be very useful to optimize the burning behavior of a polymer for particular purposes by the addition of fire-resistant materials. These possibilities are presently exploited in various ways in different technical fields, such as the building industry, transport and electric technology. They are usually dependent on the requirements of the relevant national or international safety regulations.

The authors deal mainly with the possibilities of improving the fire-resistant properties of polymers by incorporating halogen, phosphorus, and metal derivatives. They consider the effects of these additives independently of other fire risks and of applications. They, therefore, describe the effect of composition changes largely in terms of changes of flammability and combustion propagation, for well-defined samples. The evaluation is made using various techniques for the determination of the ignition temperature and the minimum oxygen concentration necessary to maintain the burning process, methods applied throughout the world although they are neglecting other features of actual risks.

The authors deal at length with the various theories for the processes, leading to ignition and burning of polymers, as well as with the conditions leading to fire extinction after removal of the ignition source.

The effort of the authors to present the state of the art from the point of view of a chemist is especially valuable because of the many literature references from socialist countries, which have been largely ignored until now in North America and Western Europe, and which are compared with the results of investigations, developments and reflections in western countries.

BASF Aktiengesellschaft, D-6700 Ludwigshafen,   Wolfram Becker
June 1984

List of Symbols

a   thermal diffusivity
c   heat capacity
D   diffusivity
De   Damköhler number

E   reaction activation energy

ΔH enthalpy

k   reaction rate constant

k_0 preexponential factor

L   thickness

LOI   limiting oxygen index (by ASTM D 2863 as volume fraction or vol %, may also be expressed in mass or molar units)

$\dot{m}$ mass rate of burning or mass loss per time unit

P   pressure

PF   parameter of flammability

$\dot{q}$ heat flux

R   universal gas constant

r   stoichiometric oxidant/fuel ratio

r'   reflection coefficient

T   temperature

V   volume

v   linear velocity

x,y,z   space coordinates

Y   concentration

α coefficient of convective heat transfer

$\bar{\alpha}$ mean coefficient of absorption (absorptivity)

δ dimension of region (of heating or reaction)

ε emissivity

λ thermal conductivity

ν reaction order

ϱ density

τ time

Subscripts

c   combustion

ck   coke

e   external

ex   extinguishant

F   fuel

f   flame

FR   flame retardant

g   gas

ign   ignition

l   losses of heat

o   initial state

ox   oxidant

p   pyrolysis

s   solid phase, surface

si   self-ignition

v   vaporization, volatilization

*   critical value,

∞ ambient

1 Introduction

Polymeric materials are today part and parcel of our everyday life. They are used in nearly every branch of industry and are responsible for rapid progress made in many of them. In view of the fact that raw materials and traditional energy resources will be soon depleted, the production of synthetic polymeric materials becomes particularly important as a way to the rational utilization of both. Today, synthetic polymeric materials are used not only as highly effective substituents for expensive and/or not readily available materials such as steel and nonferrous metals, wood and cotton, natural rubber, etc., but also as quite original materials possessing a unique set of valuable properties.

However, most polymer-based materials, especially those produced on a large commercial scale, have one important setback: they are combustible. In combustion they may release toxic and noxious gases and fumes. The problem of controlling the fire hazard of polymeric materials is a very important and urgent one. The risk and damage incurred by fire in which polymers may be involved is great, considering the modern way of construction, the erection of multi-story and high-rise buildings for various functions, the use of modern means of transportation — all involving risks of high rates of occupancy and difficulty of speedy evacuation of people from the fire zone, if a critical situation should arise.

The concept of "fire hazard" of polymer materials is a rather broad one. It includes the tendency of these materials to initiate and develop the combustion (flammability), as well as the negative effects of their exposure to high temperatures, fire or other external thermal actions.

This paper is concerned with the flammability of polymer materials. It is an attempt to review the state of the art in the field of ignition and combustion of these materials. Attention is paid to the establishment of fundamental relationships between chemical structure and flammability of polymers. The existing theories for the action mechanisms of various flame retardants are critically analyzed.

2 Ignition and Combustion of Polymeric Materials

Combustion is understood to be a fast reduction-oxidation reaction between substances, capable of spreading in space at a subsonic speed and usually accompanied by flame and light. An oxidant and reducing agent are necessary for this reaction to occur. Combustion processes are as manifold as the chemical nature of compounds having oxidizing and reducing properties.

When dealing with the combustion of standard polymeric materials it is usual to consider it as a reaction between the combustible polymeric compound and atmospheric oxygen. In this case, the polymer is in the condensed state and the oxidant is a gas.

Reactions between substances in different phase states are called heterogeneous reactions. They take place on the surface of the condensed phase. The dynamics of heterogeneous reactions depend on the rate of supply of the gaseous reactant to the surface and removal of the reaction products from the surface. Depending on whether the rate of oxygen supply is higher or lower than the rate of the reaction of oxygen with the polymer compound, the heterogeneous reaction is either activation- or diffusion-controlled.

It is well known that under normal conditions polymeric materials may be in contact with atmospheric oxygen for very long periods without being affected considerably. Only at elevated temperatures the oxidation rate becomes important. The oxidation affects the physico-chemical and physico-mechanical properties of polymers. However, at moderate temperatures, phenomena typical of a combustion process are not observed.

2.1 Ignition Phenomena

The combustion is always preceded by a situation promoting its origination. Essentially, this unsteady process called ignition arises from conditions in the system which cause self-acceleration of the oxidation. This may be caused by the progressive accumulation of heat, or by active particles initiating the chain reaction of oxidation. Heat generation in the system is the dominant factor responsible for polymer ignition. Real processes are never isolated from the environment. Therefore, heat generation depends on the heat exchange between reactants and environment.

A prerequisite of the ignition of a polymeric material is that the rate of heat accumulation due to the exothermal oxidation reaction (heat generation) is higher than the rate of heat dissipation to the environment (heat removal).

To stimulate a self-enhancing reaction between such a relatively weak oxidant as molecular oxygen and the polymer it is necessary that the system is preheated to a certain temperature. The most frequent polymer ignition stimulators are: a stream of a hot oxidizing gas (convective heat exchange), an incandescent plate or coil (conductive heat transfer), a radiating panel (radiative heat transfer), or a flame (combined heat transfer). Frank-Kamenetskiy [1] proposed a convenient diagram of the heterogeneous oxidation of a condensed substance (Fig. 1). The thermal conditions

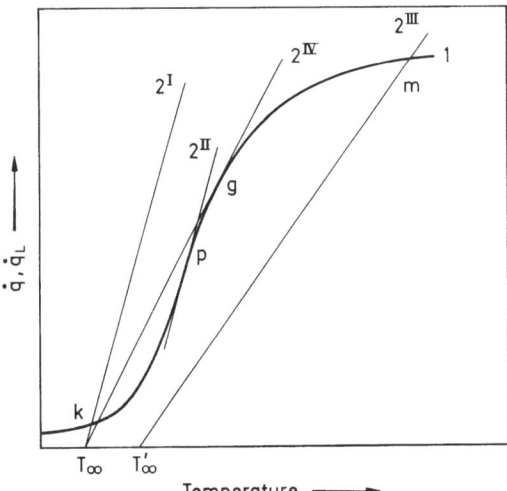

Fig. 1. Diagram of heterogeneous oxidation reaction. Heat generation (1) and heat removal (2^I–2^{IV}) curves [1]

prevailing on the surface depend on the relative positions of the heat generation and heat removal curves (Curves 1 and 2). The heat buildup rate is proportional to the oxidation rate which increases exponentially with temperature according to the Arrhenius law:

$$\dot{q} = \Delta H \, k_0 \exp{(-E/RT)} \cdot Y_{ox} \tag{2.1}$$

where $\dot{q}$ is the amount of heat produced per unit surface area per unit of time, ΔH the enthalpy of reaction, k_0 the preexponential factor, E the activation energy, R the universal gas constant, T the absolute temperature, and Y_{ox} the concentration of the oxidant.

The heat removal rate $\dot{q}_L$ is proportional to the temperature difference between the reaction zone (T) and the ambient medium (T_∞) to which the heat is transferred:

$$\dot{q}_L = \alpha(T - T_\infty) \tag{2.2}$$

α is the coefficient of heat transfer.

From Fig. 1 it is seen that the oxidation reaction can take place under various conditions. The abscissa represents the temperature of the reacting substance. The positions of the curves 2^I–2^{IV} correspond to the different conditions of heat removal from the solid surface. The points of crossing of 2^I–2^{IV} with the abscissa characterize the temperature of the environment in which the heat transfer occurs.

The relative position of the curves defines the regime of the oxidation reaction on the surface. Point k is characterized by a weak autoheating since $\dot{q}_L > \dot{q}$ with temperature increase. Here, the oxidation reaction is an activation-controlled steady-state reaction. Ignition of the system occurs at the point of contact of the two curves, i.e. at point p ($\dot{q}_L = \dot{q}$). It indicates the point where the activation-controlled reaction becomes diffusion-controlled with a weaker dependence on temperature and stronger dependence on the gas flow rate. When the heat dissipation is not sufficient, the autoheating of the system results in a steady-state diffusion-controlled combustion (point m). The reversal from the steady-state diffusion-controlled combustion to activation-controlled combustion is no longer sensitive to the ignition conditions. As the heat loss is increased and the reaction zone cools down, the reaction first smoothly changes from a state indicated by point m to a point g ($\dot{q}_L = \dot{q}$) corresponding to extinction of combustion followed by transition to the activation-controlled state.

From Fig. 1 follows the important conclusion that the conditions under which the diffusion combustion process is initiated and extinguished are not identical. The heat exchange between the system and the environment controls the ignitability and combustion stability. Hence, "flammability" is not an absolute but a relative term.

The reaction of gaseous oxygen with a polymer is an extremely complex process. Its behavior largely depends not only on the polymer nature but also on the physical characteristics of the environment and of the ignition stimulator. These factors also determine the localization of the self-enhancing exothermal reaction which controls the inflammation and combustion of the condensed phase. Such a reaction may, in fact, take place either on the polymer surface or in the gas phase adjacent or some distance away from the surface.

In an overwhelming majority of cases, heated polymers decompose into combustible gaseous products. The latter diffuse into the oxidative environment, mix and react with oxygen. Therefore, the self-enhancing exothermal reaction responsible for the heat generation and initiation of polymer combustion is in most cases a homogeneous gas phase reaction [2]. In order to stress this particular feature of the reactions between condensed systems and gaseous oxidants, the process is sometimes called quasi-heterogeneous.

However, truly heterogeneous reactions may also play an important role in preparing the materials for ignition and combustion in the gas phase [3-5].

In the case of polymers exhibiting a tendency to undergo carbonization during pyrolysis, the heterogeneous ignition and combustion may either precede the gas-phase processes, or occur simultaneously, or follow them. The smouldering of cellulosic materials accompanied by a bright glow of the surface layers is actually a heterogeneous combustion of the carbonized product from pyrolysis.

One of the authors of this review observed heterogeneous ignition and combustion in air of the highly reactive polyvinylene, a polymer with an acyclic system of conjugated $-CH=CH-$ bonds. After only minor heating, this originally black polymer becomes red and finally transforms into a light-yellow nonvolatile oxidized product.

Under conditions where it is possible to increase the rate of oxidant supply to the polymer surface and adequately remove the combustible gases and vapors near the hot surface, the heterogeneous mechanism may be responsible for the ignition of polymers which otherwise do not form a carbonized product on pyrolysis. For example, heterogeneous ignition and combustion of poly(methyl methacrylate) (PMMA) was observed in oxygen heated to 480–550 K at 50–80 atm. Under these conditions, the methyl methacrylate concentration above the polymer surface was much lower than the lower concentration limit at which its ignition may occur [6]. It has been shown experimentally [7] that at normal pressure the location of the exothermal reaction zone may vary, depending on the power of the radiative heat flux stimulating ignition of PMMA. If the heat flux power is below 60 W $\cdot$ cm^{-2}, ignition occurs far away from the polymer surface. For powers above 60 W $\cdot$ cm^{-2} the reaction zone moves closer to the surface. The ignition process becomes unsteady.

Experimentally, the moment of ignition is recorded according to the appearance of light emission or a rapid rise of the temperature in the oxidation zone. The ignition is considered established as soon as the combustion process becomes steady. Short-term glow is regarded as a flash. In the steady-state combustion the heat flux from the exothermal reaction zone to the "fresh" polymer surface makes up for all the heat necessary for the self-sustaining reaction.

All currently available theories explaining polymer ignition in terms of localization of the reaction controlling the combustion process are classified into gas-phase, heterogeneous and solid-phase approaches. The latter are left out here because ignition in the condensed phase does not require a gaseous oxidant and occurs chiefly in polymers containing oxidizing groups in the molecule (nitro groups in nitrocellulose, for example).

A theory is always expected to predict the behavior of a material (product) under the effect of an external power source and defined conditions. Of greatest practical importance for the characterization of the ignitability of polymer materials are the energy flux required for the stimulation of ignition ($\dot{q}_0$), and the ignition time, i.e. the time elapsed before ignition takes place (τ_{ign}). The minimum energy E_{ign} required for the ignition of a polymeric material may then be expressed as the product of these two parameters

$$E_{ign} = \tau_{ign} \cdot \dot{q}_0 \qquad (2.3)$$

This Equation describes the stimulation of ignition, both by a small-power source acting for a long time, and by a stronger but short heat pulse. The energy flux $\dot{q}_0$ supplied to the material surface controls the development of the heat wave. This, in turn, has an effect on τ_{ign}.

At some minimum, critical energy flux the ignition time approaches infinity (the ignition of the polymer does not occur). At any energy flux above the critical value the relationship between the ignition time and heat flux is given by [8]

$$\tau_{ign} \simeq 1/\dot{q}_0^n \tag{2.4}$$

in which the exponent n varies from 0 to 2.

If the time required for the diffusion of the reactant and for the combustion is considerably longer than the time necessary for heating the reacting material, $n \to 0$, meaning that the ignition time becomes independent on $\dot{q}_0$.

The heterogeneous and gas-phase ignition theories are based on a number of assumptions and approximations. It is difficult to take into account the entire unsteady development of the combustion process in space and time and the complex hydrodynamic and diffusion phenomena in the different phases of the system. Considerations are therefore usually restricted to a one-dimensional heat wave propagation scheme. The major objective of a theoretical analysis of the ignition of polymers is to determine the temperature and concentration profiles in the system at different periods, as a function of the initial physico-chemical parameters of the polymer material and environment, as well as of the conditions and mechanisms of heating. If the ignition criterion has been selected properly, the ignition time τ_{ign} and ignition energy E_{ign} may be calculated.

2.2 Ignition Criteria

The choice of the ignition criterion is one of the most important and difficult problems in any theoretical treatment of the ignition of condensed systems. Transition to steady combustion is an asymptotical process and the moment of ignition is a mathematically indeterminate value.

To simplify the analysis, the entire period up to the establishment of a steady combustion process is divided into several steps: heating of the material, decomposition and gasification of the polymer, mixing of the volatile products with oxygen, and oxidation of the fuel. The ignition includes the period from the moment of the attack of the ignition stimulator to the development of a fast chemical reaction. Under the real conditions of ignition of polymers, by means of a hot oxidative gas, a flame, or a radiating panel, most of this period is required for heating the material. In this case, the methods of calculating ignition characteristics involve the solution of the heat conductivity equation for a chemically inert body

$$\frac{\partial T}{\partial \tau} = a \frac{\partial^2 T}{\partial x^2} \tag{2.5}$$

where a is the thermal diffusivity.

Provided that the initial and boundary conditions have been properly selected, Eq. (2.5) describes the effect of an ignition stimulant and of the heat generation from a chemical reaction. The maximum temperature of the inert body ($\hat{T}_m$), corresponding to the beginning of the autoheating of the system as a result of the ignition reaction, is defined as the ignition temperature:

$$T_{ign} \equiv \hat{T}_m \qquad\qquad (2.6)$$

In order to determine the moment of ignition, an ignition criterion must be formulated on the basis of the particular behavior of the process under given conditions.

Another approach involves the analysis of equations corresponding to the discrete steps of combustion. The ignition parameters are calculated from the coupling condition of the steps by computerized numerical, or approximate analytical methods. In Ref. [9] where the heterogeneous ignition of a condensed system was analyzed, a sharp transition from activation to diffusion control was assumed as ignition criterion:

$$\left.\frac{\partial^2 T}{\partial \tau^2}\right|_{\tau = \tau_{ign}} = 0$$

Considering the oxidant supply to the surface, it has been shown that there is an oxidant gas concentration limit below which no heterogeneous ignition can take place.

On the basis of theoretical treatments of the experimental relationships $\tau_{ign} = f(T_{ign})$ obtained for variable ambient temperatures and pressures and different initial sample temperatures, the kinetic parameters of the heterogeneous reaction between PMMA and oxygen were determined: $E = 50$ kJ/mol, $\Delta H k_0 = 8.4 \times 10^{10}$ J cm^{-2}s^{-1}. T_{ign} was assumed to be equal to the temperature at the interface [6]. Treatment of the heterogeneous reaction as an activation-controlled process allows to disregard the diffusion factors and to apply the classical condensed system ignition theory in order to find the features specific to heterogeneous ignition [10-12].

In the gas-phase ignition theories different ignition criteria were proposed. All of them more or less reflect that the principal condition $\dot{q} \geq \dot{q}_L$ must be met. The rise in temperature to a certain critical ignition temperature, $T_{ign} = const$, has been one of the most frequently used criteria. In theory, T_{ign} is identified either by the critical surface temperature T_s^* at which active vaporization or gasification of the polymer begins [13, 14], or by the maximum gas phase temperature close to the adiabatic flame temperature [15]. In practice, ignition and self-ignition temperatures are often used as polymer ignition characteristics. However, it has been stressed many times that these temperatures cannot be considered as physico-chemical material constants. Only if care is taken to keep the conditions, under which these temperatures are measured, identical or on a standard basis [16,17], they can be useful parameters in comparative polymer ignition estimates. Kashiwagi [18] who studied the effect of heat flux (radiative) on ignition surface temperature has shown that whether the criterion $T_s^* = const$ is valid, or not, depends on the nature and decomposition mechanism of the polymer material. For PMMA, for example, this condition is satisfied within a broad range of $\dot{q}_0$. According to Librovich [13], the gas-phase ignition of a solid fuel is possible if the following two criteria are met: 1) the surface temperature must be equal to the gasification temperature, and 2) the temperature gradient near the surface must be below a certain value depending on the rate of reaction in the flame and the rate of oxidant supply to the fuel surface. Deverall and Lai [19] characterized the moment of ignition of cellulose materials by conditions necessary for the attainment of a critical gas liberation rate and a critical temperature gradient near the surface on the side of its contact with the gas phase. The first condition is actually the criterion of the minimum pyrolysis product mass flow to the gas phase [20]. According to Ref. [20] this criterion is closely related to the lower concentration limit of ignition in the boundary layer of combustible gases mixed with air. In Ref. [15], a numerical analysis is given of the gas-phase ignition of polymer materials under the effect of radiative heat flux. Six criteria are considered in that work:

1) The total rate of the gas-phase reaction must be equal to or greater than a certain constant c_1

$$\int_0^\infty (\text{Reaction Rate}) \, dy \geq c_1$$

2) The maximum local rate of gas-phase reaction must be equal to or greater than a certain constant c_2

$$(\text{Reaction Rate})_{max} \geq c_2$$

3) The maximum local gas phase temperature must be equal to or greater than a certain constant c_3;

$$(T_g)_{max} \geq c_3$$

4) The total rate of heat generation due to the gas-phase exothermal reaction is equal to or greater than the rate of radiative power absorption by the solid phase:

$$\Delta H_c \int_0^\infty (\text{Reaction Rate}) \, dy \geq \dot{q}_0(1 - r')$$

(r': reflection coefficient)

5) The acceleration of the total rate of gas-phase reaction is equal to or greater than a constant c_4:

$$\partial[(\text{Reaction Rate})dy]/\partial\tau \geq c_4$$

6) The temperature gradient near the surface on the gas-phase side becomes equal to zero or positive:

$$(\partial\theta/\partial\eta)_\eta \geq 0$$

where θ and η are nondimensional terms of the temperature ($\theta = T/T_0$) and space coordinate ($\eta = x/x_0$).

Using given numerical values of the reaction activation energies in the condensed and gas phases, the heat flux to the surface, and other physico-chemical characteristics of a polymer system, Kashiwagi [15] calculated τ_{ign} as a function of the initial oxygen concentration. A system of complex differential equations describing the ignition process is used for this calculation.

Figure 2 shows the effect of selected ignition criteria on $\tau_{ign} = f(Y_{ox})$. Although the overall behavior is the same in all cases, the figure reveals that the quantitative result, namely the ambient oxygen concentration limit, is sensitive to the criterion selection. The author discovered the existence of an upper and a lower ignition limit related to the fuel pyrolysis activation energy for a constant gas-phase reaction energy, and vice versa. These ignition limits are due to the fact that in order for a gas-phase ignition to occur the following conditions must be satisfied:
1) sufficient amount of fuel and oxygen present in the gas phase,
2) temperature high enough to sustain the self-accelerating gas phase reaction.

It is quite clear that these limits become broader as the oxygen concentration and the incident heat flux increase. For the selected system parameter values the exponent of $\dot{q}_0$ in Eq. (2.4), for $\dot{q}_0$ up to 200 W cm^{-2}, was found to be 1.9, which is close to n = 2 predicted by theory [8] for small $\dot{q}_0$.

Williams et al. [21-24] developed an asymptotic method of analysis of condensed system ignition. Using this approach they solved the energy and mass conservation equations describing the time-dependent behavior of the solid and gas phases at the different stages of the ignition process. They spent considerable effort on studying the solid fuel gasification under the influence of a constant heat flux and radiation.

The gasification process may take place either in the surface layer or in the bulk of the solid fuel. For a theoretical treatment it is subdivided into three steps: 1) inert heating of the substance, 2) transition stage during which the temperature rises rapidly, and 3) transport-controlled stage during which the temperature remains virtually constant. The latter stage may become steady for a long period. Depending

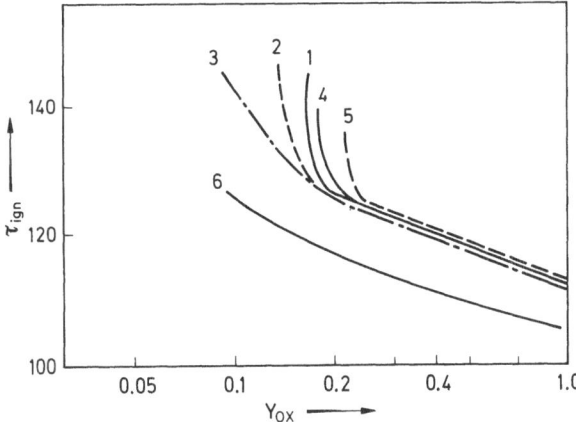

Fig. 2. Calculated ignition time (in dimensionless units) as a function of atmospheric oxygen concentration for six different ignition criteria (see text); 1) $c_1 = 2 \times 10^{-5}$ g $\cdot$ cm^{-2}s^{-1}; 2) $c_2 = 10^{-5}$ g $\cdot$ cm^{-2}s^{-1}; 3) $c_3 = 2.5$; 4) $\dot{q}_0 = 8.36$ W $\cdot$ cm^{-2}; 5) $c_4 = 10^{-3}$ g $\cdot$ cm^{-2}s^{-1}; 6) $(\partial\theta/\partial\eta)_n = 0$
Conditions: dimensionless activation energies for pyrolysis $\Theta_p = 66.7$; for gas-phase reaction $\Theta_g = 30$
(After Ref. [15] with permission)

on the relative rates of the endothermal gasification and gas-phase ignition reactions, ignition occurs either during the transition or the transport-controlled stage of the gasification process. If it occurs at the latter stage, the ignition time is independent of the absorptivity $\bar{\alpha}$ of the material. When the gasification becomes steady it does no longer affect neither the surface temperature nor the gasification rate. When ignition takes place during the transition stage, τ_{ign} depends on $\bar{\alpha}$. As the absorption coefficient decreases, τ_{ign} increases. The agreement of the approximate asymptotic method with Kashiwagi's numerical solution for τ_{ign} is better for high heat flux than for small $\dot{q}_0$.

Concerning the activation energy limits of the gas-phase ignition, there is a considerable uncertainty associated with the choice of the ignition criterion. In theory, the chemical processes accompanying ignition are described only by the effective rate constants of the pyrolysis and gas-phase oxidation reactions. It is apparent that, by using an appropriate combination of these parameters, it is possible to converge the ignition limits and thereby reduce the flammability of polymeric materials. Another way of controlling the ignition limits of polymers is by suitable selection of the gas-phase diluent. Inert gases which have higher thermal diffusivity and heat capacity increase the ignition time, limiting the oxygen concentration at which ignition is still possible.

To date, there have been quite few experimental studies on polymer ignition that might be used for verifying theoretical predictions. Most endeavors have been directed towards the determination of the self-ignition (T_{si}) and ignition (T_{ign}) temperatures. T_{si} is the lowest gaseous oxidant temperature (e.g. air temperature) at which spontaneous flame (or surface) combustion may still take place. T_{ign} is the flow temperature at which the same combustion phenomena are observed, but with the help of an external ignition energy source (spark, flame) brought close to the specimen surface.

The temperatures of polymer self-ignition and ignition in air are reported by Hilado [25]. In Ref. [26], the ignition time of polymers in contact with hot air is related to the air temperature. The $\tau_{ign} = f(T_\infty)$ curves have allowed to calculate the apparent self-ignition activation energies E_{si}. The rather low values of E_{si} (33 to 42 kJ/mol) indicate a strong influence of the diffusion factors. As established by Korshak et al. [27], heat resistant polymers have slightly higher values of E_{si}. It is interesting to note that the introduction of bromine into the poly(phenylene quinoxaline) molecule increases E_{si} from 56 to 77 kJ/mol, independent of the bromine content. An increase of the oxygen concentration in the oxidizing medium reduces both the self-ignition temperature and self-ignition time of polymers. In Ref. [28] cross-linked polymers obtained by polymerization of dimethacrylates of different glycols were studied. It was found that, in pure oxygen at 1 atm and at a gas flow temperature above 430 °C, the self-ignition times were much shorter. The E_{si} values increase from 33–42 kJ/mol at low temperatures to about 90 kJ/mol for temperatures above 430 °C. This fact is related to a change of the self-ignition mechanism. At low self-ignition temperatures, the role of the chain oxidation reactions which proceed practically without any heat release, is more important. These are the so-called isothermal or cold-flame reactions. It has been noted [29] that certain polymers go through all the stages of reaction with oxygen, from the slow oxidation reaction to the single-stage high-temperature self-ignition. For polyolefins, for example, the pressure and temperature limits have been determined, within which slow oxidation, cold-flame reaction, two-stage and, finally, single-stage high-temperature self-ignition can take place (Fig. 3). Secondary "multi-

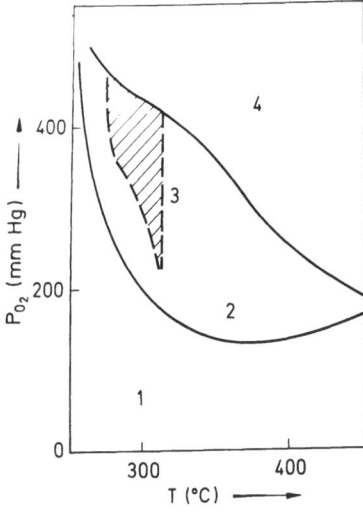

Fig. 3. Self-ignition of polypropylene (PP) in oxygen; 1) slow oxidation, 2) cold flame, 3) "multiplet" secondary cold flames, 4) two-stage self-ignition. The single-stage self-ignition is not shown (After Ref. [29], with permission)

plet cold flames" arise as a result of the ignition of the products of heterogeneous polymer oxidation reaction, and of periodic variations of the temperature gradient in the surface layer [29,30]. Polystyrene and poly(vinyl chloride) were not observed to take part in cold flame reactions [31]. According to the investigators, cold flame reactions are specific of those polymers which decompose into products able to support a cold flame.

Like gas systems, polymers were also found to have concentration limits of ignition and self-ignition. The polymer "concentration" was represented as the ratio of the mass of a polymer sample charged into the reaction vessel to the sum of masses of polymer and oxygen in the closed system [29]. The analysis of the chemical composition and concentration of gases liberated until ignition takes place is of special interest for the understanding of the polymer ignition mechanism, and of the effect of various additives on the initial and later stages of combustion. Martin [2] was the first to relate the polymer heating prehistory (decomposition, concentration and composition of combustible gases) with polymer ignition characteristics. He noted that the liberation rate of the decomposition products of cellulose materials, at moderate heat fluxes to the surface, is directly proportional to the heat flux intensity $\dot{q}_0$, whereas the integral quantity of the volatile components is inversely proportional to $\dot{q}_0$. In the $\dot{q}_0$ range from 20 to 90 W cm^{-2}, $\tau_{ign} \sim 1/\dot{q}_0^2$. As $\dot{q}_0$ increases the coke residue yield decreases and the amount of resinous liquid products increases.

Heterogeneous ignition usually occurs under conditions of slow heating of a cellulose sample at lower temperatures than the gas-phase ignition does. Martin conjectured that many light products ignitable in the gas phase are formed via secondary reactions in the carbonized layer. As they filter through the carbonized layer the heavier products undergo additional cracking.

The works of French scientists [32–34] demonstrate considerable potential in the study of the chemical aspects of ignition. It has been shown in Ref. [34] that the addition of ferrocene to poly(vinyl chloride) strongly affects the self-ignition behavior of the polymer. Various glow phenomena are then observed which usually do not occur in the pure polymer. Weak glow changes to normal self-ignition (Fig. 4) under specific conditions (temperature, oxygen pressure). Addition of ferrocene affects the product composition, decreases the content of aromatic compounds that are conventionally regarded as soot precursors (a state-of-the-art reviews is given in Ref. [35]). It has been noted that other additives (e.g. V_2O_5) cause the PVC ignition limits in oxygen to become narrower.

At higher temperature and normal or elevated pressure, polymers usually ignite via a single-stage process. The so-called pilot ignition of polymers under the effect of a benzene flame and a high-temperature tungsten lamp was studied in Ref. [36]. The ignition energy sources used in this work allowed the heat flux to be varied from 2.5 to 14.5 W cm^{-2}. It has been found that the ignition time depends on the nature of the ignition stimulant, the heat flux intensity and the optical and thermophysical properties of the material. The ignition time satisfies the Equation:

$$\tau_{ign} = \frac{160(\lambda \varrho c)^{0.75} (T_{ign} - T_\infty)}{(\bar{\alpha} \dot{q}_0)^2}, \tag{2.7}$$

where $\lambda, \varrho, c, \bar{\alpha}$ are the thermal conductivity, density, heat capacity and mean coefficient of absorption of the polymer, respectively.

According to an approximate Equation given in Ref. [8]

$$\tau_{ign} = \tau_{CD} + \frac{1}{\bar{\alpha}} \left[\frac{\varrho c (T_{ign} - T_\infty)}{\dot{q}_0 (1 - r')} \right] + \frac{\pi}{4} a \left[\frac{\varrho c (T_{ign} - T_\infty)}{\dot{q}_0 (1 - r')} \right]^2 \tag{2.8}$$

where a is the thermal diffusivity, r' the reflectivity of the surface of the solid sample, τ_{CD} the time necessary for reactant diffusion and gas-phase reaction.

The time τ_{CD} may control τ_{ign} in the case of ignition by a very large radiative heat flux or a shock wave. For small heat fluxes or convective heat transfer the last term of Eq. (2.8) is essential.

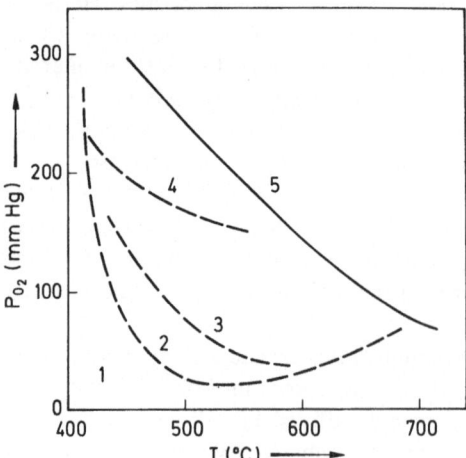

Fig. 4. Self-ignition in oxygen of poly-(vinyl chloride) containing 1.5% of ferrocene. Regions: 1) slow oxidation, 2, 3) weak glow, 4) secondary flames, 5) single-stage ignition
(After Ref. [34], with permission)

Equation (2.7) derived by Hallman et al. [36] is, in fact, a certain superposition of the second and third terms of Eq. (2.8) but involves also the relationship (see Eq. 2.4):

$$\tau_{ign} \sim \frac{1}{\dot{q}_0^2} .$$

By laser techniques it has become possible to vary the radiative heat flux intensity $\dot{q}_0$ in very broad ranges; Ohlemiller and Summerfield [37] used a monochromatic CO_2 laser (10.6 μk) for investigating the flammability of polystyrene and an epoxy polymer. They varied $\dot{q}_0$, oxygen concentration, ambient pressure and polymer absorptivity (by adding different amounts of soot). For heat flux $\dot{q}_0 = 384.5$ W cm^{-2}, τ_{ign} decreases with increasing oxygen concentration in the system, but remains practically constant at higher concentrations. This means that τ_{CD} in Eq. (2.8) becomes small compared to the other terms and can be ignored in polymer tests performed in pure oxygen. The authors observed a certain divergence and retardation of τ_{ign} in samples where soot had been added, compared with Eq. (2.8), taking into account the second and third terms.

Niioka and Williams [38] used the experimental results in Ref. [37] for assessing the principal tenets of the gas-phase ignition theory, taking into account the effective reaction rates in the condensed and gas phases. For polystyrene the effective activation energy of surface gasification, as well as the preexponential factor, were determined in an independent way (136 kJ/mol and 1.4×10^6 cm/s, respectively). The observed relationship between τ_{ign} in pure oxygen and ambient pressure is possible only if the ratio of the activation energies of the gas-phase and condensed phase reactions $E''/E' = 2$. From the effect of the oxygen partial pressure $\dot{P}_0$, at a fixed ambient pressure, on the ignition time it follows that the preexponential factor in the Arrhenius equation for the gas-phase reaction is inversely proportional to $P_0^{1/2}$. If we also take into account the polymer surface reflectivity in samples containing different amounts of soot, the experiment agrees with the theory (Fig. 5).

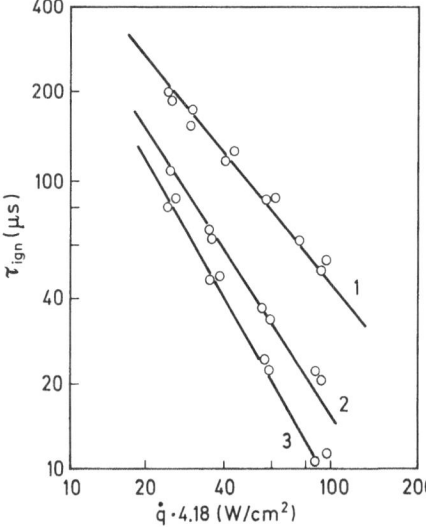

Fig. 5. Relationship between the ignition time of polystyrene in oxygen and radiative heat flux; absorptivity varied by the addition of soot: 1) 0% soot, 2) 4% soot, 3) 20% soot. Circles: experiment; lines: calculated (After Ref. [38], with permission)

Thus, the theory opens up new possiblilities for determining more details about polymer flammability by using data obtained from other independent experiments.

An analysis of the temperature and concentration profiles in the gas phase during ignition of PMMA leads to the conclusion that the process is localized in the region where the temperature necessary for the spontaneous ignition of the combustible gases and vapors is attained and the concentration of the latter is within the ignition range [7].

Much information about polymer ignition may be drawn from literature devoted to ignitability of gas mixtures and from the data on polymer pyrolysis products at temperatures developed on the surface during their combustion.

Using the Le Chatelier principle, one may calculate the lower (L) and upper (U) concentration limits of the ignition of polymer decomposition products in any oxidative medium [25, 39]:

$$L = \frac{100}{\sum\limits_{i=1}^{n} \dfrac{Y_i}{L_i}} ; \qquad U = \frac{100}{\sum\limits_{i=1}^{n} \dfrac{Y_i}{U_i}} ; \qquad \sum\limits_{i=1}^{n} Y_i = 100 \qquad (2.9)$$

where Y_i is the concentration in %, L_i and U_i are the lower and upper concentration limits of ignition, of the i-th component.

In practice, polymer ignitability is sometimes determined in terms of the limiting oxygen index (LOI). This index gives the minimum oxygen content in a O_2/N_2 mixture that maintains the flame. Deficiency of oxygen in a gas mixture (or dilution with an inert gas) results in a narrowing of the concentration limits of ignition of the combustible mixture. This is shown in Fig. 6 for pentane (pre-mixed gases). When dilution reaches an ultimate level, both concentration limits converge into a single extreme point of ignition or, as it is often called, a phlegmatization point [40] (point NP in Fig. 6). Under these conditions, the fuel/oxygen ratio in the system approaches the stoichiometry required for complete combustion of the fuel, and the flame temperature rises to a maximum. The nature of the inert diluent gas affects the concentration limits of ignition of combustible gases.

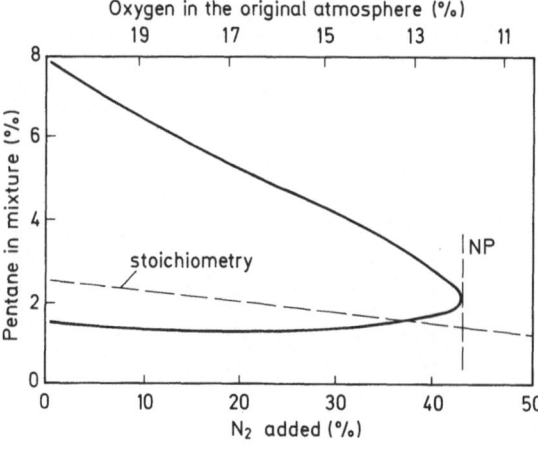

Fig. 6. Hydrocarbon ignition limits in an $O_2 + N_2$ mixture at 1 atm (After Ref. [40])

One fundamental difference between diffusion and pre-mixed gases resides in the fact that the LOI values in the first case generally correspond to a somewhat higher oxygen content (lower dilution with inert gas) compared with that at point NP for pre mixed gases.

Maček [41] believes that near the extreme point in the ignition region the chemical reaction responsible for combustion changes its direction and stoichiometry. As a consequence, the diffusion flame temperature at the extinction limit is higher than the flame temperature in the case of pre-mixed gases of the some composition, and both are smaller than the temperature corresponding to the complete combustion of the fuel.

Morimoto et al. [42] made an attempt to estimate polymer ignitability by using a limiting oxygen index of ignition (ILOI). The ILOI is the minimum oxygen content in the hot oxidizing gas ($O_2 + N_2$) flow at which spontaneous ignition of a sample is possible at a fixed ambient temperature. For many polymers a correlation between ILOI and LOI was observed (Fig. 7). It is interesting that polyoxymethylene (POM) which has the lowest LOI, ignites at temperatures above 550 °C in a practically inert atmosphere. However, a steady combustion is possible only in an oxidative atmosphere. This example is just another illustration of the discrepancy between ignition and extinction conditions of the diffusion combustion of polymers.

Chemical modification of polymers by the addition of various flame retardants may have different effects on the ignitability of polymeric materials. Information about such effects helps to explore certain features of the activity mechanism of flame retardants [34,42,43].

Thus, ignition should be regarded as the crucial stage in the initiation of the combustion of polymeric materials. The ignition stage in general includes the substages of heating, pyrolysis and gasification of polymer, depending on the heating conditions. At relatively low heat fluxes the heating controls the observed ignition times. If conditions are favorable, the ignition develops into a steady combustion process. A special feature of the latter is the movement (propagation) of the chemical reaction

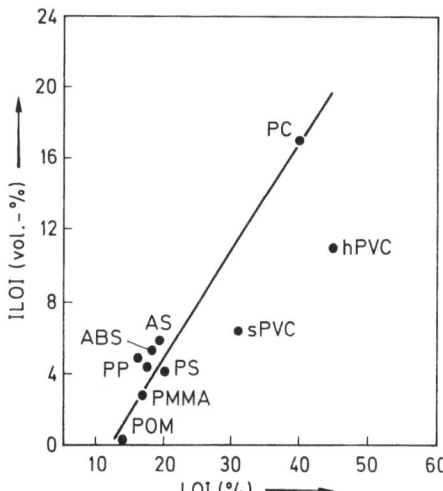

Fig. 7. Relationship between ILOI ($T_\infty = 600$ °C) and LOI of polymers (After Ref. [42], with permission)

front relative to the burning material. When the process becomes steady, an external heat flux source is no longer required. Heat, and perhaps also active particles, are transported from the reaction zone to the nearest layer of the material, causing its pyrolysis and inflammation. The type of the feedback communication between the combustion zone and a "fresh" material layer determines the combustion propagation mechanism.

2.3 Combustion Propagation

All theoretical works undertaken so far have actually been concerned with the determination of the feedback communication behavior as a function of the nature, thermophysical and physical characteristics of the polymer material, and environment (pressure, temperature, oxygen concentration, sample orientation, shape and size, etc.). Most attention in these works focused on the mechanism of heat transfer from the combustion zone to a "fresh" layer of polymer. Speed is a quantitative measure of combustion propagation in space. Accordingly, these studies are concerned with the functional relationship between the combustion rate and the above-mentioned factors.

A combustion process most frequently occurring in real fires is flame spreading along the polymer material surface. In this case, the flame front propagation along the surface involves the polymer burning in the vertical direction under the flame. In the combustion of spherical polymeric samples [44] or in the combustion of polymers in an opposed flow diffusion flame (OFDF) apparatus [45], and also in candle-like combustion, the combustion waves propagate normally to the burning surface. The polymer combustion rate may be expressed in linear or in mass units.

Mathematically, the combustion process has been modelled for the most general three-dimensional case. It is described by a sum of differential equations accounting for the heat and mass transfer in the reacting system under the assumption of energy and mass conservation laws [1]. At present, it is impossible to obtain an analytical solution for the three-dimensional form. Therefore, all the available condensed system combustion theories are based on simplified models with one-dimensional or, at best, two-dimensional heat and mass transfer schemes. In these models, the kinetics of the chemical processes taking place in the phases or at the interface is described by an Arrhenius equation (exponential relationship between the reaction rate constant and temperature), and a corresponding reaction order with respect to reactant concentrations.

The linear velocity of polymer combustion propagation (v) may be represented in terms of the ignition temperature concept (T_{ign}) by a simple fundamental formula [46]:

$$\varrho v \, \Delta H = \dot{q}_s \tag{2.10}$$

where ΔH is the difference of fuel enthalpies at the ignition and the initial temperature: $\Delta H = C_p(T_{ign} - T_0)$; $\dot{q}_s$ is the heat flux transferred from the reaction zone to a "fresh" surface with initial temperature T_0.

Eq. (2.10) indicates that one must know $\dot{q}_s$. Since the intensity of $\dot{q}_s$ is determined by the heat transfer mechanism, it is extremely important to know what particular heat transfer mechanism makes the greatest contribution to $\dot{q}_s$. Such mechanism is usually referred to as the leading combustion propagation mechanism. Theoretically, it is estimated by calculating $\dot{q}_s$ or v individually for different heat transfer mechanisms and then combining them. It is for this reason that the theoretical analysis of v is not as simple as in Eq. (2.10). It is rather difficult to measure a contribution to $\dot{q}_s$ experimentally. Therefore, the leading mechanism of the combustion wave propagation is identified in an indirect manner by measuring v as a function of pressure, temperature, oxygen concentration, thermochemical and thermophysical properties of the polymeric material, geometry, orientation, and other factors.

2.3.1 Flame Spread

The theories of flame spread over a polymeric material surface are reviewed in Refs. [46-48]. The theories may be divided into two groups:

1. those ignoring chemical kinetics in combustion and
2. those taking into account chemical kinetics.

The theory advanced by De Ris [49] belongs to the first group. In his model of flame spread along a horizontal surface it is assumed that the diffusion flame contacts the surface at the point where the polymer vaporization (gasification) begins. Reactant diffusion to the narrow zone of chemical reaction controls the heat generation process. If heat is transferred from the laminar diffusion flame to the surface by conduction, then the flame spread rate follows the Equations:

a) for thermally thin materials

$$v \simeq \sqrt{2}\,\lambda(T_f - T_v)/\varrho_s c_s L(T_v - T_0) \tag{2.11}$$

b) for thermally thick materials

$$v = v_a \varrho c \lambda (T_f - T_v)^2/\varrho_s c_s \lambda_s (T_v - T_0)^2 \tag{2.12}$$

Here, L is the material thickness, T_f is the flame temperature, v_a is the velocity of the incoming oxidant flow; if there is no external supply, v_a is controlled by natural convection. A solid fuel which is heated throughout its depth prior to flame arrival is said to be thermally thin. If the thickness (L) of the material is more than that of the layer heated prior to flame arrival the material is said to be thermally thick.

From Eqs. (2.11) and (2.12) it follows that the flame spread rate in the first case is inversely proportional to the material thickness and is independent of the incoming oxidant flow velocity. In the second case it is proportional to v_a and does not depend on the material thickness. The functional relationship between v and the initial material temperature T_0 is also different. Preheating of the material reduces $\Delta T = T_v - T_0$, thereby promoting flame spread.

This model assumes the critical ignition temperature to be equal to T_v, temperature at which the fuel is gasified in combustion. The temperature T_f is assumed to be equal to that of the adiabatic flame with a stoichiometric fuel/oxidant ratio. De Ris noted important features of flame spread along a material surface, but the fact that T_v must be determined experimentally reduces the potential of the method for predicting the flame spread rates.

According to Lastrina et al. [50], the assumption of the limiting influence of diffusion on the flame spread rate is valid only at locations far away from the flame front. In the flame front, the ratio between the mass transfer time and chemical reaction time is small. The gas-phase reaction kinetics must be taken into account. The authors analyzed a model of the flame spread over the horizontal and vertical polymer surface (downward propagation) with regard to reaction parameters in the gas and solid phases and the dimension of the ignition region, δ. They derived the equations:

$$v = \frac{\lambda \Delta H_c Y_{ox} F(P, Y_{ox})}{\varrho_s c_s c L(T_v - T_o)} \tag{2.13}$$

and

$$v = \frac{\{\lambda \Delta H_c Y_{ox} F(P, Y_{ox})\}^2}{\varrho_s \lambda_s c_s c^2 \delta (T_v - T_o)^2} \tag{2.14}$$

for thermally thin and thick materials, respectively, which are similar to Eqs. (2.11) and (2.12) concerning the functional relationships between the flame spread rate and material thickness and temperature difference $(T_v - T_o)$. It has been established that pressure and oxygen concentration influence the flame spread rate. However, the authors' attempts to obtain an analytic solution for the rate v failed, although they tried to evaluate the integral function $F(P, Y_{ox})$ from experimental data.

The effect of ambient pressure and oxygen concentration is given by:

$$v = (PY_{ox}^n)^m \tag{2.15}$$

where the exponents m and n depend on the nature of the polymer. The effects of P and Y_{ox} on the flame spread rate have been investigated for many polymers. It has been shown in particular [51] that the flame spread rate over the horizontal surface of thermally thick samples of certain polymers under oxidative environment pressure $P = 1$ bar and $Y_{ox} > 35\%$ is

$$v = KY_{ox}^n, \qquad cm\ s^{-1} \tag{2.16}$$

The constants K and n depend on the thermophysical and thermochemical properties of the polymers: $K = 0.14 + 7.5 \times 10^{-8} \cdot \Delta H_c/a$; $n = 1.35 + 0.21 \times 10^{-4} \Delta H_c$. Here, ΔH_c and a are the heat of combustion (kJ/kg) and the thermal diffusivity $(cm^2 s^{-1})$. It has been found that Eq. (2.16) with the above values of K and n is also valid for composite materials containing fillers.

In Refs. [52, 53], an analytical solution of the two-dimensional problem of flame spread over a horizontal surface is described. Formulas for the determination of the location of the chemical reaction, fuel flow to the latter, and temperature in the reaction zone have been derived. From these formulas it follows that the distance between the material surface and the chemical reaction zone is greater farther away from the flame front. Also the influx of fuel and oxidant to the reaction zone is smaller and the temperature is higher, and attains in the limit a level corresponding to adiabatic combustion. Thus, the temperature in the flame front may be substantially below the adiabatic temperature of combustion because of heat transfer to the combustible material. A criterion determining the effect of the gas-phase reaction rate on the combustion wave structure and the flame spread limits has been found. When the gas-phase reaction rate is high, the diffusion flame touches the surface, i.e. it exists in the entire region above the material surface. In this case De Ris's theory which takes into account only the heat and mass transfer is valid. If, on the other hand, the chemical reaction rate is small, the reaction begins some distance away from the combustible material surface, that is, there is a flame edge. To determine the flame spread rate it is then necessary to take into account the rate of the gas-phase reaction. The edge position affects the flame spread rate. In this model the heat is transferred from the flame to the polymer surface by conduction and convection across the gas phase and by conduction through the polymeric material.

For a thermally thin sample, the flame spread rate on the surface with an additional external heat flux $\dot{q}_e$ obeys the Equation:

$$V = \frac{\pi \Delta H_c \omega_{max} \delta_{rz}^2}{2 \varrho c \delta_T (T_f - \Delta H_0/c)} \cdot \left[\frac{(1 + \Omega) \cos \alpha - V_a}{1 + \theta \delta_s \cos \alpha / \delta_T} \right] \tag{2.17}$$

where

$$\Omega = \frac{4 \dot{q}_e}{\pi \Delta H_c \omega_{max} \delta_{rz}}; \qquad \theta = \frac{\varrho_s c_s (T_v - T_0)}{\varrho c (T_f - \Delta H_0/c)}$$

ΔH_c is the combustion enthalpy, ω_{max} is the maximum gas-phase reaction rate, δ_T, δ_s and δ_{rz} are, respectively, the thickness of the hot layer, heated polymer layer and half-thickness of the reaction zone in the flame; α is the angle of inclination of the reaction zone to the polymer surface; $\Delta H_0 = cT_0 + Y_F \Delta H_v$ is the heat necessary for polymer gasification and heating of combustible vapors.

Since the reaction rate is relatively low in the flame edge when there is no external heat source ($\Omega = 0$), the flame stability may be essentially affected by even small fluxes due to natural convection.

The criterion of flame spread over a polymeric material surface may be expressed as follows:

$$\frac{v^2 c \varrho^{v-2}}{\lambda k_0} \leqq F(Y_{ox}, Y_F, T_\infty, T_v) \exp(-E/RT_{xe}) \tag{2.18}$$

where $T_{xe} = (T_f - T_v)/\sqrt{2} + T_v$ is the flame edge temperature, v the overall reaction order in the flame, k_0 and E are the preexponential factor and activation energy of the gas-phase reaction, respectively. The functions of the reactant concentration and burning surface temperature are complex [53]. It is clear, nevertheless, that if Eq. (2.18) is not obeyed interruption of combustion must result.

An analysis has allowed the authors of Refs. [52, 53] to find out a number of interesting features of flame spread over a material surface and investigate the combustion flame structure in the gas and condensed phases. However, the equations obtained cöntain too many unknown parameters which may be determined only by way of complicated experiments. For this reason the theory in its present form can be used only for rough estimates (e.g. for flame spread limits).

Only a few experimental works are available devoted to the investigation of the temperature field in the flame front, its variation with atmospheric oxygen concencentration and the incoming oxidant flow rate [54–56]. Lalayan [56] showed that the maximum temperature of a laminar diffusion flame propagating along a horizontal surface of a thermally thick PMMA plate is indeed achieved far from the flame edge. As the oxygen concentration (in a $O_2 + N_2$ mixture) changes from 25 to 40%, the flame temperature increases from 1720 to 2000 K (the adiabatic flame temperature varying from 2230 to 2600 K under these conditions). At the same time the temperature of the flame edge, which determines the heat flux to the polymer surface, is several hundred degrees lower than the maximum one. In the flame front there is an extensive zone of practically constant temperature (Fig. 8). The author proposed that this situation could be due to the expansion of the incoming oxidant flow into the high-temperature region on account of density reduction and gas composition variations. However, the kinetic behavior which is different from diffusion control in the flame edge appears to be a more probable reason. The lower temperature and higher convective fluxes in this region result in a reduction of the ratio of characteristic reactant residence time in the zone (or reactant diffusion time), τ_r, to the overall chemical reaction time, τ_{ch}. This ratio, a nondimensional parameter, was introduced into the combustion theory by Damköhler (Damköhler number[1]: De $= \tau_r/\tau_{ch}$). The reaction zone becomes extended in the flame edge. It is interesting to note that an increase of the oxygen concentration in the oxidant flow and of the incoming flow velocity tends to dislocate the reaction zone in the flame closer to the polymer surface. This causes an increase of the heat flux to the polymer surface and of the flame spread rate. As

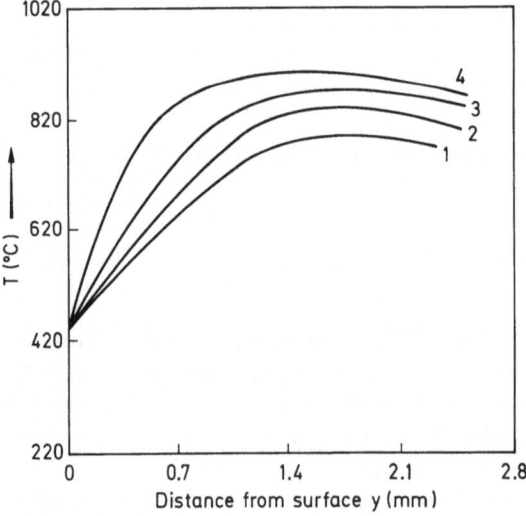

Fig. 8. Effect of the oxygen concentration on the temperature profile in the flame front spreading along a horizontal surface of a poly(methyl methacrylate) plate; 1) 25 vol%, 2) 30 vol%, 3) 35 vol%, 4) 40 vol% O_2.
(After Ref. [56])

---

[1] Damköhler, G. Z. Electrochem. *42*, 846 (1936)

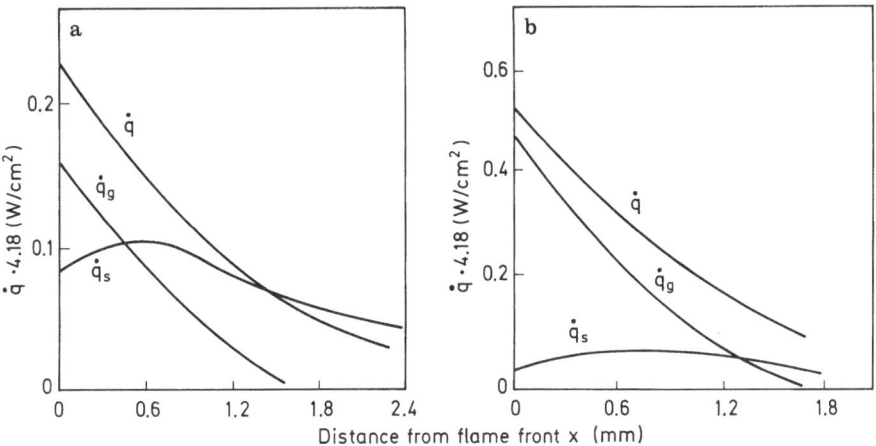

Fig. 9a and b. Effect of oxygen concentration on the total heat flux ($\dot{q}$) and individual contributions to the heat transfer from the flame to the poly(methyl methacrylate) surface through the gas ($\dot{q}_g$) and condensed ($\dot{q}_s$) phase; **a)** 25 vol %, **b)** 40 vol % O_2 (After Ref. [56])

the oxygen concentration varies in the environment, the nature of heat transfer changes too. At $Y_{ox} = 40\%$, convective heat transfer through the gas phase becomes dominant and heat conduction across the polymer sharply decreases (Fig. 9).

No doubt, an extensive investigation of the combustion wave structure under different conditions would permit to verify many conceptions of the current flame spread theories, and also to determine the applicability limits of the latter. Even now, since more experimental investigations of the rate of flame spread over polymer material surfaces as a function of various factors are being carried out, it is becoming increasingly clear that the mechanism of heat transfer from the flame to the combustible surface can change radically as the size of the combustion zone increases.

In small-scale laboratory tests (characteristic sample dimensions of less than 0.2–0.3 m), conduction or convection are the dominant heat transfer routes. It has been shown [57] horizontally or vertically, that 90 % of the heat generated in a laminar flame spreading along a thermally thick PMMA sample is transferred by conduction across the condensed phase, and only about 10 % by convection through the gas phase. In the case of thermally thin samples, on the other hand, the heat is predominantly transferred through the gas phase, independent of the flame propagation direction [58–61].

Diffusion combustion of small samples (<0.2–0.3 m) is mostly laminar, but as the combustion zone and, correspondingly, the flame height, increase the flame may become turbulent. Turbulent convective heat transfer is essential in the case of such flames. And finally, in medium and large-sized samples (>0.2–0.3 m) radiation is the prevailing heat transfer mechanism [62].

The most severe conditions are realized when the flame propagates upwards along the material surface. This type of combustion has been dealt with in a large number of theoretical and experimental works [63–67]. It has been established that the laminar flame spontaneously increases in size and may become turbulent as it moves up the surface. The flame spread is steady only until a certain moment, and then, after a critical value of the ratio of heat transfer rate to the rate of heat generation in the gas-phase reaction zone is attained, it becomes unsteady [63]. The flame spread rate begins to increase with time. For thermally thin polymeric materials the flame spread rate increase up to a certain constant value [65]. For thermally thick materials it increases exponentially with the flame height, h_f [66]. Orloff et al. [67] have shown that about 85% of the total heat flux to the polymer surface in flame propagation upward a 3.5 meter PMMA sample is transferred by radiation. In this case the flame spread rate on the surface of thermally thick materials may be approximately expressed as [46]

$$v = (\epsilon \sigma T_f^4)^2 \cdot h_f / \varrho_s c_s \lambda_s (T_v - T_o)^2 \tag{2.19}$$

where h_f is the flame height, $\sigma = 56.7 \times 10^{-8}$ Wm^{-2}K^{-4} the Stephan-Boltzman constant, and ϵ the emissivity of the flame. In the case of upward flame spread the flame may be said to "wash" the surface. Roughly, it may be assumed that the flame height is proportional to the characteristic dimension over which heat is transferred to the polymer surface, δ_f, and, accordingly, to the dimension of the pyrolysis zone, δ_p; δ_p is a function of the time and it increases according to:

$$d\delta_p / d\tau \sim \beta \delta_p^n \tag{2.20}$$

where β and n are constants. Therefore the flame spread rate, v, increases with time.

Principally different physical laws govern the heat transfer by convection, conduction and radiation. No wonder that often the flame spread rate results obtained in small-, medium- or large -scale material tests do not correlate well.

More and more attempts are made to investigate flame spreading in small-scale experiments using an external radiation source and varying the radiative heat flux [68-71]. Such works are aimed to find out how radiated heat may affect fire propagation. The increase of the flame spread rate over a material surface under the effect of an external radiation is attributed by some investigators [70,71] to the increase of the surface temperature in front of the propagating flame. The rate of heat transfer from the flame to the material increases, because the incoming flow gradient and velocity rise with $\dot{q}_e$.

As reported in Ref. [68], the spread rate of a flame moving up a vertical surface of a sufficiently thick PMMA sheet increases under the effect of an external heat radiation. Depending on the heat radiation intensity and exposure time, various effects on the flame spread rate are observed. Additional heating of the polymer surface by a radiative flux results, first of all, in a decrease of the temperature difference $(T_v - T_o)$ and, in accordance with Eq. (2.19), in an increase of v. The experimental relationship $v \sim (T_v - T_o)^{-2.2}$ at $T_v = 363$ °C is close to that predicted by theory. According to Fernandez-Pello [68], an increase of the initial polymer surface temperature, T_o, cause a parallel enhancement of the natural convection in the boundary heat layer and heat radiation by the surface, leading to its partial cooling. Therefore, when the intensity of the external radiative heat flux is low, the flame spread rate increases with time, but only up to a certain "constant" value.

At large $\dot{q}_e$ the cooling effect of the above-mentioned factors is small and the flame spread rate increases exponentially with time. In this case, polymer gasification and ignition of combustible gases by the flame occurs, the entire surface being heated by the extra source. The flame spread is jerky and unsteady. Upon additional heating of the polymeric material with a radiation panel, interesting information concerning the material flammability may be derived from simple standard flame spread tests. Thus, Quintiere [72] studied the rate of flame spread over vertical surfaces of polymer materials heated by radiation panels to provide a predetermined radiative flux distribution over the test sample surface. He showed that under thermal equilibrium conditions (i.e. when the surface temperature has become practically constant under the extra heat supply), the flame spread velocity depends on $\dot{q}_e$ in the following manner:

$$v = [c(\dot{q}_{0,ign} - \dot{q}_e)]^{-2} \tag{2.21}$$

within the range

$$\dot{q}_{0,s} \leqq \dot{q}_e \leqq \dot{q}_{0,ign}$$

where $\dot{q}_{0,s}$ is the minimum heat flux necessary for steady flame spread, $\dot{q}_{0,ign}$ the minimum heat flux necessary for material ignition, and C a coefficient depending on the nature of the polymer.

The $\dot{q}_{0,ign}$ values are obtained by extrapolation of the straight portions of the $v^{-1/2} = f(\dot{q}_e)$ curves (Fig. 10) to the abscissa. The minimum heat flux $\dot{q}_{0,s}$ may be estimated from the point where the curves start to deviate from linearity. The constant C is determined as the slope of the straight portions of the curves. For example, for unsaturated polyester and PMMA, $\dot{q}_{0,s} \simeq 0.2$ Wcm^{-2}; $\dot{q}_{0,ign}$ are 2.1 and 2.8 Wcm^{-2}, and C = 1.6 and 1.1 ((s mm^{-1})$^{1/2}$cm^{2}W^{-1}), respectively. The surface temperature may be approximated by:

$$T_0 = 80° + 110\,\dot{q}_e,\ °C \tag{2.22}$$

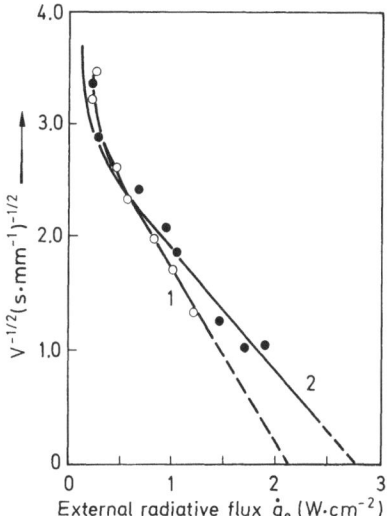

Fig. 10. Effect of external heat flux intensity on the rate of flame spread on a polymer surface; 1) poly-(methyl methacrylate), 2) polyester (After Ref. [72], with permission)

For $\dot{q}_e = \dot{q}_{0,\,ign}$, the ignition temperature is then defined as $T_{ign} = 80° + 110\,\dot{q}_{0,ign}$. Accordingly, Eq. (2.21) may be rewritten in the form

$$v = [C/110\,(T_{ign} - T_0)]^{-2},\; mm \cdot s^{-1} \tag{2.23}$$

T_{ign} for PMMA and unsaturated polyester is 310 and 390 °C, respectively. From these results it follows that PMMA and similar products are more easy to ignite and to burn than unsaturated polyesters, under a given external heat flux $\dot{q}_e$. It is interesting to note that $\dot{q}_{0,\,ign}$ values, obtained from flame spread experiments in this manner, are consistent with the $\dot{q}_{0,\,ign}$ values determined from direct ignitability tests. This is an evidence of the validity of the treatment of flame spread on a surface as a continuous gas-phase ignition of polymer decomposition products.

2.3.2 Mass Burning Rate

Detailed information about material flammability derived from macroscopic combustion parameters is given by Petrella [73.] who considered the effect of an external heat source on the mass rate of polymer combustion. From the energy conservation law it follows:

$$\dot{q}_s + \dot{q}_e = \dot{q}_v + \dot{q}_L \tag{2.24}$$

This implies that the heat fluxes to the surface from the flame ($\dot{q}_s$) and external source ($\dot{q}_e$) are counterbalanced by the heat flux for polymer gasification and fuel supply to the flame zone ($\dot{q}_v$), and losses across the polymer ($\dot{q}_L$). The heat flux $\dot{q}_v$ is given by $\dot{q}_v = \dot{m}\,\Delta H_v$, where $\dot{m}$ is the mass rate of burning, i.e. the mass loss per time unit, ΔH_v is the polymer gasification enthalpy from the initial temperature T_0, and $\dot{q}_s = \xi Y_{ox}^n$ where ξ and n are constants depending on the polymer nature. From Eq. (2.24) it follows then

$$\dot{m} = (\xi/\Delta H_v)Y_{ox}^n + (\dot{q}_e - \dot{q}_L)/\Delta H_v \tag{2.25}$$

For $n = 1$ and $\dot{q}_e = $ const, a graphical plot of $\dot{m}$ versus Y_{ox} yields a straight line (Fig. 11a). From the slope $\xi/\Delta H_v$ is obtained; the intercept with the ordinate ($Y_{ox} = 0$) gives $(\dot{q}_e - \dot{q}_L)/\Delta H_v$. For $Y_{ox} = $ const, a plot of $\dot{m}$ versus $\dot{q}_e$ yields also a straight line (Fig. 11b), from the slope of which $1/\Delta H_v$ is obtained. Thus, from the experimental data one may determine $\xi/\Delta H_v$, $\dot{q}_L/\Delta H_v$, $\dot{q}_s$, $\dot{q}_L$, ξ. For polystyrene, for example, it was found that $n = 1$, $\Delta H_v = 1680$ kJ $\cdot$ kg^{-1}, $\xi = 16.05$ W $\cdot$ cm^{-2}, $\dot{q}_L/\Delta H_v = 16.2 \times 10^{-4}$ kg cm^{-2}s^{-1}, $\dot{q}_s = 3.34$ W cm^{-2}.

Knowing these parameters, the mass rate of burning for any oxygen concentration may be determined, including combustion in air ($Y_{ox} = 0,21$), under different external heat fluxes. According to Petrella $\dot{q}_e = \dot{q}_L$ for an "ideal" case and, consequently, $\dot{m}_{ideal} = (\xi/\Delta H_v)\,Y_{ox}^n$. Three different types of polymer behavior in combustion may then be singled out:
1) in small-scale tests

$$\dot{q}_e < \dot{q}_L,\; \dot{m} < \dot{m}_{ideal}$$

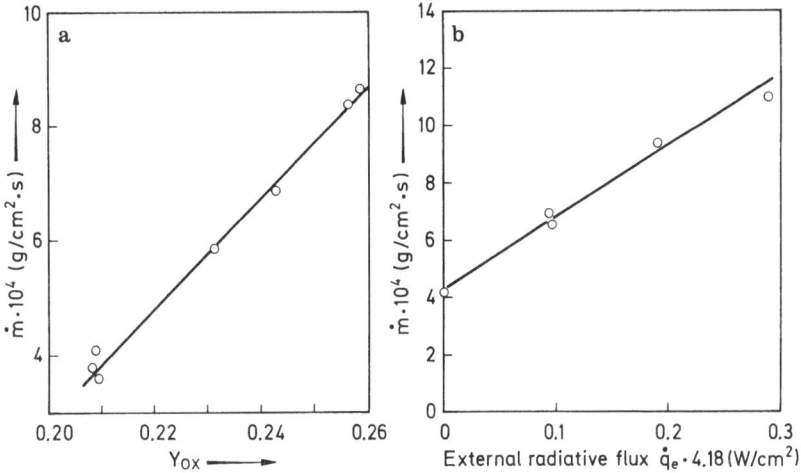

Fig. 11a and b. Mass rate of polystyrene burning as a function of oxygen concentration in the O_2/N_2 flow at $\dot{q}_e = 0$ **(a)**, and of the external heat flux intensity at $Y_{ox} = $ const **(b)** (After Ref. [73], with permission)

2) in large-scale tests

$$\dot{q}_e > \dot{q}_L, \dot{m} > m_{ideal}$$

3) the case of $\dot{q}_e = \dot{q}_L$ corresponds to enhancement of fire.

In the above-mentioned work [73] the authors recorded the total mass rate of polymer burning in a candle-like combustion rather than in flame spreading on a surface. An indirect method was proposed in Ref. [55] for measuring the mass rate of burning in flame spreading over the surface of a polymer material. It consists of measuring the burning surface profile of an extinguished sample, using a profilograph (surface roughness recorder):

$$\dot{m} = \varrho v \, tg \, \alpha \tag{2.26}$$

where v is the flame spread velocity and α the angle between destroyed and undestroyed surface.

Figure 12 indicates that the mass rate of burning ($\dot{m}$) has a maximum at $\dot{m}_1 = \varrho v \, tg \, \alpha_1$. With increasing distance from the flame edge (x), $\dot{m}$ decreases, and becomes finally practically constant, at $\dot{m}_2 = \varrho v \, tg \, \alpha_2$.

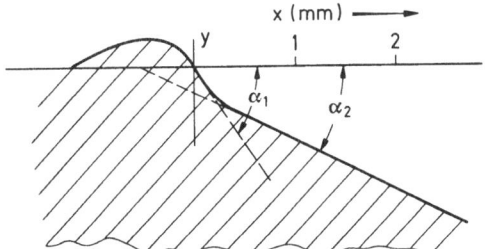

Fig. 12. Surface profile of an extinguished polymer sample. Ordinate is depth into the polymer; shaded region is the undestroyed polymer (After Ref. [55])

It has been found that when the oxygen concentration in the incoming flow is relatively low ($0.2 \leq Y_{ox} \leq 0.4$), the PMMA burning rate is practically constant over a vast region beneath the flame ($\dot{m}_1 \approx \dot{m}_2$), whereas with increasing oxygen concentration $\dot{m}_1 \sim Y_{ox}^4$, and $\dot{m}_2$ remains constant and independent of Y_{ox}. Measurements of the mass rate of burning of polymeric materials under different thermal conditions allow to establish the effects of various factors on this very complex process and, in particular, to determine the effective polymer pyrolysis rate during combustion. However, this determination requires the knowledge of the true temperature on the burning surface.

In the general case, the mass rate of burning is given by:

$$\dot{m} = \frac{\alpha}{c} \ln (1 + B) \qquad (2.27)$$

where α is the coefficient of heat transfer, c the heat capacity, and B the coefficient of mass transfer (Spalding number). The Spalding number characterizes the burning intensity of polymeric materials [63, 74–76]. It may, therefore, be conveniently used for comparative estimates of the flammability of these materials.

Under conditions of convective heat transfer, the mass transfer number B may be written as

$$B = \frac{(Y_{ox}/r) \Delta H_c - c(T_s - T_\infty)}{\Delta H_v} \qquad (2.28\,a)$$

or

$$B = (Y_{ox}/r)B_c - b \qquad (2.28\,b)$$

where

$$B_c = \frac{\Delta H_c}{\Delta H_v} ; \quad b = c(T_s - T_\infty)/\Delta H_v ; \quad r = \text{stoichiometric oxidant/fuel ratio}$$

Consequently the mass transfer rate during diffusion combustion of polymers is determined by the ratio of the heat of combustion to the effective enthalpy of polymer gasification. The lower the combustion heat and the higher the polymer gasification enthalpy or, in other words the more heat resistant the polymer, the lower is the B value. For polymer combustion in air the B value of e.g., PMMA varies between 1.3–1.4, that of polyethylene between 0.5–0.6, of phenolic resins between 0.14–0.4 [74, 75]. An increase of the oxygen concentration in the oxidative medium and of the oxidant temperature causes a rise of the mass transfer number B. Lower B values have been observed in thermally stable polymers of the carbonizable types.

As noted in Section 2.1, polymeric materials may be burned under various conditions which affect the location of the reaction responsible for heat generation during combustion. Heterogeneous combustion of polymers is also possible.

Theoretical research into the problem of combustion propagation on the surface of condensed fuels involving heterogeneous oxidation has been reported [77, 78]. It has been found that the combustion velocity depends on thermal effects, kinetic parameters of the exothermic oxidation reaction on the surface, and on the rate of oxygen supply to the surface of the condensed phase. The incoming oxidant flow velocity and oxygen concentration in the flow are important factors in this connection. The thicker the layer of a combustible material, the lower is the propagation rate of the heterogeneous combustion process.

Heterogeneous combustion (smouldering) of polymeric materials, which is accompanied by propagation of the decomposition front, charring of the polymer, as well as fuming, is a considerable fire hazard. Steadiness of smouldering of polymeric materials depends on whether or not the heat produced in the heterogeneous coke oxidation makes up for all the heat losses occurring in pyrolysis and carbonization of the polymeric material and heat losses to the environment. By investigating the smouldering of cellulosic and polyurethane foam materials, Moussa et al. [79, 80] showed that the heat flux in the coke oxidation zone and, consequently, the maximum smouldering temperature depend on the partial oxygen pressure on the coke surface and on the oxygen concentration in the environment. The individual properties of the polymer affect the coking process and the reactivity of the resultant carbonized product with respect to oxygen. This is manifested by the fact that for the same smouldering rates of polyurethane foams and cellulosic materials, the temperature at the interface between the pyrolysis and coke oxidation zone is 600–650 K in the first case and 900–1100 K in the second. The critical oxygen concentration required for a steady-state smouldering process is much lower for cellulosic materials.

Figure 13 shows how the steady-state smouldering in polyurethane foam either passes into flaming combustion or extinguishes. It is seen that flaming combustion occurs when the conditions are more severe.

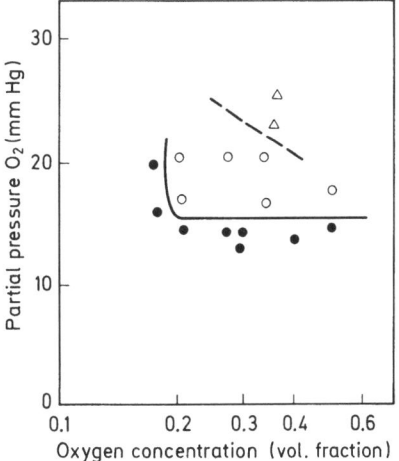

Fig. 13. Effect of partial oxygen pressure near the polyurethane foam surface and of the oxygen concentration in an oxidant flow on the conditions of combustion △: flaming combustion; ⊙: smouldering; ●: slow oxidation (After Ref. [80], with permission)

3 Extinction of Polymer Combustion

The investigation of critical conditions for polymer combustion is of great interest for the further development of the combustion theory, as well as for practical reduction of flammability of materials, fire prevention, and extinction.

   If we consider combustion as an integral process, we will find that the critical conditions of combustion are closely related to the ratio of the rates of heat transfer to and from the leading zone of the combustion reaction. This implies that to determine the critical combustion conditions one must take into account the rates of the three most important chemical and physical processes: the chemical reaction, mass transfer, and heat transfer. The various external factors such as temperature, pressure, environment, oxidizing flow rate, etc. affect primarily the rates of the above three processes. It is, therefore, possible to determine combustion limits for each of these parameters. As we have already noted (Fig. 1), the extinction of combustion occurs in the range where the rates of the chemical reaction and transport processes are commensurate or the latter become much lower than the former. The pure diffusion control of a well-developed combustion becomes a combined diffusion-activation control near the extinction limit. The extinction limit is particulary sensitive to environmental variations. The regime of exothermal reduction-oxidation is characterized by the Damköhler number, De (ratio of the residence time of the reactants in the zone of chemical reaction, τ_r, to the chemical reaction time, τ_{ch}). A steady diffusion flaming combustion of fuels is usually observed when the Damköhler numbers are high. When De decreases to a value below the critical value De_*, the combustion discontinues. The value of De may be reduced either by an increase of τ_{ch} (i.e. decrease of the chemical reaction rate) or decrease of τ_r (i.e. rise of the rate of interdiffusion of the reacting particles, reaction products and heat transfer from the reaction zone to the environment). However, the processes are interrelated to some extent. Below the critical value De_* the reactant residence time in the gas is too short, in comparison with the chemical conversion time, for significant heat release to occur. It should be kept in mind that even within one and the same system De_* is not a constant but a parameter depending on the ambient conditions.

   Diffusion flaming combustion of polymeric materials may be compared with that of unmixed gases. Zeldovich [81] related the extinction limits of the diffusion flame with the reactant concentration (mass flow rate) in the chemical reaction zone and its cooling. At the extinction limit the intensity of diffusion combustion of unmixed gases is at its maximum corresponding to the combustion intensity of a premixed stoichiometric mixture [81] of the same gases.

   A more detailed analysis [82, 83] has shown that near the extinction limit the reactant diffusion fluxes do not reach the reaction zone in a stoichiometric combination. A portion of the reactants is observed to bypass the high-temperature flame zone. However, if we consider the rate of material consumption in the reaction zone instead of its supply rate, the above statement [81] will be valid for any diffusion reaction rate. As regards the determination of the extinction limits of diffusion combustion, it is very important that the equations describing the steady-state combustion are valid even in the flame extinction stage.

   Frey and T'ien [84] estimated theoretically the extinction limits of the diffusion flaming combustion for a thermally thin fuel according to oxygen concentration, pressure, oxidant flow rate. Using predetermined kinetic parameters of the solid-phase and gas-phase reactions, thermophysical characteristics and enthalpies of the phases, they solved numerically two-dimensional differential equations for flame spread. A major conclusion is that the flame propagation velocity decreases monotonously as the Damköhler number falls to a certain critical value (ca. 10^3), whereafter flame propagation becomes impossible. A variation of the oxygen concentration in the environment affects the flame temperature which, in turn, influences the natural convection by which heat is transferred from the flame to the polymer surface, and other parameters. The estimated effects of P, V_a and Y_{ox} qualitatively agree with experiments [85]. Quantitative agreement depends on the correct choice of the gas-phase kinetic parameters which are unknown for many real polymer systems.

The available diffusion combustion theories can roughly predict the combustion limits for various parameters. Not only ambient oxygen concentration [86], pressure [87,88] and flow velocity [61,76] limits but also sample size [89-91], inertial overloads [92], ambient and material temperature [93,94] limits have been experimentally determined, and found in accordance with the theoretical predictions.

The authors of Ref. [95] generalized all the published polymer combustion limits from the viewpoint of the effect of different factors on the cooling of the reaction zone. At the extinction limit of diffusion combustion, the ratio of heat losses from the front edge of the combustion zone to the total heat generation due to the chemical reaction must be proportional to $RT_{f,*}/E$ [81]; here, $T_{f,*}$ is the flame temperature at the extinction limit and E the gas-phase reaction activation energy

$$(\Sigma \dot{q}_L / \dot{q}) \sim (RT_{f,*}/E) \tag{3.1}$$

Many factors contribute to the total heat losses and each of them may become the controlling one under certain conditions. The authors employed a semi-empirical approach to estimate the combustion limits of polymeric materials, assuming that the maximum polymer burning rate in the flame edge, at its extinction limit, is the same as the combustion rate of a stoichiometric mixture of polymer decomposition and gasification products with the oxidant. This makes it possible to calculate the maximum adiabatic combustion temperature for a standard LOI of a polymer (normal pressure and free fall acceleration of 1 G). Taking a reasonable value of the effective activation energy of the gas phase reaction (E $\approx$ 80—160 kJ/mol) and determining the mass polymer burning rate in an independent experiment, one may evaluate the preexponential factor in the Arrhenius equation for the gas-phase combustion of the polymer. Since, ultimately, all the factors influence the flame zone cooling, it is convenient to use the effect of one or another parameter on $Y_{ox,*}$ as a test of the nature of heat losses. $Y_{ox,*}$ is the critical oxygen concentration in an atmosphere necessary for the flame maintenance. An equation has been proposed relating $Y_{ox,*}$ at other than normal combustion conditions (P = 1 atm, 1 G, no forced oxidant flows) to the effective kinetic parameters of the gas-phase reaction (E_g and k_0). If these parameters are known the steady-state diffusion combustion limits may be described rather comprehensively.

Inertial overloads, for example, may enhance convective heat losses, because the buoyancy force and, therefore, the velocity of the convective rise of hot combustion products increase. At high overloads (ca. 300 G) convective heat losses become dominant. The limiting oxygen concentration is in this case proportional to the cubic root of the overload, $\bar{a}$, is weakly dependent on pressure and independent of the sample size [92]:

$$Y_{ox*} = \frac{A \cdot \bar{a}^{1/3}}{P^{0,1}} \tag{3.2}$$

where A is a constant.

Under zero-G conditions there is no natural convection and a forced oxidant flow is needed to maintain combustion. The minimum acceleration at which combustion of PMMA could still be observed in a system at rest, with $Y_{ox,*} = 17\%$, was 2×10^{-2}G. [96]. At small G, the heat is lost by convection and radiation. Information about the effect of radiation on the extinction of flame combustion of polymers may be derived from the $Y_{ox,*}$ versus pressure relationship. For extinction due to radiation, $Y_{ox*} \sim 1/P^k$; ($k \simeq 0.5$–1).

Figure 14 shows that the heat loss by radiation becomes insignificant (the oxygen index becomes independent of pressure) at $P > 0.3$ atm for PMMA, and at $P > 2$ atm and $P > 6$ atm for poly(vinyl chloride) and Teflon, respectively. A weak pressure dependence of $Y_{ox*} \simeq (1/P^{0.1})$ accounts for extinction of the flame by convective forces. Important in this case is the true velocity of the incoming oxidant flow, v_a. The latter is a sum of components due to natural and forced convection. Variations of sample dimensions and thickness of conductive walls affect the heat losses by conduction and convection. The analysis may be generalized if we consider heat losses via the condensed phase apart from the gas phase. Heat consumed by melting of the condensed phase and transferred by conduction through the condensed phase further enhances the influence of various factors on the extinction behavior. The Y_{ox*} then increases considerably. The works mentioned in Section 3 allow the polymer flammability to be predicted under different conditions. This is of great practical importance. Scarceness of our knowledge of the mechanism and kinetics of the solid- and gas-phase reactions accompanying the combustion of most polymer types pre-

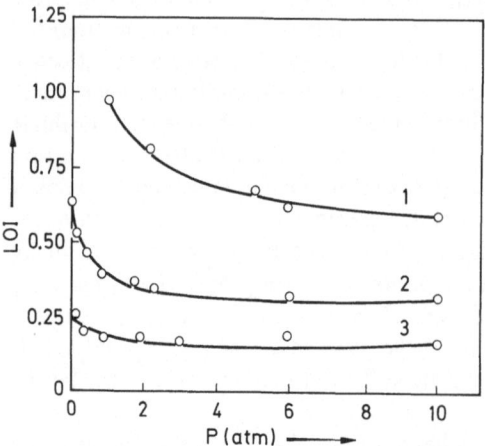

Fig. 14. Effect of pressure on the limiting oxygen index (LOI) of polymers; 1) polytetrafluoroethylene, 2) poly(vinyl chloride), 3) poly(methyl methacrylate) (After Ref. [83])

vents both an effective application of the existing theories and the development of effective ways of decreasing the combustibility of materials.

Present research efforts aim mainly at obtaining the important parameters from a study of the macrokinetics of combustion. It is important to estimate the effects of flame retardants, chemical structure of the polymer and polymer composition on variations of the solid- and gas-phase reaction kinetics.

Experiments under subcritical conditions appear to be most promising in this respect. As an example, we may cite a number of works in which different relationships at the extinction limit were used for the determination of the effective activation energy and preexponential factor of the gas-phase combustion reaction. In particular, Krishnamurthy [87] calculated the kinetic parameters of the gas-phase combustion of PMMA from the relationship between the combustion rate and the oxygen pressure and concentration at the extinction limit ($E_g = 88$ kJ/mol; $k_0 = 3 \times 10^{12}$ cm^3/mol $\cdot$ s). Other authors [76, 94] did the same by analyzing the relationship between the extinguishing oxidant flow velocity and oxidant concentration, with the help of an opposed flow diffusion flame (OFDF) apparatus. A similar relationship between flow velocity and oxidant temperature was suggested, since preheating of the oxidant was found to immediately affect the flame temperature. For PMMA, PE and polyoxymethylene (POM) $E_g = 98.5$, 140 and 121 kJ/mol, respectively, were reported.

The theoretical analysis of extinction of the diffusion flame of combustible liquids with flat flame geometry which may be established with the OFDF apparatus [97] was extended to polymer flames of similar geometry [98]. Heat losses by radiation from the polymer surface were taken into account in estimates of the flame temperature at the extinction limit. The kinetic parameters of the gas-phase reaction (E_g, kJ/mol and k_0, cm$^3 \cdot$ mol$^{-1} \cdot$ s^{-1}) have been determined at $Y_{ox} = 0.16 - 0.20$ for PMMA, POM, PE and PS, respectively: 167.2 and 1.2×10^{16}; 183.9 and 2.6×10^{16}; 167.2 and 2.2×10^{15}; 234.0 and 1.7×10^{17}. From a comparison with other data it follows that, as Y_{ox} increases, the effective energy of the gas-phase reaction decreases.

Thus, the flammability of polymeric materials is determined not only by the individual properties of the materials but also by the nature of their interactions with the environment. Moreover, size and orientation of the polymeric sample are highly important for initiation, development and extinction of the diffusion combustion.

4 Chemical Nature and Flammability of Polymeric Materials

Naturally, the effect of the chemical nature of polymers on the flammability may be established only in tests under identical conditions. At present, many procedures are available for estimating material flammability, each reflecting one or another aspect of the complex combustion process.

Usually, different methods are recommended for each particular type of material (film, fabric, carpet, rigid plastic foam, elastoplastics, etc.) to characterize its flammability adequately. Test procedures are divided into small-, medium-, large- and real-scale experiments, depending on the sample size.

Large- and real-scale tests provide most comprehensive information about the fire hazard of materials. But even such tests cannot fully reflect all the varieties of situations that may lead to fire with its undesirable consequences. Large-scale tests are costly and hard to control. Medium- and small-scale tests are more practicable and are particularly useful at the development stage of flame-retardant materials. However, neither of these alone can yield complete information about the fire hazard of a material. Therefore, the fire hazard potential is estimated from a series of experiments on ignition, flame propagation, heat generation during combustion, fire resistance, amount of fumes, and toxicity of the pyrolysis and combustion products.

A number of fire tests for polymeric materials have been developed, during the past several years, by the International Standards Organization (ISO). Hopefully, these tests will replace the present national test methods, which often correlate badly with each other. The development at ISO is aimed at describing the fire properties of polymeric materials comprehensively, with test methods chosen so as to be applicable to all types of samples. At present, the ISO fire test methods are published as standards (ISO R 1182-79 for non-combustible materials, ISO R 1326-70 and ISO R 1210-70 for flame spread, etc.), or as draft for development (DP 5657, ISO/TC 92 N 531-79 as an ignitability test). One can only sympathize with using certain complex fire hazard indices for describing material behavior in fire [99].

There is no doubt that a scientifically correct comparison of laboratory and real-scale test results will be helpful towards the development of analytical models for the objective prediction of material behavior in fire.

4.1 Limiting Oxygen Concentration as Flammability Index

What flammability criteria should be considered in order to determine the effect of the polymer nature?

Although the analysis of the relation between the chemical nature and the flammability of polymeric materials is extremely complex, we will try to carry out this analysis by focusing our attention on one of the major aspects of the combustion process, viz. the chemical reaction responsible for the combustion. Only in this aspect the material itself, i.e. its composition, structure and thermochemical properties, is fully relevant.

The combustion is a reaction between the combustible and oxygen. Oxygen consumption is one of the indicators of the amount of generated heat [100]. The limiting oxygen concentration is a convenient material flammability index. Practice has shown that the standard LOI procedure (ASTM D 2863-76) is the most reproducible test, and most sensitive to any compositional or structural variations in the material.

Van Krevelen noted that the flammability of polymers containing C, H, N, O, and halogens, depends on a composition parameter [101] CP:

$$CP = (H/C) - 0.65 (F/C)^{1/3} - 1.1 (Cl/C)^{1/3} - 1.6 (Br/C)^{1/3} \qquad (4.1)$$

where H/C, F/C, Cl/C and Br/C are element/carbon atomic ratios. The smaller the H/C ratio, the more aromatic is the polymer and the higher the LOI it exhibits. At $CP \geq 1$, polymers have LOI = 0.175; at $CP < 1$, LOI = 0.60 − 0.425. According to Moos and Skinner [102], the flammability is more a function of the hydrogen content in the polymer composition than of flame-retarding elements (P, Cl). The hydrogen content of the macromolecules is decisive both for thermal stability and flammability of polymers. On the other hand, weak bonds in the polymer, offering the possibility of radical formation and chain transfer, as well as of disproportionation involving hydrogen, reduce the thermal stability of polymers. In other words, not only the mere atomic composition but also the chemical structure of a material counts.

Flaming combustion of polymeric materials inevitably involves liberation of gaseous fuels. No wonder, therefore, that the interest in the thermal properties of polymers and the mechanisms and kinetics of polymer decomposition and gasification has been high.

4.2 Oxygen in Condensed Phase Burning

In the scientific literature the question of the role of oxygen in condensed phase decomposition during fire has been discussed repeatedly. Currently, it is assumed that polymer decomposition in flaming combustion is purely thermal. This assumption is based on the analysis and a comparison of the decomposition rates during combustion under various conditions simulating real heating rates in fires [103-105].

The pressence of oxygen near a burning polymer surface has been recorded by many investigators [3, 106-108]. But the oxygen concentration was observed to decrease with the distance from the front edge of the flame, as the ambient oxygen concentration increased [106]. These facts are perfectly consistent with the theoretical conceptions of diffusion flame combustion of materials discussed in Sect. 3 [82-84].

In the case of an intensive laminar flame combustion, oxygen penetration to the surface, across the high-temperature reaction zone, is hindered due to the high rate of its reaction in the zone. A thermal decomposition of the polymer may be expected to take place under these conditions. Near the combustion limit, oxygen is able to penetrate the zone near the surface underneath the flame edge. However, its concentration (1–2%) is insufficient for a profound oxidation of the condensed phase. The probability of oxygen penetration to the surface increases when the flame becomes turbulent and when the condensed phase decomposition products are less volatile [88].

4.3 Chemical Structure of Polymer and Susceptibility to Heat

According to their susceptibility to heat, polymers may be roughly divided into two groups [109, 110]. The first includes polymers which undergo degradation under the effect of heat, involving breaking of bonds in the main chain and formation of low molecular weight gases and liquids. At high temperatures (300 to 800 °C) polymers of this group are almost totally destroyed or leave a minute amount of a nonvolatile residue (coke). Examples of polymers of this group are: poly(methyl methacrylate), polyoxymethylene, poly-α-methylstyrene, polytetrafluoroethylene, all of which depolymerize with an almost 100% yield of the monomer. Other examples are polystyrene, polyethylene, poly(methyl acrylate) and other polymers which undergo an almost complete degradation, although with a relatively small monomer yield.

The second group includes polymers showing a tendency for inter- and intramolecular abstraction of certain atoms or groups, cyclization, condensation, recombination or other reactions which ultimately yield carbonized products. Examples of such polymers are poly(vinyl alcohol) and its derivatives, Cl-containing vinyl or diene polymers, polyacrylonitrile, and many aromatic and heterocyclic polymers. Common to the pyrolysis of all these polymers is the formation, in the macromolecules, of conjugated multiple bonds, transition from a linear to a spatially cross-linked structure, and increase of aromaticity of the nonvolatile pyrolysis product. In contrast to polymers of the fist group, pyrolysis of the latter is usually an exothermic process.

If we consider, as examples of the second group, polymers containing aromatic carbon- and/or heterocyclic links in the main chain of the macromolecules, we will notice the following general features of their pyrolysis, thermal properties and chemical structure:

1. The thermal stability of the polymers and the yield of carbonized residues in pyrolysis increase with the relative number of aromatic groups per repeat unit of the main macromolecular chain.

2. The thermal stability of heterocyclic polymers increases with the aromaticity of the heterocycles.

3. Pyrolysis begins with the breaking of the weakest bonds in the bridging groups connecting the aromatic rings or heterocycles.

4. Substituents on aromatic rings and heterocycles that take part in intermolecular cross-linking, or intramolecular cyclization reactions, reduce the rate of release of volatile products and increase the yield of the carbonized residue. Otherwise, the contribution of the substituents to the coking process is small. The substituents are often split off and removed as volatile products, the thermal stability of the polymers being thereby reduced.

5. Reduction of the number of mobile hydrogen atoms capable of taking part in chain transfer and disproportionation reactions improves the thermal stability of polymers.

6. Isomerism of the substitution of aromatic rings affects the thermal stability of the polymer and the formation of carbonized residue in pyrolysis. Meta-isomers are less stable and yield less coke than para-isomers.

7. Ladder and cross-linked polymers are more heat resistant than their linear analogs.

8. Heteroatoms with unpaired p-electrons (N, O, S) are partly removed from the carbonized pyrolysis products at temperatures above 600 °C. This process of heteroatom removal from the carbon skeleton may continue up to temperatures of 1300 to 1700 °C.

In general, carbocyclic aromatic polymers yield more carbonized products than heterocyclic ones. Susceptibility of polymers to carbonization under pyrolysis conditions is characterized by the so-called coke number [111]. Van Krevelen [101] concluded that the tendency for carbonization may be regarded as an additive resultant of the contributions of the individual functional groups in the monomer units. The contribution of a group is expressed in carbon equivalents contributed by each group to the coke residue, CR, per monomer unit:

$$CR = \frac{\sum_i (CFT)_i \times 12}{M} \cdot 100 \tag{4.2}$$

where (CFT)$_i$ is the coke formation tendency of a group expressed in C-equivalents, and M the molecular mass of the monomer unit.

4.4 Flammability and Structural Characteristics of Polymers

For polymers whose pyrolysis products are not inhibitors for the flaming reaction, Van Krevelen established a simple empirical correlation between the limiting oxygen index and the coke residue:

$$LOI = 17.5 + 0.4\, CR \qquad \text{vol. \%} \tag{4.3}$$

Polymer modification by inclusion of high-(CFT)$_i$ fragments is therefore an effective method of imparting flame retardant properties to the polymer.

The relationship between other flammability characteristics and the polymer tendency for carbonization has also been studied [112–114]. Flame retardant effectiveness of coke is, primarily, related to the reduced release of fuels into the gas-phase, and to the ability of the coke to undergo heterogeneous oxidation. Therefore, not only the composition is important but also the morphological structure, porosity, specific surface area and thermophysical properties of the carbonaceous residue [115,116].

According to Eq. (4.3), the slope of the straight line obtained from a plot of the LOI versus the coke yield during pyrolysis or combustion of a polymer may be a measure of the ability of coke to improve the flame retardancy of polymers. Fig. 15 demonstrates that, when siloxane monomers are introduced into aromatic polycarbonates, the flame retardant effect of the coke increases considerably as compared with polymers not containing such monomers. It has been noted, however [116],

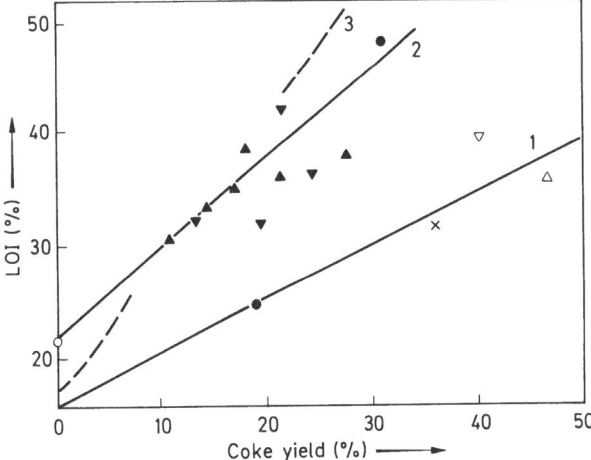

Fig. 15. LOI as a function of coke yield for different polycarbonates; 1) polycarbonates including C, H, O, S atoms; 2) polycarbonates additionally containing Si atoms; 3) line 2, recalculated taking into account SiO$_2$ in coke; different signs relate to polymer samples of different structure (After Ref. [116] with permission)

that the enhancement of the activity of the coke is not common to all Si containing polymers. The chemical nature of the polymeric material affects the course of the pyrolysis process, as well as the structure and properties of the carbonized product. Gasification and burning of the latter may occur as a result of the heterogeneous reaction with the oxidant. Impurities of metals with varying valency states are known to play an important role in the heterogeneous oxidation of carbonized products. The amount of impurity is usually estimated from the ash content. Thus according to Ref. [117], for phenolic fibers LOI = 36 %, whereas the LOI of carbon fabric made therefrom is 74 to 81 %. The presence of as little as 0.28 % ash reduces the LOI of the carbon fabric to 32 %. A similar effect may be obtained by oxidation of carbon fabric and an increase of its specific surface to 350–750 m^2g^{-1}. In this case, the LOI is 41–31 %.

Stuetz et al. [118] attempted to correlate the flammability of organic polymers with their thermooxidative stability in an atmosphere consisting of 1 % oxygen and 99 % nitrogen. The temperature (T_{od}) at which the polymer decomposes at a rate of 3.3 % per minute was used as the thermooxidative stability index. A straight-line relationship was shown to exist between the limiting oxygen index and $1/T_{od}$. Polymers for which T_{od} is above 525 °C are regarded as non-combustible in air according to the accepted testing standards. Stuetz et al. assume that the apparent relationship between flammability and the chemical nature of the polymer ensues from the thermodynamic conditions required for a steady-state combustion process, and the decomposition involving formation of volatile fuels. It can be easily seen, however, that the order in which T_{od} varies for the polymers considered by Stuetz et al. [118] practically follows the order of their thermal stabilities. Apparently, the features specific to the high-temperature decomposition of polymers in the presence of oxygen have much in common with the high-temperature oxidative decomposition of low molecular weight hydrocarbons and other organic compounds. It has been shown for many polymers that the high-temperature oxidation mechanism of hydrocarbons is substantially different from the low-temperature oxidation mechanism. Even in systems with a relatively high oxygen concentration the ratio between the products of the purely thermal cracking and the oxidation products rapidly increases with temperature (above 300 °C). Oxidation of condensed matter is then particularly unfavorable. Actually, the cracking of this type is no longer an oxidation process. Nevertheless, oxygen has a useful effect in that it promotes thermal dissociation of weak bonds or decreases the temperature limit of dissociation. The further thermal conversions (decay, structurization) are totally controlled by the fragment structure. The mechanism and kinetics of the high-temperature decomposition of polymers in the presence of oxygen have received little attention thus far [119]. Studies are necessary to understand the true role of oxygen in polymer decomposition during combustion and the extent of its participation in preflame reactions.

Important in combustion is not so much the thermal stability of the material itself but rather the amount and nature of the decomposition products. It is sufficient to compare the LOI of poly(vinyl chloride), whose thermal decomposition begins at 160–175 °C with that of heat resistant phenol-formaldehyde fibers (Kynol). The thermodynamic approach to the problem seems to be most reasonable. It allows to consider the polymer structure to explain the details of the combustion reactions and to estimate the heat of combustion of polymers.

Heat of combustion is one of the objective characteristics of material structure. Johnson [120] has published a simple empirical formula relating LOI to the specific heat of combustion of polymers:

$$LOI = 7.95 \cdot 10^5/\Delta H_c \qquad (4.4)$$

where ΔH_c is in $kJ \cdot kg^{-1}$. In many cases, however, Eq. (4.4) is not sufficiently accurate. Japanese scientists [121] proposed a more sophisticated correlation equation:

$$LOI = 0.00307(\Delta H_c \cdot MW/m_p + m_n) - 0.044 \qquad (4.5)$$

Here, ΔH_c is in $kJ \cdot kg^{-1}$, MW is the molecular mass of the monomer unit, m_p is the number of moles of products formed in the stoichiometrically complete combustion of the polymer, and m_n is the number of nitrogen moles in the O_2/N_2 atmosphere, per 1 mol of monomer unit. LOI is given as molar oxygen fraction in the atmosphere.

This equation has been derived from the assumption that the combustion heat is spent to heat the inert gas coming from the environment to the flame temperature, and to increase the enthalpy of the combustion products. The larger the bracketed term in Eq. (4.5), the higher is the limiting oxygen index of the polymer.

To demonstrate more clearly the effect of structural factors on polymer flammability, we considered the relationship between the limiting oxygen index and the specific

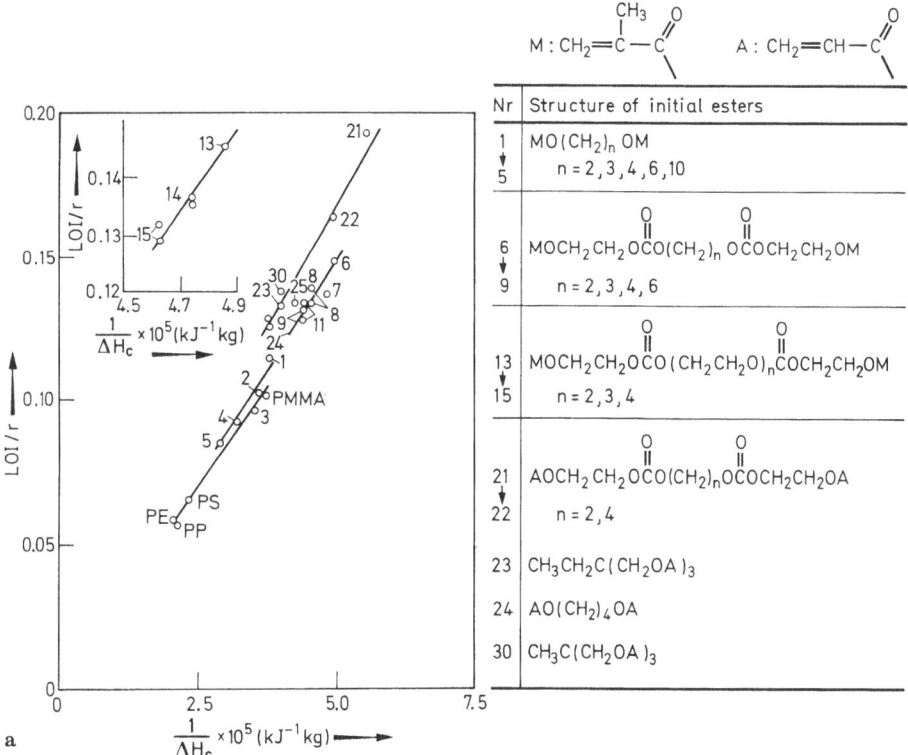

Nr	Structure of initial esters
1 ↓ 5	$MO(CH_2)_n OM$ $n = 2,3,4,6,10$
6 ↓ 9	$MOCH_2 CH_2 OCO(CH_2)_n OCOCH_2 CH_2 OM$ $n = 2,3,4,6$
13 ↓ 15	$MOCH_2 CH_2 OCO(CH_2 CH_2 O)_n COCH_2 CH_2 OM$ $n = 2,3,4$
21 ↓ 22	$AOCH_2 CH_2 OCO(CH_2)_n OCOCH_2 CH_2 OA$ $n = 2,4$
23	$CH_3 CH_2 C(CH_2 OA)_3$
24	$AO(CH_2)_4 OA$
30	$CH_3 C(CH_2 OA)_3$

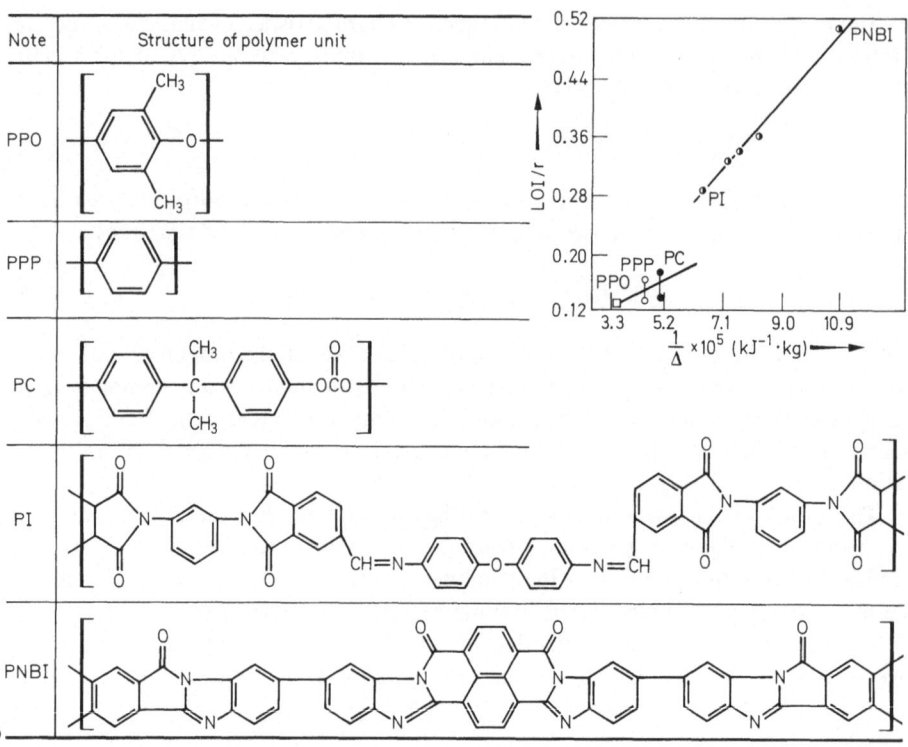

Fig. 16a and b. LOI/r as a function of the combustion heat of linear and cross-linked polymers of the noncarbonizable (**a**) and carbonizable (**b**) type (After Ref. [122])

heat of combustion, making allowance for the stoichiometric oxidant/fuel ratio, r [122]. For polymers producing a carbonized residue, its amount and the calorific power $\Delta H_{c, ck}$ are considered. For polymers which undergo almost complete degradation during pyrolysis in an inert medium and in combustion, LOI/r was plotted versus $1/\Delta H_c$ (Fig. 16a). For carbonizable polymers LOI/r was plotted versus $1/\Delta$ where $\Delta = \Delta H_c - Y_{ck} \Delta H_{c, ck}$ (Fig. 16b). LOI is taken as the mass fraction of O_2 in the $(O_2 + N_2)$ mixture. The figures show that the relationships are linear. The slopes of the straight lines give the amount of heat spent for the liberation of a unit mass of fuel to the gas phase and for the increase of the enthalpy of the decomposition product [122]. The flammability of crosslinked polymers based on acrylates and methacrylates of different glycols increases with the number of repeating groups in the starting glycol, i.e. the LOI of the polymers decreases in the same order as the thermodynamic stability of the monomers [123].

The nature and number of the terminal groups, the type of bonds, the presence of aromatic cycles and carbonate groups in the unsaturated ester molecule, as well as other structural features affect polymer flammability. The slope of the straight lines increases from methacrylate to acrylate polymers. Thus, for polymers of alkylene glycol dimethacrylates it is $2{,}82 \times 10^3$ kJ/kg, for polymers with carbonate bonds 2.86×10^3–2.9×10^3 kJ/kg; for acrylic polymers it is somewhat higher, 3.27×10^3 kJ per kg. Linear polymers, e.g. PMMA, PE, etc. have a smaller slope corresponding to

2.68×10^3 kJ/kg. Thermostable cross-linked polymers and copolymers of aromatic oligoimides and oligonaphthoylene-bis-benzimidazol are more difficult to gasify than aromatic linear polymers of the carbonizable type.

Thus, considering the combustion heat and the oxygen consumption for the combustion it is possible to study the influence of certain structural features on polymer flammability.

This explains the fact that the flammability tests based on estimates of the thermochemical parameters of polymeric systems (by direct calorimetric or indirect measurements) have been gaining more and more preference recently.

5 Retardation of Polymer Combustion

An analysis of the wealth of scientific and patent publications shows that all the efforts of affecting the combustion of polymeric materials are ultimately aimed at reducing the rate of heat release and increasing the rate of heat losses from the zone of the combustion reaction. In the light of the concept of the diffusion combustion process as a multistage process, extinction or retardation of combustion may generally be achieved by using active physical or chemical means at each stage.

Physical means of affecting the combustion process include: 1) retardation of heat supply to the polymer (e.g. heat insulation of its surface); 2) cooling of the combustion zone by increasing the physical drain of heat to the environment (for example, heat removal from a polymeric coating through a thermally conductive substrate, losses due to vaporization of components, removal of heat with the downward flowing melt); 3) hindrance of the transport of reactants to the flame zone (creating a physical barrier between the polymeric material and the oxidizing medium, slow-down of the diffusion of combustible ingredients in composite materials, material removal by dispersion); 4) extinguishing the flame by the incoming gas flow or explosion wave; 5) application of acoustic, magnetic, gravitational etc. fields.

Chemical means include: 1) modification of polymer morphology and structure; modification of composition and relative amounts of material components, causing a variation of condensed- and gas-phase reaction kinetics and mechanisms, at their interface; 2) affecting the flame with various chemical agents (gas-phase combustion inhibitors).

The externally applied physical and chemical measures of affecting a polymeric system are used in practice to extinguish a fire already under way. However, the problem of flame retardance of polymers actually implies using any measures to preclude or retard the development of fire. In other words, the physical and chemical factors must be effective within the system itself.

The leading mechanisms of flame retardance of polymers may be related to physical or chemical effects at any stage of the combustion. As a rule, the chemical influences (characterized by the rate constants of the respective reactions) are closely interrelated with the physical ones (characterized by heat- and mass-transfer parameters). Establishing the role of each factor and estimating its individual contribution to the overall effect is important for the development of ways of reducing the flammability of polymeric materials.

The methods currently used for improving flame retardance of polymeric materials are as follows:

1) preparation of polymers with a minimum content of combustible organic components or preparation of carbonizable thermally stable polymers;

2) chemical modification of polymers by introducing appropriate fragments, atoms or groups of atoms into the macromolecules, or by transformation of the polymers to more thermally stable types;

3) compounding with chemical agents having flame retarding activity.

The latter two methods are most frequently used at present. Since polymeric materials are produced on a very large scale, rendering them less combustible is of paramount importance.

Flame retardants are widely used for the purpose. The particular choice is based on technological, service, sanitary, hygienic, and other factors. Whether a flame retardant is to be used at all depends on its effectiveness and on the economy of the production and use of the polymeric material.

Studies on the development of flame retardant polymeric materials have been reviewed very extensively and discussed at many national and international conferences and symposia and in state-of-the-art reports and monographs [25, 35, 117, 124–128]. But, as it has been rightly noted by Tesoro [127], the problem is scientifically very complex and includes many unsolved aspects. This field of science is developing very rapidly thanks to the concentrated efforts of many specialists.

We considered it most rational to begin this paragraph with a discussion of the mechanistic concepts of the effects of flame retardants on the combustion of polymeric materials.

Flame retardant is a term usually applied to organic or inorganic substances which contain halogenes, phosphorus, nitrogen, boron, metals, groups containing these elements in various combinations, chemically bound water or carbon dioxide. Attempts have been made to classify flame retardants by dividing them into groups having common features [124–128]. Classifications were made according to: 1) presence of an element or a combination of elements having flame retarding activity; 2) main zone of the action, i.e. condensed- or gas-phase; 3) activity mechanism (physical or chemical). 4) other material factors (use of fillers, plasticizers, hardeners, blowing agents, surfactants, etc.); 5) intended use.

The activity mechanism of most flame retardants currently used has not been established unambiguously so far. Therefore, the division of flame retardants according to mechanistic principles (2,3 above) can be only tentative, and is often based on very general concepts and conjectures about the role of the flame retardant. Actually, the effect of a flame retardant on the combustion process is frequently related both to the condensed-phase and to the gas-phase processes.

The folowing features are used as macroscopic criteria of the activity mechanisms of flame retardants [129]:

1) variations of the composition of the volatile polymer decomposition products in the presence of a flame retardant;

2) variation of the nonvolatile residue (coke) yield;

3) ability of the flame retardant to be released from the polymer into the gas phase during combustion;

4) effect of a flame retardant on the nature of the gaseous oxidant and polymer substrate.

Moreover, the sensitivity of the effect of a flame retardant to the ambient pressure should also be taken into account. Flame retardants that are active only in the gas phase usually fail to affect the composition of the volatile pyrolysis products and the coke yield. In this case whatever the nature of the polymer, the flame retarding element is released into the gas phase during combustion; the type of oxidant (O_2/N_2, N_2O/N_2) strongly affects the flammability. On the other hand the effect of flame retardants active in the solid phase depends on the polymer nature, but is independent of the nature of the oxidant. Variations of the pressure of the oxidative environment affect the rates of gas-phase as well as heterogeneous (interfacial solid-gas) reactions.

The manner in which flame retardants affect the flammability at ambient pressure may obviously indicate which spatial region of the combustion event is most affected by the flame retardant. As the environment pressure increases, one may expect an increase of the contributions of the gas-phase and heterogeneous reactions and a decrease of the effectiveness of the flame retardant [130]. Only detailed studies of the effects of flame retardants on all steps of the complex combustion process allow the true activity mechanism to be determined. Thus, further investigations of the effects of flame retardants on pyrolysis, ignition, steady-state combustion and extinction, and on the thermochemical characteristics of the combustion of polymeric materials are being made. The behavior of the polymer and the chemical additive during decomposition influences all other flammability characteristics of a polymeric material. Synchronism of decomposition and gasification of the polymer substrate and flame retardant is considered as one of the major factors having a positive influence on most parameters of flammability. The effectiveness (a) of a flame retardant is estimated from the change of any parameter of material flammability, (PF), per unit concentration of the flame retardant (Y_{FR}):

$$a = \frac{(PF) - (PF)_0}{Y_{FR}} = \frac{\Delta(PF)}{Y_{FR}}$$

or from the relative change of the parameter of flammability per unit concentration of the flame retardant:

$$a = \frac{(PF) - (PF)_0}{(PF)_0 Y_{FR}} = \frac{\Delta(PF)}{(PF)_0 Y_{FR}}$$

Three groups of macroscopic parameters of flammability are used. The first group of parameters characterizes the critical ignition and extinction conditions (τ_{ign}, LOI, etc.), the second group characterizes the steady-state combustion (v, ṁ), and the third group characterizes the combustion heat released by the material under different conditions. The concentration of flame retardant in a system is expressed either by the concentration of the flame retardant itself, or by that of the element exibiting the flame retarding activity. The above parameters relate to the different aspects of the complex combustion process. Application of various criteria and the nonlinear variation of the flammability parameters with the concentration of the flame retardant are the most frequent causes of artifacts occurring in estimates of the flame retardant efficiency. Nevertheless, a comparative analysis of flame retardant effects on different flammability parameters is useful for the understanding of their activity mechanisms and of their role in each particular stage of the combustion process [117, 131].

Experience shows that there are no "all-purpose" flame retardants available. This is because the complex decomposition patterns of polymer and flame retardant at high temperature depend on the nature of these materials and their composition.

5.1 Mechanism of the Effect of Halogen-Containing Flame Retardants

The halogen-containing compounds most commonly used as flame retardants in polymeric materials are generally classified into three types: derivatives of aliphatic, cycloaliphatic and aromatic compounds. The compounds concerned may have high or low molecular weights and be of the reactive or additive type with respect to the polymer. Depending on the structure, the halogen-containing compounds either decompose in the condensed phase or evaporate and decay in the gas phase. Primary pyrolysis of halogen-containing compounds yields HX, X_2 and RX products where X is a halogen atom. Condensed-phase reactions succeeding the abstraction of HX, X, or RX from macromolecules may result in the formation of a nonvolatile carbonized residue. This means that less fuel will reach the flame. The flame retarding effectiveness of halogen-containing compounds of the same structure increases in the order $F < Cl < Br < I$.

Aliphatic compounds (RX_n) are more effective than aromatic compounds. There is sufficient evidence that in one form or another the halogen is released into the gas phase during combustion. Halogen-containing compounds behave in an overwhelming majority of cases as gas-phase flame retardants [124-126]. However, each particular compound may retard flaming combustion of polymers in its own special way.

In analogy with the manner in which flame inhibitors of gas systems are studied, the effects of flame retardants in polymer systems are analyzed by introducing them into the flame either on the fuel side [132] or on the oxidant side [133]. Flames of monomers (polymer decomposition products) are used as models [134].

The retarding effect of HX, X_2, RX_n introduced into the flame may be of either physical or chemical nature (in either case, the residue R in RX_n is an extra fuel source). In the first case, the flame retardant reduces the oxygen concentration in the combustible mixture in the flame reaction zone by mere dilution, in the same way as added carbon dioxide or nitrogen. The heat capacity of the resultant mixture determines the amount of heat drained off for its heating. In the second case, the flame retardant directly participates in the flame reactions and affects the complex combustion process kinetics. The HX molecule is the main inhibiting particle. When RX_n is released into the gas phase of fuel-rich flames, HX is formed predominantly via $RX_n + H \rightleftharpoons HX + RX_{n-1}$; when X_2 is released, HX is formed via $X_2 + H \rightleftharpoons HX + X$.

The inhibition of the free-radical fuel oxidation chain reaction occurs according to the following scheme:

$$H\cdot + HX \rightleftharpoons H_2 + X \tag{A}$$
$$O\cdot + HX \rightleftharpoons \cdot OH + X \tag{B}$$
$$\cdot OH + HX \rightleftharpoons H_2O + X \tag{C}$$

In fuel-rich flames such as polymer flames reaction (A) is the dominant one. The nature of the halogen atom strongly affects the equilibrium constant of these reversible reactions ($K = k_+/k_-$). Thus, for reaction (A) $K_{HBr(A)}$ increases with decreasing flame temperature and is 75, 1.6×10^3 and 4.19×10^5 at 1580, 1000 and 600 K, respectively; for $K_{HCl(A)}$ the corresponding values are 2.5, 2.9 and 3.7 [135]. From these data it is evident that HCl is a poorer inhibitor than HBr. Calculations show that the back reaction in (A), though apparently detrimental for the overall inhibiting effect of HX, is beneficial at high degrees of conversion of the fuel to the combustion products [136].

The retardation effect increases further due to regeneration of HX and formation of a secondary inhibitor X_2 [135, 136]:

$$X + X + M \rightleftharpoons X_2 + M \tag{D}$$
$$H + X + M \rightleftharpoons HX + M \tag{E}$$
$$X + HO_2 \rightleftharpoons HX + O_2 \tag{F}$$
$$X_2 + H \rightleftharpoons HX + X \tag{G}$$

Dixon-Lewis [135] found that the inhibition of hydrogen-air flames with HBr causes a considerable reduction of the heat release rate. However, one of the main draw backs of the chemical inhibition with halogens is that large amounts of the fire retardant (12–15% of Br and 20% of Cl are required to achieve a marked fire retarding effect [137].

Larsen [137] has shown for the combustion of halobenzenes and hydrocarbon systems, with the supply of halogen-containing compounds on the fuel side, that the minimum weight percentage of halogen at the combustion limit must be as high as 70–80%. The effectiveness of F, Cl, Br and I is the same if compared on a mass rather than on a volume basis. In the latter case, it is proportional to the atomic weights of the halogens.

The function of halogens in reducing the flammability of polymers consists simply in increasing the rate of the supply of matter to the flame without a parallel substantial increase of the heat of combustion and consequently of the heat flux from the flame to the fuel. In experiments involving feeding of CF_3Br into a flame of poly(methyl methacrylate) on the oxidant side it was found [133] that this compound is more active than nitrogen. The ultimate effect depends on the flame direction (a candle-like flame, or a flame which completely encompasses the sample). Thus, the minimum concentration of CF_3Br to be added to the atmosphere (O_2/N_2) at the extinction limit of PMMA is 5% and 10% by weight for candle-like and upward flames, respectively. This is almost half the extinguishing concentration of nitrogen (Fig. 17, curves 2, 3). Thus, CF_3Br is a more effective extinguishing agent than N_2. This result indirectly confirms that CF_3Br acts by chemical means in the extinction of the flame. Y_{ex} (abscissa of Fig. 17) is the mass fraction of extinguishing agent in the atmosphere. It is interesting that at constant oxygen concentration in the atmosphere the mass rate of polymer burning in a CF_3Br-containing atmosphere ($\dot{m}$) is even higher than that in a O_2/N_2 atmosphere ($\dot{m}_0$); the flame is not extinguished (Fig. 17, curve 1). This shows that CF_3Br functions both as a physical and a chemical retardant [133]. The problem is to separate the two effects. An attempt in this direction has been made recently [138]. It was based on the changing limiting oxygen index with the addition

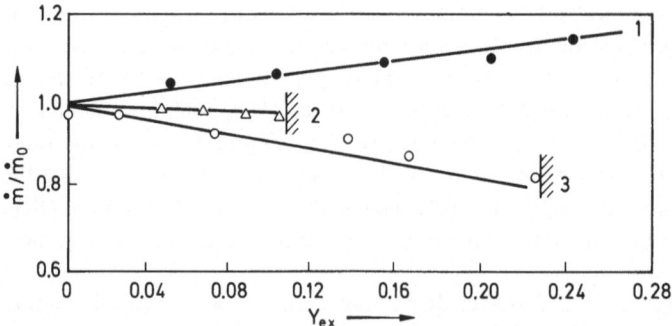

Fig. 17. Effect of the concentration of the extinguishing agent on the relative burning rate of PMMA and flame extinction: 1) $O_2/N_2/CF_3Br(Y_{O_2} = 0.233)$; 2) $CF_3Br(Y_{O_2}/Y_{N_2} = 0.233/0.767)$; 3) $N_2(Y_{O_2}/Y_{N_2} = 0.233/0.767)$; For further explanations see text. (After Ref. [133], with permission)

of a flame retardant. The change of enthalpy of the mixture containing oxygen, nitrogen, a combustible and an extinguishing agent was taken into account. It was shown that CF_3Br functions merely as a diluent, i.e. by physical means, when its concentration is high.

5.2 Mechanism of the Effect of Phosphorus-Containing Flame Retardants

Phosphorus as well as its organic and inorganic derivatives (with tri- or pentavalent phosphorus) are widely used for flame retardation in polymeric materials. Like halogen-containing compounds, phosphorus compounds (mainly organic) may be reactive or additive with respect to the polymer substrate. The introduction of phosphorus-containing fragments into polymeric macromolecules affects the decomposition process to a higher degree than mere compounding. In the latter case, the pyrolytic process in the composite is controlled by the interactions between the polymer and the additive. The activity mechanism of phosphorus-containing compounds has not been established unambiguously. Phosphorus-containing compounds may affect combustion both in the condensed and gas phase. It has been found, for example, that if phosphate groups are introduced into the phenyl groups of polystyrene, all phosphorus will remain in the nonvolatile residue during combustion. The limiting oxygen index of such polymers varies with the phosphorus concentration in the same way in experiments performed in different oxidizing atmospheres, i.e. the effect of the flame retardant is due to its activity in the condensed phase [139]. On the other hand, if p-tert-butyl diphenyl phosphate is used as an additive of PVC in plastisol adhesives, 95% of the total phosphorus is transferred to the gas phase and only 5% remains in the coke. Thus, this phosphate ester is a gas-phase flame retardant. An analysis of a similar coke residue from a formulation containing PVC-plastisol and isodecyl-diphenyl phosphate, however, shows that up to 50% of the phosphorus is retained by the condensed phase [140]. The volatility of the starting P-compounds and of the P-containing products formed as a result of the interaction with the polymer, and of subsequent decomposition, determines the penetration of phosphorus into the gas phase. It is evident that the temperature on the burning surface of the polymer is extremely important. The organic groups attached to the phosphorus, especially if they contain only a small number of carbon atoms, have a favorable effect on the volatility of the P containing flame refardant. The boiling point (T_b) of aliphatic esters decreases in the order: phosphates > phosphonates > phosphites. For instance, the T_b values of $(C_2H_5O)_3PO$, $(C_2H_5O)_2P(O)C_2H_5$ and $(C_2H_5O)_3P$ are 215, 198 and 158 °C respectively. The boiling points of the phosphine analogs R_3P are even lower. Aromatic substitutents raise T_b (for $(C_2H_5O)_2P(O)CH_2C_6H_5$, $T_b = 270$ °C and for $(C_6H_5O)_3P$, $T_b = 370$ °C). The volatility of incompletely sub- stituted phosphorus compounds drops rapidly with the number of free acid groups. These compounds are in the associated state and easily react with each other upon heating, yielding polyphosphorus compounds. The volatility also decreases with increasing molecular mass. The more complicated the structure of the organic com- pounds, the more difficult it is to directly transfer them into the gas phase and the more probable becomes the decomposition of these substances in the condensed phase before they can be gasified. Phosphorus containing flame retardants are effective both according to chemical as well as physical mechanisms, both in the condensed and gas phase. The chemical activity mechanism of P-containing flame retardants is assumed to be related to a variation of the pyrolysis process by enhancing car- bonization, to the reduction of the amount of the volatile gases reaching the flame zone, and to the inhibition of the heterogeneous oxidation of the coke residue and of flame reactions. The physical effects include simple dilution, flame cooling, and the formation of a layer of polymetaphosphoric acid on the polymer surface. This layer serves as a physical barrier to the heat transfer from the flame to the polymer, and also to the diffusion of the reactants. The coke layer is also regarded as a barrier of this kind which reduces the heat transfer from the flame and the mass transfer of fuel and oxidant. Recently, some evidence has been obtained supporting earlier

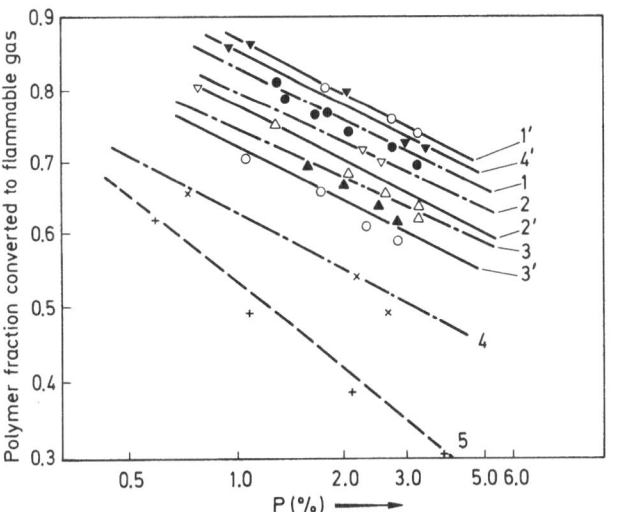

Fig. 18. Effect of flame retardants on the yield of volatile components in the combustion of cotton fabric: 1, 1') Ph_3P and Ph_3PO; 2, 2') $Ph_2P(OPh)$ and $Ph_2P(O)(OPh)$; 3, 3') $PhP(OPh)_2$ and $PhP(O)(OPh)_2$; 4, 4') $P(OPh)_3$ and $P(O)(OPh)_3$; 5) H_3PO_4 (After Ref. [143] with permission)

conjectures about the activity mechanism of P-containing flame retardants. Calorimetry was used to evaluate the effect of phenyl derivatives of tri- and pentavalent phosphorus on the flammability of cotton farbrics [141]. The addition of P-compounds reduces the amount of flammable gases released during the combustion (see Fig. 18). However, it is not only the volatility of the products what determines the effect of the flame retardant, but also their reactivity with respect to the cellulosic polymer. Triphenyl phosphite is much more active than triphenyl phosphate. The latter is inferior to its phosphoramide analog. The effectiveness of phosphoramide-containing flame retardants increases with the number of phosphoramide groups [142].

For cellulose it has been shown quite convincingly that P-containing flame retardants promote polymer dehydration and thus increase the coke yield in combustion and pyrolysis. Polymer dehydration includes a cellulose phosphorylation stage [142]. Phosphorus-containing flame retardants have stronger inhibiting effect on polymers which exhibit a tendency for carbonization in pyrolysis. They promote dehydrohalogenation, dehydrocondensation, cyclization, crosslinking, etc., i.e. reactions leading to the formation of a carbon skeleton. The coke yield of such polymers is much higher. However, in most cases the decomposition is accelerated at the initial stages, or the temperature at which decomposition starts is decreased. At later stages, the generation of volatile substances is slowed down. Treatment with P-containing flame retardants of polymers displaying no tendency for carbonization also accelerates decomposition. However, at later stages, carbon skeleton formation seems to compete with volatilization of the low molecular weight products [143]. Direct analysis of the carbonized residues of burned or pyrolyzed polymers by X-ray electron spectroscopy (ESCA) has verified the formation of polyphosphoric acid-related PO_4 fragments from the P-containing flame retardants. The acid is localized in the surface layer and carbon is accumulated on the surface itself. Phosphorus migration is observed under certain conditions [144]. Scanning electron spectroscopy [145] failed to distinguish between a carbonized layer and any kind of a surface layer that might look like a film of polyphosphoric acid. In carbonized silicon polycarbonates a certain surface layer has been detected [116]. Gas-filled agglomerates and inclusions (nodules) were discovered in the surface layer of carbonized cotton fibers treated with phosphoramide fire retardants. Their size decreases when more effective flame retardants are used. It has been hypothesized that this might be due to a decreasing amount of volatile polymer decomposition products captured by the softened surface layer. In this case, a foamed carbonized layer is formed which may contain up to 75% of phosphorus.

A number of nitrogen derivatives of phosphoric and polyphosphoric acid (ammonium polyphosphate, melamine pyrophosphate) are used for improving the flame retardance of polyurethanes and other polymers. In thermal decomposition these compounds produce ammonia and the corresponding phosphoric acids which catalyze dehydration and other reactions, causing polymer dehydration during combustion. The coke produced in this process is more or less foamed. Ammonium polyphosphate and melamine pyrophosphate are added to compositions of intumescent coatings used for fire protection of various structural elements in construction [146, 147].

The effectiveness of coke as a heat barrier retarding heat transfer to the undecomposed polymer depends primarily on the thermophysical properties of coke. Semiempirical models of intumescent coatings [148, 149] allow the coke effectiveness to be estimated as a function of such factors as layer thickness, thermal conductivity of the carbon foam, enthalpy of coking, convective heat transfer by the decomposition products, heat radiation, or reflection by a surface.

Gibov et al. [150] found that liquid pyrolysis products and molten polymers are capable of ascending to the surface through the pores in the carbonized layer, by the action of surface tension forces (capillary effect). In this manner, the flaming combustion of carbonizable polymers may be sustained. The velocity of ascension of the liquid which affects the polymer combustion rate follows Darcy's law and depends on the coke porosity, layer thickness, viscosity and surface tension of the liquid. Treatment of a coke layer with phosphorus-containing compounds reduces the surface tension and the velocity of liquid flow through the coke layer; as a consequence, retardation of the combustion occurs. A highly foamed carbonized layer however, falls short of exhibiting the expected heat insulating properties; oxygen easily penetrates into the polymer surface and the combustion rate is not reduced.

The effect of volatile phosphorus-containing compounds in polymer flames has not yet been studied. For gas systems it has been found [151] that the flame retarding effect of the additives depends on the concentration (trimethyl phosphate, phosphorus halides and thiohalides). HPO fragments were detected in the radiation spectrum of a H_2/O_2 flame, upon the introduction of trimethyl phosphate [152]; P_2, PO, PO_2 and HPO, P and PH fragments in low concentrations were discovered in a CH_4/O_2 flame inhibited with triphenylphosphine oxide. As regards the flame retardation effect, phosphorus in triphenylphosphine occupies an intermediate position between Sb and As. Triphenyl phosphate is less effective than $SbCl_3$. Most effective are $POCl_3$, PCl_3 and PBr_3.

According to Hastie [129], flame retardation is due to the following reactions:

$$H + PO + M \rightleftharpoons HPO + M$$
$$HPO + H \rightleftharpoons H_2 + PO$$

In the flame front where the concentrations of O and OH radicals are commensurate with or greater than that of H, the following reactions are possible [153]:

$$PO + OH \rightleftharpoons HPO + O$$
$$PO + O (+ M) \rightleftharpoons PO_2 (+ M)$$
$$PO + OH (+ M) \rightleftharpoons HOPO (+ M)$$

The maximum flame temperatures of polymers to which P-containing flame retardants have been added are normally lower than those of unmodified polymers. The low LOI values (18–26 %) of the volatile low molecular weight organophosphorus compounds (phosphates, phosphonates, phosphites) indicate their insufficient activity in the gas phase [154, 155].

5.3 Mechanism of the Effect of Metal-Containing Flame Retardants

More than 4/5 of the elements of the Periodic Table are metals or metalloids. If a substance contains metal atoms, it is impossible to state a priori, from this fact alone, whether or not it will exhibit the desired effect of flame retardance in a particular polymer system. The metal-containing compound and the polymer must be considered as one system. The possible participation of the metal-containing compounds in chemical reactions is related to the electronic structure of the constituent atoms and the bonds between them. Among metal containing compounds (hydroxides, oxides, inorganic and organic acid salts, complexes), those with electron density in the outer s- and p-orbitals of the metal atom have received the widest application as flame retardants of polymeric materials. They include alkaline earth metal hydroxides and carbonates, aluminum hydroxides, oxides and other compounds of antimony and tin, as well as borates. In recent years, there has been also interest for compounds of d-type metals and of some f-type metals. The d- and f-type atoms actively participate in electron-transfer reactions and donor-acceptor complexing. Depending on the type of the system and the conditions, ions of these metals function either as oxidants or reducing agents. Many factors control the effect of these compounds in polymer combustion: The nature of the metal, degree of oxidation, type of compound, manner of introduction into the polymer, concentration, conditions of combustion and pyrolysis and the nature of the polymer are important. Metal compounds may affect processes taking place in the condensed phase, gas phase and at the interface. Aluminum hydroxide is by far the most commonly used inorganic metal compound. It has been found that $Al(OH)_3$ reduces the flammability of polyolefins, polyacrylonitrile and polystyrene, as well as epoxy- and polyester-based materials [156, 157]. The similar response of aluminum hydroxide-filled epoxy resins in O_2/N_2 or N_2O/N_2 atmosphere led to the conclusion that the flame retardant is mainly active in the condensed phase [157]. In order to achieve a tangible flame-retardant effect, metal hydroxides and carbonates have to be added to the polymers in large amounts, which is an indirect indication of the predominantly physical mechanism of their effect. It has been established that metal hydroxides and carbonates ($Al(OH)_3$ and $CaCO_3$) undergo endothermal decomposition producing water or carbon dioxide. Their effect is attributed to the large amount of heat required for the decomposition of the flame retardant and the vaporization of the decomposition products (surface and flame cooling). Recently, it has been shown that the addition of 50 to 80% of aluminium hydroxide to ultrahigh molecular weight polyethylene permits to obtain plastics with an LOI of up to 45% and with other flammability and smoke generation parameters close to those of thermally stable aromatic polymers [158].

It is now conventional to use antimony oxides, or other Sb- compounds, as synergists with halogen-contaning flame retardants. The synergistic effect of the antimony compounds is correctly attributed to the formation of volatile antimony trihalides SbX_3 which affect the flame reactions [159, 160]. The more $SbCl_3$ is released into the gas phase, the higher is the LOI of a polymeric material [4].

The maximum flame-retardance is obtained with an antimony to halogen atomic ratio of 1:3 [161]. An intermediate product of the reaction of Sb_2O_3 with a halogen-containing polymer substrate, or with a flame retardant, is an antimony oxyhalide. Pitts found that other oxides may either promote (Fe_2O_3, CuO, TiO_2) or hinder (ZnO, CaO, MgO) the decomposition of antimony oxyhalides. This, in turn, enhances or diminishes the overall flame-retardant effect [161].

It has been shown [162] that volatile metal halides with relatively weak Me-X bonds are extremely effective flame retardants. Di- or trivalent iron halides in very low concentrations (1–5 ppm) considerably reduce the flammability of ABS plastics (the LOI is increased from 18.5 to 30). In combination with halogen-containing compounds, oxides and salts of Mn^{2+}, Sn^{4+}, Mo^{6+} and other metals also reduce the flammability of ABS plastics. The activity of $FeCl_3$ is much higher than that of $SbCl_3$. Metal compounds are also used as the basic components of effective fire extinguishing powders [163]. Complex as their activity mechanism is, there is no doubt that many compounds may become involved in the inhibition of flame reactions. Lead tetraethyl, ferrocene, iron and manganese carbonyls are also known to inhibit flame reactions [164].

The promotion of the decay of active particles (H atoms and OH-radicals) in the presence of various metal compounds has been repeatedly verified [165]. Which metal particles do actually react with the atoms and radicals propagating the chain reactions in a flame? Hastie [129] studied flames of $SbBr_3$-inhibited methane and discovered CH_3Br and HBr in the preflame zone, and Sb and SbO in the high-temperature reaction zone. Thus, halogen as well as metal are supplied to the gas phase. Both are capable of reacting with the active particles in the combustion reaction, embracing extensive flame areas. According to Ref. [129], the metal reacts via the following route:

$$Sb + O + M \rightleftharpoons SbO + M$$
$$Sb + OH + M \rightleftharpoons SbOH + M$$
$$SbO + H \rightleftharpoons SbOH$$
$$SbOH + H \rightleftharpoons SbO + H_2$$

Volatility and reactivity of the metal-containing compound determine the probability of its participation in flame reactions. Since metal halides are easily hydrolyzed they are of no practical importance.

An alternative activity mechanism of metal compounds is their participation in pyrolytic reactions in the condensed phase, i.e. promotion of carbonization and reduction of the yield of volatile products.

It has been shown that it is possible to effectively reduce the flammability of many polymeric materials by using metal compounds in very low concentrations. Metals and their compounds may be included in the polymeric macrochains, either coordinatively bonded to functional groups of the polymer or used as additives. For example optically transparent, low-flammability polycarbonates have been obtained by adding as little as 0.001 to 2 % by weight of alkali or alkaline earth metal sulfonates [166].

The flammability of wool products is reduced by modification with K_2ZrF_6 and K_2TiF_6 in combination with alkali metal isopolymolybdates, vanadates, tungstates, and complexes of these alkali metal salts with organic and inorganic acids. Such a modification increases the LOI from 24.5 to 31.5 %. It has been found that the flammability of wool is the lower the smaller the radius of the metal ion and, correspondingly, $(MeF_6)^{-2}$ ions are effective in the series W > Mo > V [167]. Judging by macroscopic criteria, titanium and zirconium complexes are active in the condensed phase. They catalyze the decomposition of the natural polymer during the combustion [168]. The effect of other metal salts on the flammability of wool has been considered in Friedman's review [117, p.229].

Certain polymers are able to bind metal ions by chelating. The heterogeneous oxidation during combustion is thereby affected. An example is the polymer formed by polycondensation of oxalic acid bis(amidohydrazone) and terephthalic acid dichloroanhydride [169].

The amino groups as well as oxygen and hydrogen are involved in the chelating of the metal ions. During combustion, this polymer produces N_2, CO_2, H_2O, metal oxides, and a coke residue. The flammability of fibers from this polymer is reduced by chelated Ca, Zn, and Zr ions (LOI $\simeq 40\%$), whereas Cu and Fe ions catalyze the complete combustion of the fibers, under the effect of a 1500 °C flame.

The flammability of silicone elastomers is considerably decreased if various platinum complexes are applied to the filler surface. Nonvolatile Pt compounds in catalytic concentrations (1–30 ppm) increase the yield of the coke and affect its structure and properties [170].

McCarter [171] studied the activity of 185 inorganic salts and oxides in flaming combustion and smouldering of cellulosic fabrics. Powders of these substances were dispersed onto the fabric surface. The effect of the nature of the metal (alkali and alkaline earth metals, p- and d-type metals in different states of oxidation) and anion (sulfates, nitrates, halides, borates, phosphates, acetates) has been reported. Smouldering is inhibited by salts whose anions contain B, P, S, $(SiF_6)^{2-}$, halides. Flaming combustion is retarded by salts of Fe. Alkali metals, Mg, Zn, Pb, Al, Ca and Ba salts are less effective. About 80 compounds inhibit both flaming combustion and smouldering. A difference has been noted in the activity of compounds applied as a powder or as an impregnating solution. Solutions of dichromates, permanganates and perchlorates even enhance the flaming combustion of cellulose. Košik et al. [172] found that impregnation of cellulose with $ZnCl_2$ increases the LOI from 19 to 35%. $ZnCl_2$ promotes the thermooxidative decomposition of the polymer at low temperatures. At 300–400 °C the mass loss is practically the same in an inert or oxidative atmosphere. At higher temperatures cellulose carbonization is enhanced.

Various organic compounds (phenolates, acetylacetonates, salts of organic acids, complexes) of d-type metals, e.g. Zn, Mo, Fe, promote the coking of clorine- and bromine-containing polymers, inhibiting flamming combustion and smoking. Thus, the addition of zink pyromellitate (1–5 ppm) to poly(vinyl chloride) increases the coke yield by almost a factor of two, reduces the yield of combustible volatile products by 39% and brings the LOI from 50 to 72–75. The coking promotion effect correlates better with the boiling point than with the strength of the Lewis acids of the metal halides produced.In other words, the less volatile metal halides retained in the condensed phase promote the cross-linking by dehalogenation of the polymer [173]. It is interesting to note that the use of a combined system of oxidant/Cl-containing product/metal oxide did not cause any tangible reduction of polystyrene flammability despite the almost two fold increase of the coke yield. Most of the Fe^{3+} (though not as $FeCl_3$) remained in the condensed phase [173]. This fact highlights the importance of the type of metal compound reacting with the polymer, as well as the nature of the polymer. With PMA minor amounts of $ZnCl_2$ reduce the decomposition rate at 250–300 °C both in air and in the absence of oxygen, because $ZnCl_2$ promotes cyclization and cross-linking in the case of this polymer, whereas it accelerates considerably the degradation of PMMA under the same conditions [174].

Small amounts of Co^{2+}, Ni^{2+}, Mn^{2+} and Fe^{2+} promote the thermooxidative decomposition of polystyrene and reduce its ignition time in air (Fig. 19) [175]. The effect increases with the redox potential of Me^{2+}. Only Ni^{2+} does not fit neatly into this relationship. It was found that at high self-ignition temperatures of PS, cobalt salicylate is converted to Co_3O_4. It is assumed that metal complexes with varying valence of the metal atom increase the rate of processes involving electron transfer:

$$Me^{2+} \dashrightarrow Me^{3+} + e^-$$

$$\sim\!\!\!\sim Re + e^- \dashrightarrow \sim\!\!\!\sim R^{\cdot}$$

$$Me^{3+} + \sim\!\!\!\sim R^{\cdot} \dashrightarrow Me^{2+} + \sim\!\!\!\sim R$$

Several Co^{2+}, Sn^{2+}, Pr^{3+}, Cu^{2+}, Cu^{1+}, and Fe^{3+} compounds were used for flame retardation of epoxy-based polymers [176]. Zarkhina et al. [177] have shown that Mn^{2+}, Cr^{3+} and Ce^{3+} compounds reduce the temperature of initiation of the thermooxidative decomposition of the same polymer by 80–100 °C, but the rate of the decomposition itself is lower than that of the unmodified polymer. Thermogravimetric and differential thermal analysis have shown that Me^{n+} have no influence on the thermal decomposition of the polymer in the absence of an oxidative atmosphere.

The available data concerning the effect of metal compounds on the pyrolysis and combustion of polymeric materials show that there is a great potential in the control of these materials. This field is still in its embryonic stage. Much is yet unclear. In many cases the analysis is rendered difficult because of the dual function of metal compounds (especially of the d- and f-types) in redox reactions. Their ability not only to inhibit but also to promote polymer decomposition manifests itself most clearly in concentration effects [178]. The specific effects of certain metal compounds on the ignition and combustion rate of polymers indicate that heterogeneous oxidation

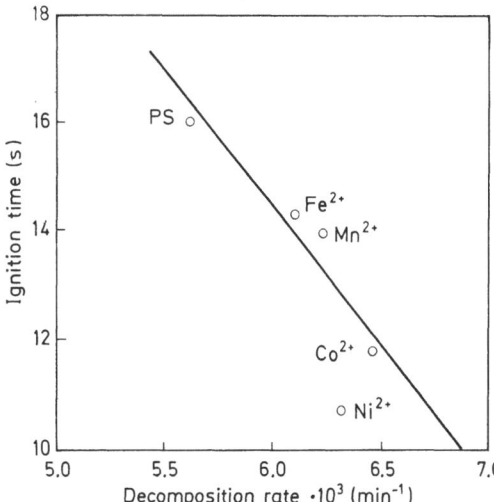

Fig. 19. Ignition time as a function of the rate of decomposition of polystyrene in the presence of transition metal salicylates (After Ref. [175] with permission)

reactions do play a certain role. Variable valence metal ions are known to promote hydroperoxide decomposition by a radical or molecular mechanism involving the formation of intermediate complexes: [178,179]

$$nROOH + mMe^{(n+1)+} \rightarrow [nROOH...mMe^{(n+1)+}] \rightarrow RO_2^\cdot + H^+$$

$$+ Me^{n+} + (n-1)\,ROOH + (m-1)\,Me^{(n+1)+}$$

$$nROOH + mMe^{n+} \rightarrow [nROOH...mMe^{n+}] \rightarrow RO^\cdot + OH^- + Me^{(n+1)+}$$

$$\downarrow \quad + (n-1)\,ROOH + (m-1)\,Me^{n+}$$

$$[Me^{(n+1)+}, OH^-, RO^\cdot] \rightarrow Me^{n+} + R'C(O)\,R'' + H_2O$$

In the highest valence state $Me^{(n+1)+}$ may react with radicals R, RO, $RO_2^\cdot$, and promote the decomposition of aldehydes, acids, alcohols, and ketones.

The retarding effect of metal compounds on the flaming combustion of polymers must be primarily attributed to the decrease of the yield of flammable volatile pyrolysis products. There are two alternative paths: 1) catalysis of the formation of noncombustible products (CO_2, H_2O); 2) inhibition of polymer decomposition at the high temperatures developed during steady-state combustion (high-temperature stabilization) [180].

6 Conclusions

The large volume of production and the versatile use of polymer materials calls for a classification of these materials according to fire hazards.

The combustion is an extremely complex process including many chemical and physical phenomena of transformation of matter. The need and desire to know and control this process urges man to study its various aspects. Organic polymers are but one example of the multitude of materials used by man. They possess peculiar features and properties which individually affect the material behavior in a critical fire situation. It is, therefore, important to study the flammability characteristics of polymeric materials and the factors affecting them.

The general aspects of ignition, combustion and extinction in polymeric materials have been considered. There is a relationship between the chemical nature of the polymer and its flammability, which can be estimated from the influence of the polymer structure on the stoichiometry and the specific heat of the combustion reaction.

This review demonstrates that only theoretical investigations supported by experiments can open up scientifically based ways of creating flame retardant polymeric materials. This field of materials science is still in its early development stage. More profound investigations of the mechanisms and kinetics of high-temperature polymer decomposition processes, formation of a new phase on the burning surface (e.g. a carbonized layer) and of the effect of this phase on the heat and mass transfer in combustion are necessary. Evidently, the question of oxygen participation in polymer decomposition during combustion is not only of theoretical but also of practical importance. There is virtually no information available about the effective kinetic parameters of the gas-phase flame reactions of polymers and the effects of flame retardants on these parameters. Extension of work in this direction and investigations of the structure of inhibited polymer flames will help towards the understanding of the activity mechanisms of flame retardants.

Halogen-, phosphorus- and metal-containing substances as flame retardants of polymers may exert their action at the different stages and zones of combustion, by chemical as well as by physical mechanisms.

Metal-containing compounds appear to be the most promising flame retardants. Research in this direction is still in the nascent state but the very first results demonstrate the high effectiveness of this type of flame retardants and their ability to control the combustion process on a catalytic level.

7 References

1. Frank-Kamenetsky, D. A.: Diffusion and Heat Transfer in Chemical Kinetics. Moscow: Nauka 1967 (in Russian)
2. Martin, S.: 10th Internat. Symp. on Combustion, p. 877, Pittsburgh: The Combustion Institute 1965
3. Stuetz, D. E. et al.: J. Polym. Sci., Pol. Chem. Ed. *13*, 585 (1975)
4. Fenimore, C. P., in: Flame Retardant Polymeric Materials. Lewin, M., Atlas, S. M., Pearce, E. M. (eds.), p. 371, New York, London: Plenum Press 1975
5. Jakes, K. A., Drews, M. J.: J. Polym. Sci., Polym. Chem. *19*, 1921 (1981)
6. Isakov, G. N., Grishin, A. M.: Fizika Goreniya i Vzryva, *10*, 191 (1974)
7. Muton, N., Hirano, T., Akita, K.: 17th Internat, Symp. on Combustion, p. 1183, Pittsburgh: The Combustion Institute 1979
8. Ohlenmiller, T. J., Summerfield, M.: AIAA Journal *6*, 878 (1968)
9. Grishin, A. M., Ignatenko, N. A.: Fizika Goreniya i Vzryva, *7*, 510 (1971)
10. Merzhanov, A. G., Averson, A. E.: Combustion and Flame *16*, 89 (1971)
11. Averson, A. E., Barzykin, V. V., Martemyanova, T. M.: Fizika Goreniya i Vzryva *10*, 498 (1974)
12. Averson, A. E., in: Heat- and Mass-Transfer in Combustion, Stesik L. N. (ed.), p. 16, Chernogolovka: Branch of the Institute of Chemical Physics Academy of Sciences of the USSR 1980 (in Russian)
13. Librovich, V. B.: Zhurn. Prikl. Mekh. i Tekhn. Fiz. N2, 36, 1968
14. Kanury, A. M.: Combust. Sci. Technol. *8*, 171 (1974)
15. Kashiwagi, T.: Combust. Sci. Technol. *8*, 225, (1974)
16. Setchkin, N. P.: J. Res. NBS *43*, 591 (1949)
17. Fire Hazard of Materials Used in Chemical Industry, Ryabov I. V. (ed.), Moscow: Khimiya 1970
18. Kashiwagi, T.: Combust. Sci. Technol. *20*, 225 (1979)
19. Deverall, L. I., Lai, W.: Combustion and Flame *13*, 8 (1969)
20. Kanury, A. M.: Combust. Sci. Technol. *16*, 89 (1977)
21. Kindelan, M., Williams, F. A.: Combust. Sci. Technol. *10*, 1 (1975)
22. Kindelan, M., Williams, F. A.: Acta Astronautica *2*, 955 (1975)
23. Kindelan, M., Williams, F. A.: Combust. Sci. Technol. *16*, 47 (1977)
24. Niioka, T., Williams, F. A.: Combustion and Flame *29*, 43 (1977); Combust. Sci. Technol. *18*, 207 (1978)
25. Hilado, C. J.: Flammability Handbook for Plastics Stanford, Connecticut. Technomic Publishing Co. 1969
26. Miller, B., Martin, I. R., Meiser, C. H.: J. Appl. Polym. Sci. *17*, 629 (1973)
27. Korshak, V. V. et al.: Vysokomol. Soed. *23*, 789 (1981)
28. Aseeva, R. M., Serkov, B. B., Zaikov, G. E.: Vysokomol. Soed. *23*, 868 (1981)

29. Baillet, C., Delfosse, L., Lucquin, M. in: The Combustion Institute European Symposium, p. 148, London: Academic Press 1973, Europ. Polym. J. *17*, 779, 787, 791 (1981)
30. Rychlý, J., Matisová-Rychlá, L., Špilda, J.: Europ. Polym. J. *15*, 565 (1979)
31. Delfosse, L.: J. Macromol. Sci. Chem. *11A*, 1491 (1977)
32. Lecomte, L. et al.: J. Macromol. Sci. Chem. *11*, 1467 (1977)
33. Bert, M., Michel, A., Guyot, A.: Fire Research *1*, 301 (1977)
34. Descamps, J. M., Delfosse, L., Lucquin, M.: Fire and Materials *4*, 37 (1980)
35. Aseeva, R. M., Zaikov, G. E.: Combustion of Polymeric Materials, Moscow: Nauka 1981 (in Russian)
36. Hallman, J., Welker, J. R., Sliepcevich, C. M.: SPE Journal *28*, 43 (1972)
37. Ohlemiller, T. J., Summerfield, M.: 23th Internat. Sympo. on Combustion, p. 1087, Pittsburgh: The Combustion Institute 1971
38. Niioka, T., Williams, F. A.: 17th Internat. Symp. on Combustion, p. 1163. Pittsburgh: The Combustion Institute 1979
39. Hilado, C. J.: J. Fire and Flammability *6*, 130 (1975); *10*, 227 (1979)
40. Monakhov, V. T.: Methods of Investigating Fire Hazard of Materials. Moscow: Khimiya, 1979 (in Russian)
41. Maček, A.: Combust. Sci. Technol. *21*, 43 (1979)
42. Morimoto, T., Mori, T., Enomoto, S.: J. Appl. Polym. Sci. *22*, 1911 (1978)
43. Miller, B., Martin, J. R. in: Flame-Retardant Polymeric Materials. Lewin, M., Atlas, S. M., Pearce, E. M. (eds.), vol. 2, p. 63. New York, London: Plenum Press 1978
44. Weibel, R. T., Essenhigh, R. H.: 14th Internat. Symp. on Combustion, p. 1413, Pittsburgh: The Combustion Institute 1973
45. Holve, D. J., Sawyer, R. F.: 15th Internat. Symp. on Combustion, p. 351, Pittsburgh: The Combustion Institute 1974
46. Williams, F. A.: 16th Internat. Symp. on Combustion, p. 1281, Pittsburgh: The Combustion Institute 1976
47. Magee, R. S., McAlevy, R. F. III: J. Fire Flammability *2*, 271 (1971)
48. Sirignano, W. A.: Combust. Sci. Technol. *6*, 95 (1972)
49. De Ris, J. N.: 12th Internat. Symp. on Combustion, p. 241, Pittsburgh: The Combustion Institute 1969
50. Lastrina, F. A., Magee, R. S., McAlevy, III, R. F.: 13th Internat. Symp. on Combustion, p. 935, Pittsburgh: The Combustion Institute 1971
51. Lalayan, V. M., Khalturinsky, N. A., Berlin, A. A., Vysokomol. Soed. *21 A*, 11 (1977)
52. Rybanin, S. S.: Dokl. AN. SSSR *235*, 1110 (1977)
53. Rybanin, S. S., Sobolev, S. L., Stesik, L. N.: Chemical Physics of Combustion and Explosion: Combustion of Condensed Heterogeneous Systems, p. 32, Chernogolovka: Branch of the Institute of Chemical Physics Academy of Sciences of the USSR 1980 (in Russian)
54. Sibulkin, M., Hansen, A. G.: Combust. Sci. Technol. *10*, 85 (1975)
55. Lalayan, V. M. et al.: Vysokomol. Soed. *22B*, 150 (1980)
56. Lalayan, V. M.: Experimental Study of Laminar Flame Propagation on a Polymer Surface. Moscow: Institute of Chemical Physical Academy of Sciences of the USSR 1980
57. Fernandez-Pello, A., Williams, F. A.: 15th Internat. Symp. on Combustion, p. 217, Pittsburgh: The Combustion Institute 1974
58. Sibulkin, M. et al.: Combust. Sci. Technol. *14*, 43 (1975)
59. De Ris, J.: 17th Internat. Symp. on Combustion, p. 1003, Pittsburgh: The Combustion Institute 1979
60. Hirano, T., Noreikis, S. E., Waterman, T. E.: Combustion and Flame *22*, 353 (1974)
61. Hirano, T., Noreikis, S. E., Waterman, T. E.: Combustion and Flame *23*, 83 (1974)
62. Fernandez-Pello, A. C., Santoro, R. J.: 17th Internat. Symp. on Combustion, p. 1201, Pittsburgh: The Combustion Institute 1979
63. Sibulkin, M., Kim, J.: Combust. Sci. Technol. *17*, 39 (1977)
64. Fernandez-Pello, A. C.: Combustion and Flame *31*, 135 (1978)
65. Markstein, C. H., De Ris, J.: 14th Internat. Symp. on Combustion, p. 1085, Pittsburgh: The Combustion Institute 1973
66. Orloff, L., De Ris, J., Markstein, G. H.: 15th Internat. Symp. on Combustion, p. 183, Pittsburgh: The Combustion Institute 1975

67. Orloff, L., Modak, A. T., Alpert, R. L.: 16th Internat. Symp. on Combustion, p. 1345, Pittsburgh: The Combustion Institute 1977
68. Fernandez-Pello, A. C.: Combust. Sci. Technol. *17*, 1 (1977)
69. Kashiwagi, T., Newman, D. L.: Combustion and Flame *26*, 163 (1976)
70. Hirano, T., Tazawa, K.: Combustion and Flame *32*, 95 (1978)
71. Kashiwagi, T.: 15th Internat. Symp. on Combustion, p. 255, Pittsburgh: The Combustion Institute 1974
72. Quintiere, J.: Fire and Materials *5*, 52 (1981)
73. Petrella, R. V.: J. Fire and Flammability *11*, 3 (1980)
74. Kanury, A. M.: 15th Internat. Symp. on Combustion, p. 193, Pittsburgh: The Combustion Institute 1974
75. Holve, D. J., Sawyer, R. F.: 15th Internat. Symp. on Combustion, p. 351, Pittsburgh: The Combustion Institute 1974
76. Matthews, R. D., Sawyer, R. F.: J. Fire and Flammability *7*, 200 (1976)
77. Feng, C. C., Sirignano, W. A.: Combustion and Flame *29*, 247 (1977)
78. Rybanin, S. S., Sobolev, S. L.: Combustion Wave Propagation in Heterogeneous Reaction, a preprint, Chernogolovka, Branch of the Institute of Chemical Physics Academy of Sciences of the USSR 1981 (in Russian)
79. Moussa, N. A., Roong, T. Y., Garris, C. A.: 17th Internat. Symp. on Combustion, p. 1447, Pittsburgh: The Combustion Institute 1977
80. Ortiz-Molina, M. G. et al.: 17th Internat. Symp. on Combustion, p. 1191, Pittsburgh: The Combustion Institute 1979
81. Zeldovich, Ya. B.: Zh. Teor. Fiziki *19*, 1199 (1949)
82. Margolin, A. D., Krupkin, V. G.: Dokl. AN SSSR *242*, 1326 (1978)
83. Margolin, A. D., Krupkin, V. G.: Fizika Goreniya i Vzryva, *17*, 3 (1981)
84. Frey, A. F., T'ien, J. S.: Combustion and Flame *36*, 263 (1979)
85. Frey, A. F., T'ien, J. S.: Combustion and Flame *33*, 55 (1978)
86. Fenimore, C. P., Martin, F. J.: Combustion and Flame *10*, 135 (1966)
87. Krishnamurthy, L.: Combustion Sci. Technol. *10*, 21 (1975)
88. Bolodian, I. A. et al.: Fizika Goreniya i Vzryva *15*, 63 (1979)
89. Komamiya, R.: J. Fire and Flammability *4*, 82 (1973)
90. Margolin, A. D., Krupkin, V. G.: Fizika Goreniya i Vzryva, *14*, 56 (1978)
91. Aldabaev, L. I. et al.: Chemical Physics of Combustion and Explosion. Combustion of Heterogeneous Gas Systems, p. 45, Chernogolovka: Branch of the Institute of Chemical Physics Academy of Sciences of the USSR 1980
92. Margolin, A. D., Krupkin, V. G.: Fizika Goremiya i Vzryva *16*, 47 (1980)
93. Perrins, S., Pettet, I.: J. Fire and Flammability *5*, 85 (1974)
94. Holve, D. J., Sawyer, R. W.: Polymer Flame Retardant Mechanisms; Report ME-75-2, Berkeley: Colledge of Engineering Department of Mechanical Engng. 1975
95. Margolin, A. D., Krupkin, V. G.: Dokl. AN SSSR, *257*, 1369 (1981)
96. Melikhov, A. S., Potyakin, V. I.: Chemical Physics of Combustion and Explosion. Combustion of Heterogeneous and Gas Systems, p. 48, Chernogolovka, Branch of the Institute of Chemical Physics of Academy of Sciences of the USSR, 1980 (in Russian)
97. Krishnamurthy, L., Williams, F. A., Sechadri: Combustion and Flame *26*, 363 (1976)
98. Sohrab, S. H., Williams, F. A.: J. Pol. Sci., Pol. Chem. Ed. *19*, 2955 (1981)
99. Kanury, A. M., Martin, S. B., Alvares, N. J.: Fire and Materials *1*, 141 (1976)
100. Krause, R. F., Gann, Jr. and R. G.: J. Fire and Flammability *12*, 117 (1980)
101. Van Krevelen, D. W.: Polymer, *16*, 615 (1975): J. Appl. Polym. Sci., Appl. Polym. Symp. *3*, 269 (1977)
102. Moos, E. K., Skinner, D. L.: J. Cell. Plast. *13*, 276 (1977)
103. Blazovski, W. S., Cole, R. B., McAlevy, III, R. F.: 14th Internat. Symp. on Combustion, p. 1177, Pittsburgh, The Combustion Institute 1973
104. Shteinberg, A. S. et al.: Combustion and Explosion, p. 124, Moscow: Nauka 1972 (in Russian)
105. Brauman, S. K.: J. Polym. Sci., Polym. Chem. Ed. *15*, 1507 (1977)
106. Burge, S. J., Tipper, C. F. H.: Combustion and Flame *13*, 495 (1969)
107. Zhubanov, B. E., Davlichin, T. Kh., Gibov, K. M.: Vysokomol. Soed. *17B*, 746 (1975)

108. Efremov, V. D. et al.: Chemical Physics of Combustion and Explosion. Combustion of Hetero-geneous and Gas Systems, p. 38, Chernogolovka: Branch of the Institute of Chemical Physics Academy of Sciences of the USSR 1980
109. Grassie N.: Chemistry of High Polymer Degradation Processes, London: Butterworths Sci. Publ. 1956
110. Aseeva, R. M. et al.: Structural Chemistry of Carbon and Coal, p. 161, Moscow: Nauka 1969 (in Russian)
111. Aseeva, R. M.: Encyclopedia of Polymers, vol. 1, p. 2061, Moscow: Sovetskaya Entsiklopedia 1972 (in Russian)
112. Brauman, S. K.: J. Fire Retardant Chem. 6, 249 (1979)
113. Parker, J. A.: Fire and Flammability 6, 435 (1975)
114. Hilado, C. J. et al.: J. Fire and Flammability 9, 367, 553 (1978)
115. Brauman, S. K.: J. Fire Retardant Chem. 6, 266 (1979)
116. Kambour, R. P., Klopfer, H. J., Smith, S. A.: J. Appl. Pol. Sci. 26, 847 (1981)
117. Economy, J.: Flame Retardant Polymeric Materials, Lewis M., Atlas, S. M., Pearce, E. M. (eds.), vol. 2, p. 213. Mew York, London: Plenum Press 1978
118. Stuetz, D. E. et al.: J. Polym. Sci., Polym. Chem. Ed. 18, 967, 987 (1980)
119. Shlyapnikov, Yu. A.: Uspekhi Khimii 40, 1105 (1981)
120. Johnson, P. R.: J. Appl. Polym. Sci. 18, 491 (1974)
121. Ohe, H., Matsuura, K.: Textile Res. J. 45, 778 (1975)
122. Aseeva, R. M., Ruban, L. V., Zaikov, G. E., in: Flame Retardant Polymeric Materials: Pro-perty Estimation Problems. Conference of October 19–21, 1981. Report Theses, p. 19, Tallin: Institute of Chemistry, AN ESSR, 1981; Plastmassy (1983), p. 34
123. Asejeva, R. M., Lomakin, S. M., Zaikov, G. E., in: Natürliche und künstliche Alterung von Kunststoffen, Woebcken, W. (ed.), Kunststoffe-Fortschrittsberichte, vol. 7, p. 109, München, Wien: Hanser Verlag 1981
124. Lyons, J. W.: The Chemistry and Uses of Fire Retardants, New York: Wiley Intersci. 1970
125. Flame Retardance of Polymeric Materials, Kuryla, W. C., Papa, A. J., vols 1–3, New York: Marcel Dekker 1973–1975
126. Flame Retardant Polymeric Materials, Lewin, M., Atlas, S. M., Pearce, E. M. (eds.), vol. 1, New York, London: Plenum Press 1975
127. Tesoro, G. C.: J. Polym. Sci. Macromol. Rev. 13, 283 (1978)
128. Kodolov, V. I.: Flame Retardants for Polymeric Materials. Moscow: Khimiya 1981 (in Russian)
129. Hastie, J. W.: J. Res. NBS 77A, 733 (1973)
130. Bolodian, I. A., Denisiuk, A. P., Zhevlakov, A. P.: Fizika Goreniya i Vzryva 17, 146 (1981)
131. Kishore, K., Das, K. M., Kakodhar, S. P.: Colloid and Polymer 258, 51 (1980)
132. Gibov, K. M. et al.: Izv. AN Kaz. SSR, Ser. Khim. 1978, 60
133. Sibulkin, M., Little, M. W., Kulkarni, A.: Fire and Flammability 10, 263 (1979)
134. Petrella, R. V.: J. Fire and Flammability 10, 52 (1979)
135. Dixon-Lewis, G., Simpson, R. J.: 16th Internat. Symp. on Combustion, p. 1111, Pittsburgh: The Combustion Institute 1976; Combustion and Flame 36, 1 (1979)
136. Lovachev, L. A.: Combust. Sci. Technol. 25, 49 (1981)
137. Larsen, E. R., Ludwig, R. B.: J. Fire and Flammability 10, 69 (1979)
138. Tücker, D. M., Drysdale, D. D., Rasbaseh, B. J.: Combustion and Flame 41, 293 (1981)
139. Camino, G. et al.: Fire Safety J. 2, 257 (1980)
140. Keeney, C. N.: Adhesives Age 25, p. 578
141. Drews, M. J., Yeh, K., Barker, R. A.: Textilverdlg. 8, 180 (1973)
142. Yurchenko, V. M., Zubkova, N. S., Tyuganova, M. A.: Khimiya Drevesiny 4, 37 (1980)
143. Day, M. et al.: J. Appl. Polym. 26, 277 (1981); 27, 575 (1982)
144. Lipanov, A. M. et al.: Chemical Physics of Combustion and Explosion. Combustion of Con-densed and Heterogeneous Systems. p. 85, Chernogolovka, Branch of the Institute of Chemical Physics Academy of Sciences of the USSR 1980
145. Pandya, H. B. et al.: J. Fire Retardant Chem. 7, 27 (1980); 5, 86 (1978)
146. Vandersall, H. L.: J. Fire and Flammability 2, 97 (1974)
147. Kay, M., Price, A. F.: J. Fire Retardant Chem. 6, 69 (1979)
148. Ellard, J. A.: Am Chem. Soc. Div. Org., Coat. Plast. 33, 531 (1973)
149. Cagliostro, D. E., Riccitiello, S. R.: J. Fire and Flammability 6, 205 (1975)

150. Gibov, K. M., Zhubanov, B. A., Shapovalova, L. N. in: ref. [122], p. 93
151. Lask, G., Wagner, H. G.: 8th Internat. Symp. on Combustion, p. 432, Baltimore: Williams-Wilkins Co. 1962
152. Fenimore, C. P., Jones, G. W.: Combustion and Flame 8, 133 (1964)
153. Davis, P. B., Thrush, B. A.: Proc. Roy. Soc. A302, 243 (1968)
154. Nelson, G. L., Webb, J. L.: J. Fire and Flammability 4, 210 (1973)
155. Arney, W. C., Kuryla, W. C.: J. Fire and Flammability 3, 183 (1972)
156. Sobolev, J., Woycheshin, E. A.: J. Fire and Flammability/Fire Retardant Chem. 1, 12 (1974); 2, 224 (1975)
157. Martin, F. J., Price, K. R.: J. Appl. Polymer Sci. 12, 143 (1968)
158. Glazar, B. L., Howard, E. G., Collatte, J. W.: J. Fire and Flammability 9, 430 (1978)
159. Fenimore, C. P., Jones, G. W.: Combustion and Flame 10, 295 (1966)
160. Brauman, S. K.: J. Fire Retardant Chem. 3, 66, 117, 138, 225 (1976)
161. Pitts, J./J.: J. Fire and Flammability 3, 51 (1972)
162. Whelan, W. P.: J. Fire Retardant Chem. 6, 206 (1979)
163. Baratov, A. A.: Zh. Vs. Khim. Obshch. im. D. I. Mendeleeva 19, 531 (1974)
164. Yantovsky, S. A.: Zh. Prikl. Khimii 40, 1856 (1967)
165. Iya, K. S., Wollowitz, S., Kaskan, W. E.: 15th Internat. Symp. on Combustion, p. 329, Pittsburgh: The Combustion Institute 1974
166. US Patent No. 4,067,846, 4,073,768 (General Electric Co.)
167. Benisek, L. in: Proceedings 5th Internat. Wool Textile Res. Conf., Ziegler, K. (ed.). German Wool Research Institute, Aachen, vol. 5, p. 31 (1976)
168. Beck, P. J. et al. in: ref. [167], vol. 2, p. 549 (1976); Text. Res. J. 46, 478 (1976)
169. Van Berkel, F. G. A., Grotjahn, H.: J. Appl. Polym. Sci., Appl. Polym. Symp. 21, 67, 1973
170. McLaury, M. R.: J. Fire and Flammability 10, 175, (1979)
171. McCarter, R. J.: Fire and Materials 5, 66 (1981)
172. Košik, M. et al.: Fire and Materials 1, 19 (1976)
173. Brauman, S. K.: J. Fire Retardant Chem. 7, 119, 154, 161 (1980)
174. Kochneva, L. S. et al.: Europ. J. 15, 575 (1979)
175. Kishore, K., Prasad, G., Nagarajan, R.: J. Fire and Flammability 10, 296 (1979)
176. Ksandopulo, G. I. et al. in: Inhibition of Chain Gas Reactions. Ksandopulo, P. I. (ed.), Alma-Ata: Kaz. State University Press 1971
177. Zarkhina, G. S. et al.: Vysokomol. Soed. 22B, 690 (1980)
178. Skibida, I. P.: Uspekhi Khimii 54, 1729 (1975)
179. Blyumberg, E. A.: Dokl. AN SSSR 242, 358 (1978)
180. Shustova, O. A., Gladyshev, G. P.: Uspekni Khimii 55, 1695 (1976)

G. Henrici-Olivé, S. Olivé (Editors), W. Becker (Guest Editor)
Received January 1, 1984

Author Index Volumes 1–70

Subject Index

A. Knop, L. A. Pilato

Phenolic Resins

**Chemistry, Applications and
Performance Future Directions**

1985. 109 figures, 114 tables. Approx. 350 pages
ISBN 3-540-15039-0

From the preface: "… Since their introduction in
1910, the highly versatile family of phenolic resins has
demonstrated an important role in the developments
in electrical, automotive, construction, and appliance
industries. In the 80's the wave of high technology has
fostered their active participation in "high tech" areas
ranging from electronics, computers, communication,
outer space/aerospace, biomaterials, biotechnology
and advanced composites. Many phenolic resin
systems are actively involved in the "leading edge" of
these innovative technologies. Thus, they demonstrate an uncanny versatility to be adaptable to
prevailing times as the today's society is transforming
from an industrial to an information/communication
society.

The excellent participation at the recent scientific
symposia and the acceptance of the early edition
"Chemistry and Application of Phenolic Resins" by
A. Knop and W. Scheib – including the Japanese and
Russian translation – by the industrial and chemical
community demonstrated a high level of interest in
the broad subject of phenolic resins and has provided
the stimuli of this present publication.

This volume covers fundamentals, the chemical and
technological progress, and new applications including
the literature generally up to July 1984. Special
emphasis was assigned to advanced instrumental and
analytical techniques and environmental aspects..."

Springer-Verlag
Berlin
Heidelberg
New York
Tokyo

Characterization of Polymers in the Solid State I:

Part A: **NMR and Other Spectroscopic Methods**

Part B: **Mechanical Methods**

Editors: **H. H. Kausch, H. G. Zachmann**

1985. 135 figures, 16 tables. XI, 222 pages.
(Advances in Polymer Science/Fortschritte der
Hochpolymeren-Forschung, Volume 66)
ISBN 3-540-13779-3

Contents:

Springer-Verlag
Berlin
Heidelberg
New York
Tokyo